Algorithms and Combinatorics 10

János Pach (Ed.)

New Trends
in Discrete and
Computational
Geometry

With 46 Figures

Springer-Verlag
Berlin Heidelberg GmbH

János Pach

Mathematical Institute of the
Hungarian Academy of Sciences
H-1364 Budapest, PF-127, Hungary

and

Courant Institute
New York University
251 Mercer Street
New York, NY 10012, USA

Mathematics Subject Classification (1991):
52-XX, 68-XX, 05-XX, 14-XX, 57-XX, 60-02

ISBN 978-3-642-63449-9   ISBN 978-3-642-58043-7 (eBook)
DOI 10.1007/978-3-642-58043-7

Library of Congress Cataloging-in-Publication Data
New trends in discrete and computational geometry / János Pach (ed.).
p. cm. – (Algorithms and combinatorics; 10)
Includes bibliographical references and index.
ISBN 978-3-642-63449-9
1. Geometry – Data processing. 2. Combinatorial geometry. I. Pach, János. II. Series.
QA448.D38N49 1993 516-dc20 92-23684 CIP

© Springer-Verlag Berlin Heidelberg 1993
Originally published by Springer-Verlag Berlin Heidelberg New York in 1993
Softcover reprint of the hardcover 1st edition 1993

Typesetting: Springer TEX in-house system
41/3140-543210 – Printed on acid-free paper

Table of Contents

Chapter III. Epsilon-Nets and Computational Geometry
Jiří Matoušek

Chapter IV. Complexity of Polytope Volume Computation
Leonid Khachiyan

Chapter V. Allowable Sequences and Order Types
in Discrete and Computational Geometry
Jacob E. Goodman and Richard Pollack

Chapter VI. Hyperplane Approximation and Related Topics

Nikolai M. Korneenko and Horst Martini

Chapter VII. Geometric Transversal Theory

Jacob E. Goodman, Richard Pollack and Rephael Wenger

Chapter VIII. Hadwiger-Levi's Covering Problem Revisited

Károly Bezdek

Contributors

Imre Bárány

Mathematical Institute of the
Hungarian Academy of Sciences
Pf. 127
H-1364 Budapest, Hungary
e-mail: h2923bar@ella.hu

Károly Bezdek

Department of Geometry
Eötvös University
Múzeum krt. 6–8
H-1088 Budapest, Hungary
e-mail: h1515bez@ella.hu

Gábor Fejes Tóth

Mathematical Intitute of the
Hungarian Academy of Sciences
Pf. 127
H-1364 Budapest, Hungary
e-mail: h1160fej@ella.hu

Jacob E. Goodman

Department of Mathematics
City College, CUNY
New York, NY 10031, USA
e-mail: jegcc@cunyvm.cuny.edu

Leonidas Guibas

Department of Computer Science
Stanford University
Stanford, CA 94305, USA and
DEC Systems Research Center
Palo Alto, CA 94301, USA
e-mail: guibas@crl.dec.com

Leonid Khachiyan

Department of Computer Science
Rutgers University
New Brunswick, NJ 08903, USA
e-mail: leonid@aramis.rutgers.edu

Péter Komjáth

Department of Computer Science
Eötvös University
Múzeum krt. 6-8
H-1088 Budapest, Hungary
e-mail: h825kom@ella.hu

Nikolai M. Korneenko

Institute of Mathematics
Belorussian Academy of Sciences
Surganova 11
220604 Minsk, Belorussia

Wlodzimierz Kuperberg

Department of Mathematics
Auburn University
Auburn, AL 36849-5310, USA
e-mail: wkuperb@ducvax.auburn.edu

Horst Martini

Großmannstraße 13
O-8512 Großröhrsdorf, Germany

Jiří Matoušek

Department of Applied Mathematics
Charles University
Malostranské nám. 25
11800 Praha 1, Czechoslovakia
e-mail: matousek@earn.cspguk11

William Moser

Department of Mathematics
McGill University
805 Sherbrooke St. West
Montreal, P.Q., H3A 2K6, Canada
e-mail: moser@gauss.math.mcgill.ca

János Pach

Mathematical Institute of the
Hungarian Academy of Sciences
Pf. 127
H-1364 Budapest, Hungary
h1158pac@ella.hu
and
Courant Institute, NYU
251 Mercer Street
New York, NY 10012, USA
e-mail: pach@cims6.nyu.edu

Richard Pollack

Courant Institute, NYU
251 Mercer Street
New York, NY 10012, USA
e-mail: pollack@geometry.cs.nyu.edu

Raimund Seidel

Computer Science Division
University of California at Berkeley
Berkeley, CA 94720, USA
e-mail: seidel@cs.berkeley.edu

Micha Sharir

School of Mathematical Sciences
Tel-Aviv University
Tel-Aviv 69978, Israel
and
Courant Institute, NYU
251 Mercer Street
New York, NY 10012, USA
e-mail: sharir@taurus.bitnet

Rephael Wenger

Department of Computer Science
Ohio State University
Columbus, OH 43210, USA
e-mail: wenger@lasagna.cis.ohio-state.edu

Introduction

János Pach

Willst du ins Unendliche schreiten,
Geh im Endlichen nach allen Seiten.
J. W. Goethe

§1

Had you asked a layman in the fifties what mathematics was about, he would have thought "of dull uninspiring numbers, of a lifeless mechanism which functions according to laws of inescapable necessity" [1]. Coming from a family of many mathematicians, I learned in my younger days how to refute the widely held opinion that mathematicians are mainly concerned with counting, calculating and computing. Euclid's famous proof of the existence of infinitely many primes by *reductio ad absurdum* was the standard example of a "philosophical" argument using no calculation: Assume, for the sake of contradiction, that there were only finitely many primes $p_1, p_2, \ldots, p_n$. Then $p_1 p_2 \cdots p_n + 1$ could not have any prime divisor, which is clearly impossible! I was trained to believe that mathematics, a "gigantic, bold venture of the mind", was "die Schule des Denkens" (the school of thinking) [2, 3]. For me it had much more to do with infinity, philosophy and puzzles, than with counting or tedious computation.

The insufficiency of this "Weltanschauung" had, however, become apparent by the early seventies. The new situation is amply demonstrated by the titles of two extremely influential mathematical works which appeared in that period: The Art of Counting (by P. Erdős) and The Art of Computer Programming (by D. E. Knuth) [4,5]. The propagation of computers led to the upgrading of finite (i.e., combinatorial or discrete) mathematics, whose possibly most significant prophet was Erdős. It has also raised many exciting new problems. For example, how fast can we decide if a given integer is a prime? A wide class of questions of this kind has suddenly become tractable by computers. Knuth's books constituted the first serious modern attempt to create a coherent theory of algorithms.

A classical prototype of an exact mathematical algorithm is the so-called Euclidean algorithm, whose discovery is usually attributed to the Pythagoreans. Given three segments s_1, s_2 and s, we say that s is a *common measure* of s_1 and s_2 if the lengths of s_1 and s_2 are integer multiples of the length of s. To find the largest common measure of s_1 and s_2, determine how many times the shorter segment (say, s_2) goes into the other. If it does not leave any remainder, then $s = s_2$. Otherwise, the remainder, s_3, is shorter than s_2, and we check how many times s_3 goes into s_2. If there is no remainder this time, then $s = s_3$. Otherwise, we continue this procedure. It is not hard to show that, if this algorithm does not terminate in a finite number of steps, then s_1 and s_2 are incommensurable. For instance, this is the case when s_1 and s_2 are the diagonal and the side of a square, respectively. This was the first proof of the existence of an irrational number.

§2

Computer science has engendered a permanent revolution in mathematics. Among the fields whose sytematic revision, from the algorithmic point of view, proved to be the most fruitful, are algebra (arithmetic), combinatorics (graph theory, flows in networks, combinatorial optimization) and geometry. The spectacular initial results of computational geometry have already been summarized in three monographs [6,7,8]. They cover, among others, the following basic problems.

(A) Construction of Convex Hulls and Voronoi Diagrams. The problem of determining the convex hull of n points in the plane plays a similar role in computational geometry to what sorting does in many other areas of computer science. Moreover, it can be reduced to a sorting problem in linear time. There are many different methods to solve this problem in (optimal) $O(n \log n)$ time; e.g., by Graham's scan, by Prune and Search techniques, by the gift-wrapping algorithm (Jarvis march), etc. They illustrate a wide range of basic concepts, techniques and paradigms of the field, and may therefore serve as the perfect material for an introductory course. The higher dimensional versions of this problem lead naturally to the discussion of the combinatorial structure (the face lattice) of *convex polytopes*. In particular, McMullen's Upper Bound Theorem implies that the convex hull of n points in d-space can have as many as $c\,n^{\lfloor d/2 \rfloor}$ facets ($d-1$ dimensional faces), which provides a lower bound for the time complexity of any algorithm for the construction of the convex hull. On the other hand, it is not hard to see that any such algorithm can be used to find the Voronoi diagram of a set of points in $(d-1)$-space. (Recall that the *Voronoi diagram* of a set S is a cell decomposition of the space such that the cell assigned to any $p \in S$ consists of all points which are closer to p than to any other element of S.)

(B) Proximity Problems for Point Sets, Geometric Optimization. Given a set S of n points in d-space, find a pair of points in S which are closest (farthest). More generally, for every element of S, find a point that is nearest to it (farthest away from it). Find a spanning tree of S of minimum (maximum) total length. Find a spanning cycle of minimum length, i.e., an optimal *traveling salesman tour*. Find a ball of minimum radius enclosing all elements of S.

The traveling salesman problem is known to be NP-complete already for $d = 2$. The planar versions of all the other problems mentioned above can be solved in $O(n \log n)$ time by reducing them to the construction of a (suitably modified) Voronoi diagram. In particular, to find a spanning tree T of minimum length for a point set S in general position in the plane, we can use the fact that every edge of T must belong to the *Delaunay triangulation* of S, i.e., to the dual graph of its Voronoi diagram. Some of these techniques also work in higher dimensions. For example, the closest pair of points can be determined in $O(n \log n)$ time for any n element point set in d-space. However, we do not know the optimal time complexity of an algorithm to find the *diameter* (i.e., the farthest pair of points) for $d \geq 3$. Since Khachiyan's seminal discovery of the so-called "ellipsoid algorithm", whose running time is polynomial in the number of variables, in the number of constraints and in the size of the input, *linear programming* problems are usually discussed in the framework of computational geometry. The recent developments in this field suggest that every efficient approach to linear programming is based on new geometric ideas and leads to the discovery of many new geometric algorithms.

(C) Point Location, Range Search, Visibility. Assume that a city has n post offices, and that an operator has to answer a large number of telephone calls from customers asking which office is nearest to them. After constructing the Voronoi diagram of the set of post offices in $O(n \log n)$ time, the operator will face the problem of locating a query point in a planar map drawn by $O(n)$ straight-line segments. By some further preprocessing, one can build a tree-like data structure of depth $O(\log n)$ that enables us to answer any such query in $O(\log n)$ time, by simply tracing a path along this tree. The classical *range search* problem can now be paraphrased as follows. Suppose that the operator has to give the complete list of all post offices belonging to a certain region, where the query regions are taken from a fixed class of so-called ranges, e.g. from the class of halfplanes, triangles, circles, etc. Any efficient algorithm for the solution of this problem is based on some kind of balanced partition scheme of the plane which can be used to define a suitable data structure. The same technique can be adapted for the solution of a large variety of similar problems. For example, given n segments (triangles, circles, etc.) in the plane, provide the list of those which intersect a query segment (line, circle). One can further modify the questions above by requiring that our data structures be dynamical, i.e., that they allow fast deletion and insertion of new elements. Similar methods can be used to determine the visibility relations among n geometric objects (segments in the plane, triangles in 3-space). This can be applied in scene analysis to reconstruct the view of a collection of figures from a given point.

§3

In 1986, when Goodman and Pollack launched their journal "Discrete and Computational Geometry" at Springer-Verlag, this choice of title looked somewhat questionable. It seemed unlikely that one could fill a volume of 400 pages per year with first-rate results in the newborn field of computational geometry, whereas discrete geometry had been an established, active mathematical discipline for over a century. By now, however, it has become clear that these two subjects are deeply intertwined and their marriage promises to be long-lasting.

The aim of this book is to summarize some of the most recent developments in discrete and computational geometry, with special emphasis on those questions where the interaction between the two disciplines is the strongest.

(1) Arrangements and Combinatorial Methods. An arrangement of sets is said to form a *packing* (a *covering*) of a region R if no two of them overlap (if every point of R is contained in at least one of them). The fundamental problem in the theory of packing and covering is to determine the largest (smallest) number of congruent copies of a given body which can be packed (which are needed to cover) a region R. Originally, Voronoi diagrams were invented to tackle the special case in which the given body is a sphere. L. Fejes Tóth, the author of the first monograph devoted to this subject, initiated the investigation of the following question: find the shortest path between two points avoiding all members of a given packing [9], [10]. Independently, a more general problem was raised by theoreticians and practitioners in robotics: design an algorithm to move an object continuously from one position to another, avoiding (if possible) collision with a given collection of obstacles. By some routine transformations, this task can be reduced to a problem of the following type. Given an arrangement of not necessarily disjoint bodies $B_1, \ldots, B_n$ in d-dimensional Euclidean space, decide whether two points belong to the same connected component C of the complement of their union, and — if the answer is in the affirmative — connect the two points by a (short) curve within C. The time requirement of such an algorithm turns out to be roughly proportional to the number of "corners" on the boundary of $\cup_{i=1}^{n} B_i$ (where a corner belongs to at least d different B_i's). To find an upper bound on this quantity, one has to generalize and improve some old and deep results of Erdős, Beck, Szemerédi and Trotter on the distribution of distances among n points and on the length of Davenport-Schinzel sequences [11,12,13,14]. The unfolding intimate relationship between computational and combinatorial geometry, which this exemplifies, is perhaps the most exciting recent development in our field.

Chapter I of this volume (by Guibas and Sharir) focuses on the motion planning problem and on the combinatorial structure of arrangements of bodies and surfaces. Many related new results in combinatorial geometry are discussed in Chapter XI (by W. Moser and Pach). Chapter X (by G. Fejes Tóth and W. Kuperberg) concentrates on density problems and on some computational aspects of packings and coverings, while Chapter VIII (by K. Bezdek) surveys the state of affairs regarding the Hadwiger-Levi conjecture, one of the oldest unsolved problems about coverings.

(2) Random Methods. In the last fifty years probabilistic methods and ideas have permeated many branches of mathematics, including combinatorics and some areas of geometry (such as the theory of packings and coverings in higher dimensions). However, it has only been recognized recently that these techniques are applicable to a wide range of problems in combinatorial geometry. As was pointed out by Haussler and Welzl, many set-systems constructed by geometric means have small Vapnik-Chervonenkis dimension. (The *Vapnik-Chervonenkis dimension* of a set-system $\mathcal{F}$ is the size of the largest set A such that for any $B \subseteq A$ there exists $F \in \mathcal{F}$ with $F \cap A = B$.) On the other hand, it had been well known by statisticians that any set-system $\mathcal{F}$ of small fixed Vapnik-Chervonenkis dimension can be well represented by a small random sample taken from its underlying set X. In particular, one can pick $O(\frac{1}{\epsilon} \log \frac{1}{\epsilon})$ elements of X, so that any $F \in \mathcal{F}$ with $|F| \geq \epsilon|X|$ will contain at least one of them. These elements are said to form an *ϵ-net* for $\mathcal{F}$ [15,16]. In the solution of numerous problems in combinatorial and computational geometry (e.g. bounding the number of incidences between points and surfaces, or range searching), the probabilisitic arguments can be replaced by using the existence of a small ϵ-net in a suitably defined set-system. Moreover, in most cases such an ϵ-net can be constructed by an efficient deterministic algorithm. However, from the practical point of view, randomized algorithms are often as good as deterministic ones. In fact, in many cases they are better. The most celebrated efficient deterministic algorithms for linear programming, triangulating a polygon, finding the convex hull of a point set in higher dimensions, etc., are of theoretical interest only, because they are extremely complicated. In the last couple of years, inspired by some pioneering results of Clarkson, many elegant, easily implementable randomized algorithms have been found to solve these problems in almost optimal expected running time. There are some other important problems (e.g., the approximation of the volume of a convex body in n-dimensional space with given relative accuracy) for which it can be shown that the efficiency of the best randomized algorithm cannot be matched by using deterministic methods.

Chapter III of this volume (by Matoušek) gives a concise introduction to the theory of ϵ-nets with many geometric applications. Chapter II (by Seidel) describes some of the most important randomized algorithms in computational geometry, and offers a surprisingly elegant uniform approach to the analysis of their expected performance. Chapter IV (by Khachiyan) focuses on the problem of computing the volume of a convex body.

(3) Topological Methods. A surprising lesson one can draw from a lot of recent research in discrete and computational geometry is that the main obstacle to further progress in many areas is our ignorance of simple geometric properties of the Euclidean plane, let alone higher dimensional spaces. We have a rich arsenal of powerful combinatorial methods and involved data structures, but most of the time we operate with a handful of nontrivial topological facts such as the Borsuk-Ulam (or Ham-Sandwich) theorem and Helly's theorem. For example, until recently we did not know the answer to the following elementary question. Is it true that, if two finite point sets in the plane have the same order type, then one of them can be continuously transformed into the other without changing its order type? (Two point sets in Euclidean space have the same *order type* if there is a one-to-one correspondence between their elements so that any simplex induced by one set has the same orientation as the corresponding simplex in the other set.) Another annoyingly simple unsolved problem was to decide whether any arrangement of (say) 8 *pseudolines* (two-way infinite plane curves crossing each other pairwise at one point) is isomorphic to some arrangement of 8 straight lines. In the last decade several important tools have been developed to address problems of this type, including oriented matroids and allowable sequences [17]. Using a theorem of Milnor and Thom, Goodman and Pollack were able to estimate the number of different order types of n points in d-space with remarkable precision. It turned out that the concept of order type plays a central role in many seemingly unrelated branches of discrete and computational geometry. In particular, when combined with the Borsuk-Ulam theorem, it allows us to obtain some far-reaching generalizations of Helly's theorem for plane transversals, and to establish good upper bounds on the number of k-*sets* (i.e., those k-element subsets of an n-element set in d-space which can be cut off by a hyperplane).

Chapter IX of this book (by Bárány) focuses on recent generalizations and applications of the Borsuk-Ulam theorem. Chapters V and VII (by Goodman, Pollack and Wenger) offer comprehensive surveys on order types and geometric transversal theory, respectively. Chapter VI (by Korneenko and Martini) reviews a number of recent algorithms for the approximation of a point set by planes.

The last chapter (by Komjáth) concentrates on new set-theoretic results about colorings and decompositions of geometric figures or of the entire space. Undoubtedly the most sensational recent discovery of this type was the solution to the modern version of "squaring the circle" [18]. As Laczkovich has showed, it is possible to decompose a unit disk into finitely many disjoint pieces so that they can be rearranged to form a square of area π. Surprisingly, his argument uses the so-called "marriage lemma" (König-Hall theorem for finite bipartite graphs).

We may wonder at Goethe's wit: the road to infinity may well be paved with small stones (of finite combinatorics).

References

[1] H. Rademacher and O. Toeplitz: The enjoyment of mathematics. Princeton University Press, 1957. (Translated from: Von Zahlen und Figuren: Proben Mathematischen Denkens für Liebhaber der Mathematik. Springer, Berlin 1933)

[2] D. J. Struijk: A concise history of mathematics, 2nd edn. Dover Publ., New York

[3] G. Pólya: Mathematical discovery. John Wiley, New York London 1962. (The German translation is entitled: Die Schule des Denkens)

[4] P. Erdős: The art of counting (selected writings, ed. by J. Spencer). MIT Press, Cambridge, Mass. 1973

[5] D. E. Knuth: The art of computer programming, 2nd edn. Addison-Wesley Publ. Co., Reading, Mass. 1981

[6] F. P. Preparata and M. I. Shamos: Computational geometry. An introduction. Springer, New York 1985

[7] K. Mehlhorn: Data structures and algorithms 3: Multi-dimensional searching and computational geometry. Springer, Berlin Heidelberg 1984

[8] H. Edelsbrunner: Algorithms in combinatorial geometry. Springer, Berlin Heidelberg 1987

[9] L. Fejes Tóth: Lagerungen in der Ebene, auf der Kugel und im Raum (2nd edition), Springer, New York Heidelberg Berlin 1972

[10] L. Fejes Tóth: On the permeability of a circle-layer. Studia Sci. Math. Hung. 1 (1966) 5–10

[11] P. Erdős: On a set of distances of n points. Amer. Math. Monthly 53 (1946) 248–250

[12] J. Beck: On the lattice property of the plane, and some problems of Dirac, Motzkin and Erdős in combinatorial geometry. Combinatorica 3 (1983) 281–297

[13] E. Szemerédi and W. T. Trotter, Jr.: A combinatorial distinction between the Euclidean and projective planes. European J. Combinatorics 4 (1983) 385–394

[14] S. Hart and M. Sharir: Nonlinearity of Davenport-Schinzel sequences and of generalized path compression schemes. Combinatorica 6 (1986) 151–177

[15] V. N. Vapnik and A. Chervonenkis: On the uniform convergence of relative frequencies of events to their probabilities. Theory Probab. Appl. 16 (1971) 264–280

[16] D. Haussler and E. Welzl: ϵ-nets and simplex range queries, Discrete Comput. Geom. 2 (1987) 127–151

[17] A. Björner, M. Las Vergnas, B. Sturmfels, N. White and G. M. Ziegler: Oriented matroids. Cambridge University Press (to appear)

[18] M. Laczkovich: Equidecomposability and discrepancy: a solution of Tarski's circle-squaring problem. J. Reine Angew. Math. 404 (1990) 77–117

Chapter I. Combinatorics and Algorithms of Arrangements

Leonidas Guibas and Micha Sharir

1. Introduction

In this chapter we study the combinatorial structure of *arrangements* of algebraic curves or surfaces in low-dimensional Euclidean space. Such arrangements arise in many geometric problems, as will be exemplified below. To introduce the class of problems we will be interested in, we begin with the following concrete example, taken from the theory of *motion planning* in robotics.

Let B be a polygonal object having u sides, which is free to translate in the plane amidst a collection of polygonal obstacles, having a total of v corners. We are interested in characterizing in discrete combinatorial terms the space of all *free placements* of B, i.e. placements in which B does not intersect any obstacle. To this end, we choose an arbitrary fixed reference point p in B, and consider the following collection of "contact curves":

(i) For each corner c of B and each obstacle edge e, let $\gamma_{c,e}$ denote the locus of all points x such that when B is placed with p at x (without changing its orientation) its corner c makes contact with the edge e.

(ii) Similarly, for each side s of B and each obstacle corner q, let $\gamma_{s,q}$ denote the locus of all points x such that when B is placed with p at x its side s makes contact with the corner q.

The number of contact curves is $O(uv)$, and it is easily checked that each of them is actually a line segment parallel to the obstacle edge e in the first case, or to the side s of B in the second case.

It should be clear that to understand the manner in which B interacts with the obstacles as it is being translated in the plane it suffices to study the arrangement $\mathcal{A}$ of the $n = O(uv)$ contact curves (see Section 2 for more details concerning arrangements of curves). For example, suppose B is initially placed so that p lies in some face f of $\mathcal{A}$, and so that B is disjoint from the obstacles at this placement (namely, at a free placement). Then, as long as B translates with p remaining in f, B remains disjoint from the obstacles, but whenever p crosses an edge of f, a collision takes place. Moreover, if B does intersect some obstacles at its initial placement, the collection of obstacles that B intersects remains the same as long as p moves within a single face of $\mathcal{A}$. See Figure 1 for an illustration.

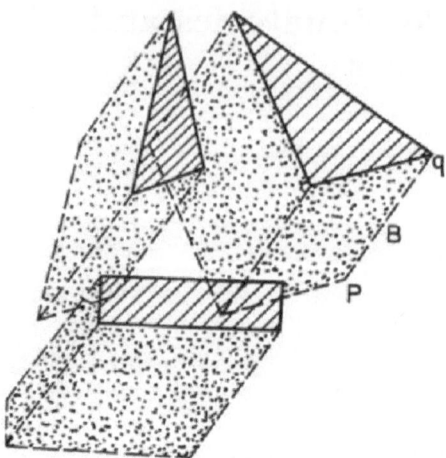

Fig. 1. The arrangement induced by placements of B

We can thus use the arrangement $\mathcal{A}$ to answer questions like the following:

1. Compute the space of all free placements of B. (This can be represented as the union of those faces of $\mathcal{A}$ that correspond to free placements.)
2. Find all placements of B that can be reached, via a collision-free translational motion, from a given initial (free) placement z. (This is just the face of $\mathcal{A}$ containing z.)
3. Find a placement of B at which it intersects the largest number of obstacles.

This reduction of placement and motion planning problems to the analysis of arrangements is fairly standard, and can plainly be generalized. For example, if B and the obstacles are bounded by curved (algebraic) edges, the contact curves will be algebraic (of degree that is generally larger than, and depends on, those of the boundary curves); see [7] for more details. Similarly, if B has more degrees of freedom, for example if it can also rotate, or if it is a rigid object moving in 3-space, then placements of B will have to be represented by points in higher dimensional space, and the contact loci become algebraic hypersurface patches. This reduction can thus create arrangements of fairly general algebraic surfaces in a high-dimensional space.

What are the geometric parameters that control the structure and complexity of the resulting arrangement? These are roughly

k – the number of degrees of freedom of the moving object;
n – the number of combinatorially distinct types of contacts between B and the obstacles (in other words, the number of contact surfaces);
d – the maximum algebraic degree of the contact surfaces.

Which of these parameters is likely to be large? In typical applications, k is small — it is two in the case of a translating object in the plane, three if we also allow rotation, and six for a general motion of a rigid object (or a manipulator arm) in 3-space. Let us consider situations in which the only parameter that is expected to be large is n — there are many possible types of contact, but each of them involves features that have low algebraic degree. This is indeed the situation in the case of the translating polygon that we started with. While this type of situation is not exclusive, it is typical enough to warrant close examination of the new kind of problems that it raises.

Specifically, we can sweep under the rug all issues involving algebraic manipulation of a fixed number of contact surfaces — since the dimension k of the parametric space in which these surfaces lie is small, and since the surfaces have uniformly bounded low degree, we can assume that any manipulation of this kind, such as finding the intersection manifold of any $j \leq k$ of these surfaces, takes constant time. For the kind of study we want to pursue in this chapter our primary focus is in the *combinatorial patterns* in which algebraic surfaces can intersect each other. Therefore we are willing to abstract away from the algebraic details of how any fixed number of bounded degree surfaces interact among themselves. The complexity we want to cope with is the combinatorial complexity arising from having to consider very many surfaces all at once.

Let us end these introductory comments by giving another example where arrangements of low-degree algebraic surfaces in a low-dimensional space arise. Let $\mathcal{L}$ be a set of n lines in 3-space. Since the space of lines in 3-space has four degrees of freedom, we can think of this space as a 4-dimensional (projective) space. For each line $\ell \in \mathcal{L}$ let Σ_ℓ denote the locus of all lines that intersect or are parallel to ℓ. It is easy to verify that Σ_ℓ is a quadratic hypersurface in 4-space.

Now consider the arrangement $\mathcal{A}$ formed by the n "contact surfaces" Σ_ℓ. This arrangement partitions 4-space into $O(n^4)$ cells, each representing an "isotopy class" of lines, that is lines that can be moved continuously within the cell from any placement to another without ever becoming incident or parallel to a line in $\mathcal{L}$. See [15] for more details.

The arrangement $\mathcal{A}$ thus enables us to solve a variety of problems involving *visibility* in three dimensions. For example, given a collection of n triangles in 3-space, does there exist a line that meets all of them? Another problem would be: find all pairs of triangles that can "see" each other, in the sense that there exists a segment connecting a point on one triangle to a point on the other and does not meet any other triangle. These problems, and many related ones, are solved by extending the edges of the given triangles into full lines, and by constructing the arrangement $\mathcal{A}$ of these $3n$ lines. By examining each isotopy class in this arrangement, we can determine the existence of a "common stabber" of the triangles, or find all "lines of visibility" between pairs of the triangles. Indeed, such properties are invariant over an isotopy class, so it suffices to check just one line in each class. See [52] for these applications.

There are many additional problems of this kind which make use of this "line-arrangement" $\mathcal{A}$, and there are many other application areas giving rise to arrangements of this sort. Generally speaking, this happens when we consider parametric representations of geometric objects (lines in space, placements of a moving system, etc.); the surfaces that arise in the parametric space correspond to loci of objects for which some critical phenomenon (coincidence, contact) occurs, and the way these surfaces partition the parametric space tells us a lot about what happens "in between" these critical events.

2. Arrangements of Curves in the Plane

We begin our study with the case of arrangements of algebraic curves in the plane. Let $\Gamma = \{\gamma_1, \ldots, \gamma_n\}$ be a collection of n such curves, where we assume that their maximum degree is bounded by some fixed constant. As a matter of fact, the results presented in this part of the chapter apply in a more general case in which the given curves satisfy the following properties:

- Each γ_i is a simple Jordan arc (which can be either closed or unbounded, thus separating the plane, or bounded and delimited by two endpoints).
- Each pair of curves in Γ intersect in at most some fixed number, s, of points.
- The shape of each γ_i is sufficiently simple, so that certain operations on a curve or on a pair of curves in Γ can be assumed to take constant time. These operations include: breaking each γ_i into maximal x-monotone pieces, finding the intersection points of a pair of curves in Γ, and testing whether a point lies above, on, or below a specific x-monotone portion of a curve in Γ.

All these assumptions hold in the algebraic case assumed above. The last assumption is important only for the algorithmic results described below, but is not needed for the combinatorial analysis of the arrangement of Γ.

Let $\mathcal{A} = \mathcal{A}(\Gamma)$ denote the arrangement of the curves in Γ. In the special case of two dimensions, $\mathcal{A}$ is a planar map, whose vertices are the endpoints and intersection points of the curves, whose edges are maximal connected portions of the curves which do not contain a vertex, and whose faces are maximal connected 2-d regions not meeting any edge or vertex. Note that the faces of $\mathcal{A}$ are open, and that the edges are relatively open. The combinatorial complexity of $\mathcal{A}$ can be measured by the number of its edges, because, by Euler's formula, the number of vertices and faces of $\mathcal{A}$ is dominated by the number of its edges.

The above assumptions make it clear that the combinatorial complexity of $\mathcal{A}$ is $O(n^2)$, and that this bound can be attained in typical cases (e.g. in an arrangement of n lines in general position in the plane). The actual complexity of $\mathcal{A}$ can be smaller than quadratic if either the curves of Γ are only sparsely intersecting, or if many of them are concurrent.

Consider first the problem of computing the entire arrangement $\mathcal{A}$. What do we want the output of this computation to be? An arrangement of curves is simply a partition of the plane as described above. We can use any one of several standard structures for representing such partitionings, for example the *quad-edge* data structure of Guibas and Stolfi [36]. In this structure the primary objects are edges. Each face is represented by the collection of the rings of edges bounding it, and each vertex by the ring of edges around it. Furthermore all these entities are cross-linked according to their topological adjacencies. For some applications it is important to further decompose the faces of the arrangement into regions such that each of them is determined by only a fixed number of the original curves defining the arrangement. There are many ways to do such "triangulations". For instance, we can break up each of our curves into arcs such that each arc is x-monotone and has no intersections with other arcs in its interior. If we then draw from the endpoints of each arc "vertical attachments", i.e. vertical line segments up and down, extended till another curve is intersected, we obtain such a refinement of the original arrangement. See Fig. 2 for an illustration. It should be clear that these triangulations leave the asymptotic space complexity of the arrangement at $O(n^2)$ (caution: in higher dimensions such a refinement may increase the space complexity).

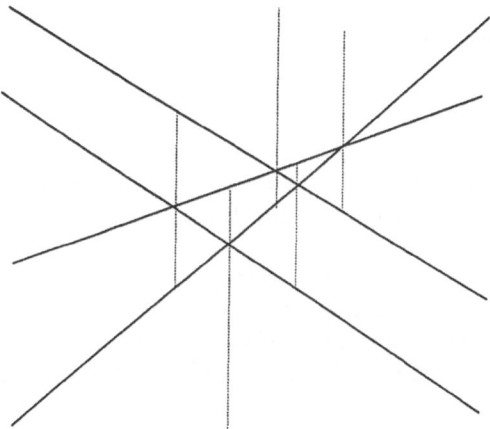

Fig. 2. A trapezoidal refinement of an arrangement

We will describe two techniques for calculating $\mathcal{A}$, or some refinement of it. The first technique, originally due to Bentley and Ottmann [9], is based on "line sweeping". That is, we take a vertical line ℓ and sweep it across the plane from left to right. As we do so, we maintain a list Λ of the intersection points of the curves in Γ with ℓ, sorted by increasing y coordinate. Since these points vary continuously, we store in Λ only pointers to the appropriate x-monotone portions of the curves of Γ that induce these intersections. We note that Λ

changes only when ℓ sweeps through an endpoint of an x-monotone portion of a curve, or through an intersection point of a pair of the curves. In the first case, we have to insert into Λ or to remove from it some curve portions. In the second case we need to swap in Λ the intersecting curves. To facilitate efficient execution of these operations, we maintain Λ as a balanced binary tree, and maintain in addition a priority queue Q of endpoints and intersection points that still lie ahead of (that is to the right of) ℓ. The endpoints can be inserted into Q ahead of the sweep, since they are all known in advance. The intersection points are added to Q as soon as the pair of intersecting curves first become adjacent in Λ. It is easily verified that this procedure can be implemented to run in time $O((n + k) \log n)$, where k is the actual number of intersection points of the curves in Γ. Thus in the worst case this runs in time $O(n^2 \log n)$. The working storage of this algorithm is largely dominated by the maximum size of the priority queue Q. This size is obviously bounded by $O(n + k)$, but this can be reduced to $O(n)$ if we are careful to delete from Q intersection points between pairs of curves that are no longer adjacent in Λ.

The sweeping procedure just described produces all intersections of the given curves, and thus all the vertices of their arrangement $\mathcal{A}$. By careful implementation, we can at the same time construct all edges and faces of $\mathcal{A}$ and find their adjacency relationships. We can thus summarize:

Theorem 2.1 *The Bentley–Ottmann procedure can compute the arrangement of n curves in the plane (satisfying the above assumptions) in time $O((n + k) \log n)$ and $O(n)$ space, where k is the size of the arrangement.*

Although the above technique is fairly general, simple and elegant, it has the disadvantage that when k is quadratic, it runs in $O(n^2 \log n)$ time. Can $\mathcal{A}$ be calculated faster than that, say in $O(n^2)$ time? This issue has been studied a lot in the last few years. For the case of lines, several algorithms were given [33], [16] for computing their arrangement in quadratic time. Especially noteworthy is the topological-sweep technique of Edelsbrunner and Guibas [28] that has numerous other applications. For the case of line segments, a recent result of Chazelle and Edelsbrunner [13] improved the Bentley–Ottmann technique and obtained an $O(n \log n + k)$ algorithm, which therefore never becomes more than quadratic. As we shall see shortly below, some of these techniques can be extended to handle the case of general curves.

In higher dimensions the situation is more complicated and fully general techniques are lacking. Under our bounded degree condition, the complexity of the arrangement of n hypersurfaces in d dimensions is $O(n^d)$, but standard triangulation techniques, like the Collins decomposition [23], may blow this up to n raised to an exponential in d. Some recent progress in this regard has been reported in [14]. There a method is given to produce a triangulation of the arrangement of n bounded-degree surfaces in d-space of size roughly $O(n^{2d-3})$, and in time roughly $O(n^{2d-1})$ (poly-logarithmic factors have been omitted in these bounds).

3. Lower Envelopes and Davenport-Schinzel Sequences

We have seen in the introduction that a typical problem that arises in applications of arrangements is to calculate a single face in an arrangement; see Figure 3. The first problem that we need to tackle is to bound the combinatorial complexity of such a face. After all, if that complexity could be quadratic in the worst case, perhaps all we should be doing is to compute the entire arrangement and extract from it the desired face. Fortunately, this turns out not to be the case. For example, in the case of lines, a single face is a convex polygon bounded by at most n edges, so it has always small complexity ($O(n)$). What happens in the case of general curves?

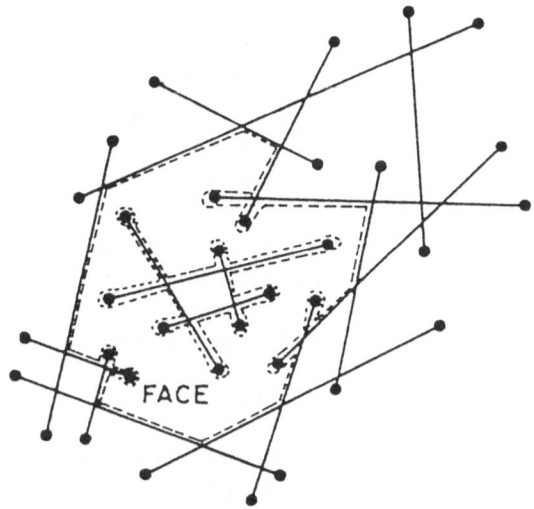

Fig. 3. A single face in an arrangement of segments

We first formulate an easier problem. Given a collection Γ of n x-monotone curves $\gamma_1, \ldots, \gamma_n$, so that no pair of them intersect in more than s points (for some fixed constant parameter s), define $\Psi = \Psi_\Gamma$ to be the *lower envelope* of the arcs in Γ. That is, viewing the arcs in Γ as the graphs of (partially-defined) functions $y = \gamma_i(x)$, Ψ is the graph of the pointwise minimum of these functions, that is, of the function

$$\Psi(x) = \min_i \gamma_i(x) \ .$$

The combinatorial complexity of Ψ can be measured in terms of the number of maximal connected pieces of the curves in Γ that appear along Ψ; alternatively we can measure that complexity in terms of the number of *breakpoints* along Ψ, where a breakpoint is an intersection point of two (or more) curves that appears along Ψ. See Figure 4 for an illustration.

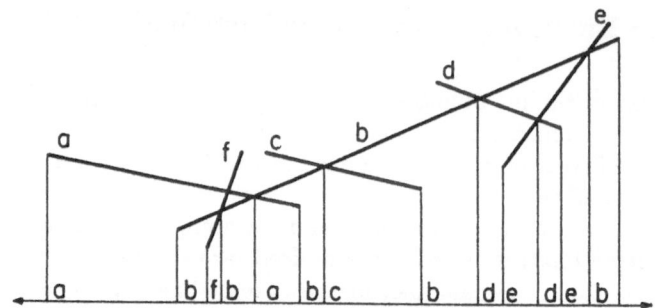

Fig. 4. The lower envelope and the lower envelope sequence in an arrangement of segments

Clearly, the lower envelope is a portion of the boundary of a single unbounded face of the arrangement of Γ. Thus combinatorial analysis and efficient construction of lower envelopes are natural precursors to the more general problems involving the whole boundary of an arbitrary face.

Given the collection Γ, consider the sequence $U = U(\Gamma)$ obtained by writing the indices of the curves γ_i in the order in which they appear along Ψ from left to right (see Figure 4 for an illustration). Assume first that the curves in Γ are all (x-monotone and) unbounded in both directions; thus they can be regarded as graphs of totally-defined continuous functions. The sequence $U = (u_1, u_2, \ldots, u_m)$ thus has the following properties:

(i) U is composed of n distinct symbols;
(ii) $u_i \neq u_{i+1}$ for all i;
(iii) U does not contain any alternating subsequence of two (distinct) symbols whose length is $s + 2$; that is, there do not exist $s + 2$ indices $i_1 < i_2 \cdots < i_{s+2}$ such that

$$u_{i_1} = u_{i_3} = u_{i_5} = \cdots = a,$$

$$u_{i_2} = u_{i_4} = u_{i_6} = \cdots = b,$$

and $a \neq b$.

The first two properties are obvious, and the third follows from the observation that such an alternation would imply that the curves γ_a, γ_b intersect in at least $s + 1$ points, contrary to assumption.

Any sequence U that satisfies these properties is called an (n, s)-*Davenport-Schinzel sequence* (a $DS(n, s)$ sequence in short). Davenport-Schinzel sequences are thus a powerful combinatorial structure that arises in many problems in discrete and computational geometry. They have been introduced by Davenport and Schinzel about 25 years ago [25], [24], and were further analyzed by Szemerédi [62], Atallah [5], Hart and Sharir [39], Sharir [56], [57], Schwartz and Sharir [55], and by Agarwal, Sharir and Shor [2].

Let $\lambda_s(n)$ denote the maximum length of a $DS(n, s)$ sequence.

It turns out that $DS(n,s)$ sequences provide a combinatorial character-ization of the lower envelope of n continuous functions, each pair of which intersect in at most s points. We have just observed that the "lower envelope sequence" $U(\Gamma)$, for any collection Γ of such functions, is a $DS(n,s)$ sequence. The converse also holds — for each $DS(n,s)$ sequence U, there exist n contin-uous functions, each pair of which intersect in at most s points, whose lower envelope sequence is equal to U (see [5] for a proof).

(Note that for a collection Γ of such functions, the total number of pairwise intersections of their graphs is at most $sn(n-1)/2$. This easily yields the fairly weak upper bound $\lambda_s(n) \leq \frac{sn(n-1)}{2} + 1 = O(n^2)$.)

If the curves in Γ are bounded, then it is not difficult to show that their lower envelope sequence is a $DS(n, s+2)$ sequence, and, conversely, for any $DS(n, s+2)$ sequence U there exists a collection Γ of n x-monotone (bounded) Jordan arcs, each pair of which intersects in at most s points, such that their lower envelope sequence is U. Thus the complexity of the lower envelope in the bounded case is at most $\lambda_{s+2}(n)$.

It is therefore of little surprise that these sequences have many applica-tions in combinatorial and computational geometry, extending beyond the few applications with which we have begun this chapter, and including problems in computer vision, computer graphics, motion planning, and other problems in robotics. Some of these applications are reviewed later in this chapter.

The major known estimates for $\lambda_s(n)$ are:

1. $\lambda_1(n) = n$; $\lambda_2(n) = 2n - 1$ (trivial; cf. e.g. [5], [25]).
2. $\lambda_3(n) = \Theta(n\alpha(n))$ (Hart and Sharir [39]). Here $\alpha(n)$ is the extremely slowly growing functional inverse of Ackermann's function. See [39] for a review of the definition and basic properties of Ackermann's function and its inverse. We note that $\alpha(n) \leq 4$ for all $n \leq 2^{2^{\cdot^{\cdot^{\cdot^2}}}}$ with 65536 2's in the exponential tower; thus we can regard $\alpha(n)$, for all practical purposes, to be a constant. Nevertheless, from a theoretical point of view, $\lambda_3(n)$ does grow faster than any linear function of n.
3. $\lambda_4(n) = \Theta(n \cdot 2^{\alpha(n)})$ (Agarwal, Sharir and Shor [2]).
4. $\lambda_{2s}(n) = O(n \cdot 2^{O(\alpha^{s-1}(n))})$, where the constant of proportionality depends on s (Agarwal, Sharir and Shor [2]).
5. $\lambda_{2s+1}(n) = O(n \cdot \alpha(n)^{O(\alpha^{s-1}(n))})$, where the constant of proportionality depends on s (Agarwal, Sharir and Shor [2]). (The last two bounds improve previous upper bounds given by Szemerédi [62] and by Sharir [56].)
6. $\lambda_{2s}(n) = \Omega(n \cdot 2^{c_s \cdot \alpha^{s-1}(n)})$ for each $s \geq 1$ and each $n \geq 1$, where c_s, as well as the constant of proportionality, depend on s (Agarwal, Sharir and Shor [2]).

In short, for any fixed s, $\lambda_s(n)$ is an almost linear function of n (and is actually ever so slightly superlinear for $s \geq 3$). Compared with the naive $O(n^2)$ upper bound noted above, these sharper bounds gain roughly an order of magnitude. Most of the geometric applications of Davenport-Schinzel sequences have indeed such a flavor — they yield algorithms whose complexity is (roughly) an order of magnitude better than what can be obtained by a naive approach.

The proofs of the above bounds are fairly complicated and are therefore omitted in this survey. Nevertheless, to give the reader some idea of what tools are used in the proofs, we will briefly sketch the analysis of the upper bound for $s = 3$.

Let U be a $DS(n, 3)$ sequence. We partition U into *blocks*, where each block is a contiguous subsequence of U in which no symbol appears more than once. It is not difficult to show that one can partition U into $2n - 1$ or fewer such blocks.

Let us denote by $\psi(m, n)$ the maximum length of a $DS(n, 3)$ sequence that consists of at most m blocks. We derive the following recurrence formula for ψ: For any partition $m = m_1 + \cdots + m_b$, where $m_1, \ldots, m_b$ are positive integers, there exists a corresponding partition of n as $n_1 + \cdots + n_b + n^\star$, where $n_1, \ldots, n_b, n^\star$ are nonnegative integers, such that

$$\psi(m, n) \leq \sum_{i=1}^{b} \psi(m_i, n_i) + \psi(b, n^\star) + c(m + n^\star) \, ,$$

for some absolute constant c. Informally, what this says is that if we partition the m blocks of an appropriate $DS(n, 3)$ sequence U into b clusters, consisting of m_1 contiguous blocks, m_2 contiguous blocks, and so on, then the n symbols forming U are partitioned into b subsets of "internal" symbols, where the i-th subset consists of n_i symbols that appear only within the i-th cluster, and a remainder set of "external" symbols, which appear in more than one cluster. The above recurrence is derived by considering the contributions to the length of U of each set of internal symbols and of the external symbols separately, and by estimating the "overhead" incurred when the internal and external symbols are merged to obtain U back. This overhead arises because two non-adjacent appearances of the same, say internal symbol a might become adjacent when we remove the external symbols, which violates the first condition that Davenport-Schinzel sequences have to satisfy. Note also that the above recurrence reflects the property that by deleting only a small number of appearances of external symbols, we can retain just a single appearance of each such symbol in each cluster, thus allowing us to regard each of these clusters as a single block for the external symbols. The proof of all these properties is not very difficult, and can be found in [2], [39].

It is an instructive exercise to solve the above recurrence. We give a few hints concerning the solution. We first note that $\psi(2, n) = 2n$. We now choose $b = 2$, and choose m_1, m_2 to be roughly $m/2$. We obtain the recurrence (in which $n_1 + n_2 + n^\star = n$)

$$\psi(m, n) \leq \psi(m/2, n_1) + \psi(m/2, n_2) + cm + (c + 2)n^\star \, ,$$

whose solution is easily seen to be

$$\psi(m, n) \le cm \log m + (c + 2)n \ .$$

We next choose $b = m/\log m$ and the m_i's to be roughly $\log m$ each. We estimate $\psi(b, n^*)$ using the bound just derived, and obtain the recurrence

$$\psi(m, n) \le \sum_{i=1}^{b} \psi(\log m, \ n_i) + 2cm + (2c + 2)n^* \ ,$$

whose solution is easily seen to be

$$\psi(m, n) \le 2cm \log^* m + (2c + 2)n \ .$$

We continue in this manner, choosing next $b = m/\log^* m$, and so on, obtaining a series of bounds of the form

$$\psi(m, n) \le ckm\alpha_k(m) + (ck + 2)n \ ,$$

where α_k is the functional inverse of the k-th "row" A_k in the Ackermann function hierarchy; see [2], [39] for more details. We finally stop this process at $k = \alpha(m) + 1$, which is shown to yield the bound

$$\psi(m, n) = O((m + n)\alpha(m))$$

which, substituting $m \le 2n - 1$, gives the asserted bound $\lambda_3(n) = O(n\alpha(n))$.

3.1 A Few Initial Applications

We begin by illustrating a simple and typical application of Davenport Schinzel sequences (which is taken from [5]). Let $p_1(t), \ldots, p_n(t)$ be n moving points in the plane; $p_i(t)$ is the position of the i-th point at time t. Assume (each coordinate of) each $p_i(t)$ to be polynomial in t of maximum degree d. At each time t we wish to find the point $p_i(t)$ nearest to the origin. More specifically, we want to partition the time axis into a minimum number of disjoint intervals $I_1, \ldots, I_m$, and obtain a corresponding sequence of indices $i_1, \ldots, i_m$ such that for all $t \in I_k$ the nearest point to the origin is $p_{i_k}(t)$.

Clearly all we need to do is to calculate the lower envelope $\min_i |p_i(t)|^2$. But by assumption, for any pair $i \ne j$ the functions $|p_i(t)|^2$ and $|p_j(t)|^2$ coincide in at most $2d$ points. Thus the number of maximal connected portions of the graphs of these functions composing their lower envelope, as well as the size of the desired time partitioning, is at most $\lambda_{2d}(n)$. Moreover, as we will show next, this partitioning can be produced in time $O(\lambda_{2d}(n) \log n)$.

As this example indicates, the first algorithmic question that comes to mind regarding Davenport Schinzel sequences is to design an efficient algorithm for calculating the lower envelope Ψ of a collection of n continuous functions $\gamma_1, \ldots, \gamma_n$, each pair of which intersect in at most s points. A simple way of doing it is to use a divide-and-conquer strategy that divides the collection of functions into two subcollections of roughly $n/2$ functions each, calculates recursively the lower envelope of each subcollection, and merges the two envelopes by a simple linear scan through both. In more details, the merge step proceeds as follows. Each subenvelope is produced with a list of its breakpoints, sorted in increasing x-order. We merge these two lists to obtain a partition of the x-axis into intervals, over each of which each of the two envelopes is attained by a single arc. We can thus in constant time compute the envelope of the whole collection over each interval, and concatenate these portions to obtain the desired envelope. Since the merge takes time linear in $\lambda_s(n)$, it follows that the entire procedure runs in time $O(\lambda_s(n)\log n)$. A cleverer divide and conquer method has been given by Hershberger [40], and runs in time $O(\lambda_{s-1}(n)\log n)$. This improvement does not amount of course to much in practice, but is interesting from a theoretical point of view.

Next consider the following special case of lower envelopes: Let $e_1,\ldots,e_n$ be n line segments in the plane, none of which is vertical. Viewing them as a collection of partially defined functions over the x-axis, and noting that each pair of them intersect at most once, we conclude, by the results stated earlier, that the complexity of their lower envelope is at most $O(n\alpha(n))$. Moreover, it was shown by Wiernik and Sharir [64] and by Shor [59] that there exist collections of n segments whose lower envelope does consist of $\Omega(n\alpha(n))$ sub-segments. This is somewhat surprising in view of the simple shape of line segments. Again, these constructions are too complicated to fit within this survey. We will see how these results are used and extended in the following sections.

3.2 Applications of Lower Envelopes and Davenport-Schinzel Sequences

To complete our detour of discussing envelopes, we mention four additional applications of them, all of which use Davenport-Schinzel sequences as a major combinatorial and algorithmic tool. These applications address problems in computer graphics, motion planning and computational geometry. This is by no means an exhaustive list; other applications, including dynamic computational geometry [5], shortest paths between two convex polyhedra [8], and finding extreme free placements of a line segment in a 2-d polygonal space [49], have also been obtained. We will also return to the topic of envelopes later, when we discuss the case of multivariate functions.

(i) **[Cole and Sharir [22]]** An algorithm is derived for preprocessing a polyhedral surface (terrain) σ having n edges, to support $O(\log n)$-time *ray shooting* queries, each of which seeks the first point on σ (if any) hit by a query ray emerging from a fixed given point a lying above σ. The preprocessing of σ requires $O(n\alpha(n)\log n)$ time and space, which is nearly an order of magnitude improvement over the worst case $\Omega(n^2)$ combinatorial complexity of the way in which σ can be visible from a.

This algorithm is closely related to (and has potential applications for) the well known hidden surface removal problem (see [48] for recent results on that problem, and [26] for some related results on visibility of a polyhedral terrain).

(ii) [**Leven and Sharir [44]**] An algorithm is derived for determining whether a given convex k-gon P can be translated and rotated to a placement in which it lies inside another given polygonal region Q (not necessarily simply-connected) whose boundary consists of n edges altogether. Chazelle [11] has studied this problem for the case in which P and Q are arbitrary simple polygons. The algorithm of [44] (for the special case in which P is convex but Q is an arbitrary polygonal region) runs in time $O(kn\lambda_6(kn)\log kn)$, and is based on the combinatorial property that (in general position) there exist at most $O(kn\lambda_6(kn))$ free placements of P at which it makes three simultaneous contacts with the boundary of Q (improving a naive $O((kn)^3)$ bound). This algorithm is a substantial improvement over the general technique of Chazelle.

(iii) [**Kedem and Sharir [43]**] An algorithm is derived for planning collision-free motion of a convex k-gon P moving (translating and rotating) amidst a collection of polygonal obstacles having n edges altogether. This problem is closely related to the polygon containment problem (ii), and the algorithm exploits the results obtained there, but also requires additional analysis of the topological and combinatorial structure of the free configuration space of P. The algorithm runs in time $O(kn\lambda_6(kn)\log kn)$, similar to the complexity of the algorithm in (ii). The algorithm is shown to be close to optimal in the worst case, and can be compared to the (nearly optimal) $O(n^2\log n)$ motion planning algorithms for the case in which B is a line segment [45], [60].

(iv) [**Chew and Kedem [18]**] The polygon containment problem is extended to the case in which P can also expand and shrink. It seeks the largest similar copy of P that can be placed inside the region Q. An algorithm is derived that runs in time $O(k^4n\lambda_4(kn)\log n)$. Thus, if k is considered to be small and fixed, the algorithm's efficiency is roughly the same as that in (ii), so adding the degree of freedom of expansion does not incur any significant increase in cost.

4. Faces in Arrangements

In this section we discuss several problems related to the structure of portions of arrangements.

4.1 The Case of a Single Face

We first return to the original problem that we faced before our digression into lower envelopes. That is, we wish to bound the combinatorial complexity of any single face in an arrangement of n Jordan arcs having the property that no two of them intersect in more than s points. Fortunately, this complexity is asymptotically the same as that of the lower envelope of such curves. That is, we have

Theorem 4.1 *The maximum complexity of a single face in an arrangement of curves as above is $O(\lambda_{s+2}(n))$ for bounded Jordan arcs, and $O(\lambda_s(n))$ for unbounded or closed Jordan curves.*

Sketch of Proof. Consider first the case of bounded Jordan arcs. Let $\gamma_1, \ldots, \gamma_n$ be the given arcs, and let f be a face in their arrangement. The idea of the proof is to trace (a single component of) the boundary of f and write down the sequence U of indices of the curves γ_i in the order in which they appear along the boundary, in the hope that U is a Davenport-Schinzel sequence of the appropriate type. This almost works, but requires a few modifications to yield the asserted bound.

One difficulty that arises is that U is a circular, rather than linear, sequence. We can of course regard U as a linear sequence, starting at some arbitrary place, but that may introduce extra alternations of pairs of symbols because of a "wrap-around" effect of symbol appearances. Another difficulty is that, unlike the case of lower envelopes, both sides of a curve γ_i may appear along the boundary of f, which again can increase the number of alternations between pairs of symbols. To overcome these difficulties, we split each curve into several symbols. First, each side of a curve γ_i is denoted by a distinct symbol. Next, we prove a "consistency" result: The circular order in which portions of (one side of) a curve γ_i appear along the boundary of f is the same as their order along γ_i; that is, there exists some place in U (which may depend on γ_i) such that if we read U from that place in circular order, we encounter portions of γ_i in the same order that they occur along γ_i. When we break U at some arbitrary place to make it a linear sequence, some curves γ_i may "wrap around" this breakpoint, which will violate the consistency property just mentioned. To retain it, we simply use two distinct symbols to denote (this side of) γ_i, so that the wrapped-around occurrences of γ_i are denoted by a new symbol.

After all these precautions, we finally obtain a linear sequence composed of at most $4n$ symbols (and having the same length as U) and one can show that this sequence is indeed a $DS(4n, s+2)$ sequence, from which the theorem follows. See [53], [37] for a more detailed proof.   □

As an easy corollary, we obtain the following result about *zones* in arrangements of curves. A zone of a curve γ_0 in an arrangement $\mathcal{A}$ of a collection Γ of n other curves is the collection of faces of $\mathcal{A}$ crossed by γ_0. If we assume, as usual, that no pair of the given curves intersect in more than s points, and also assume that γ_0 does not intersect any curve in Γ in more than (perhaps another) constant number of points, we obtain

Theorem 4.2 *The combinatorial complexity of the zone of a curve in an arrangement of n other curves that satisfy the above assumptions is $O(\lambda_{s+2}(n))$.*

Proof. Simply split each curve in Γ at each point of intersection with γ_0, and form a little gap there. This combines all the faces in the zone of γ_0 into one big face, at the cost of increasing the number of curves in Γ by a constant factor. An application of the preceding theorem completes the proof. □

An interesting application of the preceding theorem yields an efficient incremental technique for constructing an arrangement of a collection Γ of n curves as above. The idea is to add the curves in Γ in some order one by one, and to maintain dynamically the arrangement as we go. To update the arrangement when inserting the i-th curve γ_i, we trace the boundary of all the faces that it crosses. This will allow us to detect its intersections with former edges, and thus obtain the new edges, vertices, and faces that are incident to γ_i. To trace the faces crossed by γ_i we simply walk along their boundaries in linear order, looking for the next intersection of γ_i with that boundary. Since the collection of those faces is merely the zone of γ_i, their overall complexity is $O(\lambda_{s+2}(i-1)) = O(\lambda_{s+2}(n))$. Thus the entire incremental algorithm takes $O(n\lambda_{s+2}(n))$ time, which is closer to quadratic than the worst-case performance of the sweeping technique described earlier. For more details see [29]. We thus summarize:

Theorem 4.3 *The arrangement of n curves as above can be constructed in time $O(n\lambda_{s+2}(n))$.*

We note that in the case of lines the above arguments specialize to yield an $O(n^2)$ algorithm for the construction of the arrangement; see [33], [16] where this technique is first described, although the analysis in these papers is more complicated as it does not use Davenport-Schinzel sequences.

Another question that arises is how fast can one compute a single face in an arrangement, say the face containing some designated point x. This is of course what we need to solve the motion planning problem described in the beginning of this chapter. This has been studied in [30], [37], where an algorithm has been developed for this problem. It uses divide and conquer, where the merge step receives as input two faces f_1 and f_2 in two subarrangements, both faces containing x, and the task is to find the connected component of $f_1 \cap f_2$ that contains x. The algorithm achieves this by a special kind of sweeping, which takes $O((|f_1| + |f_2|) \log n)$ time, yielding an overall algorithm whose complexity is $O(\lambda_{s+2}(n) \log^2 n)$. We summarize:

Theorem 4.4 *A single face in an arrangement of n curves as above can be calculated in time $O(\lambda_{s+2}(n) \log^2 n)$.*

As noted, this yields an efficient algorithm for motion planning of fairly general systems with two degrees of freedom. As discussed above, placements of such a system can be represented in 2-d space in such a way that the collection of all free placements that can be reached from a given initial placement is a single face in an appropriate arrangement of "contact curves". We thus conclude that motion planning for such systems can be performed in close to linear time, as in the preceding theorem.

4.2 Many Faces in Arrangements and Related Problems

The preceding analysis has considered various portions of an arrangement of n curves in the plane: the entire arrangement, its lower envelope, a single face, and the zone of a curve. An intermediate problem that arises is what is the combinatorial complexity of (and how fast can one calculate) any m distinct faces of the arrangement. The hope is that their maximum complexity is considerably smaller than quadratic, when m is smaller than $\Theta(n^2)$. This indeed has been shown for several types of curves, and we will review here the main techniques and results. Consider for simplicity the case of lines. It is easily checked that any pair of distinct faces in an arrangement $\mathcal{A}$ of n lines cannot have more than four lines that are incident to both faces. Hence if one consider the bipartite graph G whose edges are all pairs (f, ℓ) of a line ℓ incident to one of the m given faces f, then this graph does not contain the complete bipartite graph $K_{2,4}$ as a subgraph. It then follows from standard results in extremal graph theory that the number of edges of G is at most $O(mn^{1/2} + n)$. If we denote by $K(m, n)$ the maximum complexity of m faces in any arrangement of n lines, then we have shown that

$$K(m, n) = O(mn^{1/2} + n) \, .$$

This bound has originally been obtained by Canham [10] more than twenty years ago.

However, we can obtain an improved bound. We apply a partitioning scheme that was given by Matoušek [46] — given n lines in the plane and a parameter $r \leq n$, it partitions the plane into $O(r^2)$ triangles such that no triangle is crossed by more than $O(n/r)$ lines. We now distribute the m given faces among these triangles. Suppose m_i faces are fully contained within the i-th triangle, which is crossed by $n_i = O(n/r)$ lines. The total complexity of these faces is

$$K(m_i, n_i) = O(m_i n_i^{1/2} + n_i) = O(m_i (n/r)^{1/2} + n/r) \, ,$$

so, summing over all $O(r^2)$ triangles, and noting that $\sum_i m_i \leq m$, we obtain an overall complexity of $O(m(n/r)^{1/2} + nr)$. This bound still does not take into account the complexity of faces that "spill" across several triangles. However, each such face must belong, for each triangle Δ that it meets, to the zone of the boundary of Δ, clipped to within Δ. It follows from the preceding results that the complexity of that zone for the i-th triangle is $O(n_i) = O(n/r)$. Thus, summing this over all triangles we finally obtain

$$K(m, n) = O(m(n/r)^{1/2} + nr) \, .$$

All that remains is to choose an appropriate value for r. We choose $r = m^{2/3}/n^{1/3}$. This is feasible as long as m is between $n^{1/2}$ and n^2, and yields $K(m,n) = O(m^{2/3}n^{2/3})$. If m is smaller, the Canham bound can be applied directly. Leading to

Theorem 4.5 *The combinatorial complexity of any m distinct faces in an arrangement of n lines in the plane is $O(m^{2/3}n^{2/3}+n)$. The bound is tight in the worst case.*

The upper bound has been proved by Clarkson et al. [20], while the lower bound follows from a construction of Szemerédi and Trotter [63].

It is interesting to note that the "many faces" problem is strongly related to the problem of bounding the number of incidences between m distinct points and n lines (or other curves) in the plane. Indeed, suppose we are given m points $p_1, \ldots, p_m$ that form many incidences with a collection of n lines. One can choose a sufficiently small $\epsilon > 0$ and replace each line ℓ by two parallel lines translated by ϵ to each side of ℓ. This creates a small face around each p_i, whose complexity is twice the number of incidences involving p_i (this argument excludes points that are incident to just one line). This implies in particular that the maximum number of incidences between m poinst and n lines is $O(m^{2/3}n^{2/3}+m+n)$, and this bound is also known to be tight in the worst case.

The results just described can be generalized to other types of curves in the plane. Clarkson et al. [20] have also shown that the maximum complexity $K(m,n)$ of m distinct faces in an arrangement of n pseudo-lines or n unit circles is $O(m^{2/3}n^{2/3} + n)$. For arbitrary circles the known upper bound is $O(m^{3/5}n^{4/5} + n)$. A similar relationship holds between these problems and the corresponding incidence problems, and the same upper bounds also apply to the maximum number of incidences between m points and n curves as above. As an interesting corollary, we note that, given a set of n points in the plane, the number of pairs of them at distance exactly 1 from each other is half the number of incidences between these n points and n unit circles, centered at each of the n given points. We thus conclude that the maximum number of unit distances among n points in the plane is $O(n^{4/3})$ [20], [61]. The known lower bound for this quantity has been given by Erdős [34], [35] and is very close to linear in n.

The bounds on the complexity of many faces and on the number of incidences for curves of the kinds mentioned above are obtained using alternative techniques for partitioning an arrangement of such curves. These techniques take a random sample of r of the curves, form and triangulate its arrangement, and then continue in much the same way as in the case of lines described above. Two issues have to be addressed. First we need to derive bounds, similar to the Canham bound given above, for the other types of curves so as to handle the base case of the recursion. Second, we need to estimate the expected value of the resulting complexity, over the random sampling. The first issue can be handled in a similar fashion to the above analysis, but it is usually an argument specific to the particular types of curves we are dealing with. The second step is handled by applying recent results of Clarkson and Shor [21] that show that the expected value of the complexity is asymptotically the same as the bound derived above; more details are given in the next subsection.

An interesting case is that of line segments. Obtaining tight bounds for the complexity of many faces in arrangements of line segments has been a more difficult task, mainly because of the difficulty of obtaining a Canham-like bound in this case. Only very recently it was proved by Aronov et al. [3] that the complexity of m faces in an arrangement of n line segments is $O(m^{2/3}n^{2/3} + n\alpha(n) + n\log m)$, which is tight except for a certain small range of values of m.

4.3 Partitioning Arrangements

In the previous subsection we mentioned Matoušek's [46] method for partitioning an arrangement of lines in the plane and the use of random sampling techniques to obtain the many faces bounds. We now further elaborate on these topics because they throw light on the structure of arrangements.

A fundamental technique for solving many combinatorial and algorithmic problems about arrangements is *divide-and-conquer*. In this approach the underlying space of the arrangement is subdivided into regions. Each region, together with all the manifolds of the arrangement crossing it, defines a sub-problem. Depending on various parameters, each subproblem is then solved either recursively, or by some more direct method. Finally, the results of the subproblems are combined to produce the result for the full problem. We saw this aproach in action when we derived the $O(m^{2/3}n^{2/3} + n)$ bound for the complexity of m faces in an arrangement of n lines in the plane.

What is needed in this approach is a way to partition the underlying space of the arrangement into regions, so that each region receives only its "fair share" of the overall complexity of the arrangement. Matoušek's method is a deterministic way of finding such a partitioning for an arrangement of lines in the plane. Specifically, given n lines in the plane, he gives a way to partition the plane into $O(r^2)$ triangles so that each triangle is cut by at most $O(n/r)$ lines of the arrangement. In a partitioning into $O(r^2)$ triangles we can argue that a typical line will intersect $O(r)$ triangles — this is another "zone theorem", such as Theorem 4.2. Thus there are overall $O(nr)$ line-triangle crossings, so on the average we expect a triangle to be crossed by $O(n/r)$ lines. In this sense Matoušek's partitioning is as fair as possible. (Notice that because geometric objects are "big", we cannot expect a divide-and-conquer step to apportion each of them to a single subproblem only. But we can expect that each geometric object will be passed to only few of the subproblems.)

Recently Matoušek has been able to extend his determinstic partitioning technique to hyperplanes in any dimension d. In [47] he gives a method for finding, given n hyperplanes in R^d, a partitioning of R^d into $O(r^d)$ simplices so that each simplex is cut by $O(n/r)$ hyperplanes. (The running time of his method is, with some recent improvements, $O(nr)$.) General, but much slower, deterministic partitioning methods were also given by Chazelle and Friedman [12]. However, no deterministic method of obtaining such a good partitioning is known for general arrangements of algebraic surfaces. For such arrangements we must resort to randomized methods to obtain good partitionings. We now explain these randomized techniques because they are quite general and informative about the combinatorics of arrangements.

How do we use randomization to obtain a good partitioning? The general idea is to select a random sample of r among our n manifolds, and then triangulate the arrangement of the r sample manifolds. By "triangulate", we mean to decompose each cell or face of the arrangement into simplices, or in general pieces of "bounded description complexity". For instance, for the case of lines in the plane we can triangulate each face by drawing all diagonals from some vertex, or trapezoidalize each face by drawing vertical lines from each vertex of the face. The crucial property is that each region thus generated defines has a simple shape fully described by a number of coordinates that is independent of r (and n, but may depend on the dimension d). Notice that for hyperplanes in d dimensions we can obtain this way $O(r^d)$ elementary regions in our partitioning, just as in the Matoušek construction. Each of these regions has the property that it is fully determined by only a constant (possibly depending on d) number of the sampled r manifolds, but does not properly intersect any of them. As we will see, probabilistically this is evidence that each region is intersected by only few of the original manifolds.

The first results in this direction were obtained by Clarkson [19], and independently by Haussler and Welzl [41]. In [19] Clarkson roughly argues as follows. Let us fix on the trapezoidalization of the sample arrangement of the r lines described above. Each trapezoid is fully determined by four or fewer of the r lines, as is easy to check. Let us call a trapezoid "fat", if it is properly intersected by at least αn of the original lines (for some α, $0 < \alpha < 1$, not necessarily a constant), but by none of the other sample lines. Each group of four or fewer of our r sample lines determine a constant number of different trapezoids, so overall the r lines define $O(r^4)$ different trapezoids. The probability that a particular such trapezoid is fat is at most $(1-\alpha)^{r-4}$ and therefore the probability that there is a fat trapezoid at all is upper bounded by $O(r^4(1-\alpha)^{r-4})$ (this argument assumes that each trapezoid is defined by exactly four lines). The critical value here is $\alpha = (c/r)\log r$, for some constant c — then the upper bound becomes less than some other constant less than 1. A careful analysis shows that:

Theorem 4.6 *For any $\epsilon > 0$ there is a constant C_ϵ such that, with probability at least $1 - \epsilon$, each of the trapezoids in which the arrangement of the r sample lines is subdivided is cut by at most $C_\epsilon(n/r)\log r$ original lines.*

Thus there is a very easy randomized algorithm to get a partitioning of an arrangement that is, with high probability, almost as good as Matoušek's (we have lost a $\log r$ factor in the bound on the number of cutting lines). In [41] Haussler and Welzl introduced a general framework for obtaining results of this kind — the theory of *epsilon nets*. In this theory we start with an undelying space and certain interesting subsets of that space we call *ranges*. In our case the underlying space X will be that of all lines in the plane, and the ranges $\mathcal{R}$ will be defined by considering the subsets of lines determined by the property that they must intersect a particular triangle (or trapezoid) in the plane. Such a range space $(X, \mathcal{R})$ has an important property known as *finite Vapnik-Chernovenkis dimension*. What this means is that there are finite subsets of this space not all of whose subsets can be extracted by considering only the allowed ranges (this is where the property that our ranges are triangles or trapezoids — objects of finite description complexity — comes in). Under this very general setting the theory of epsilon nets shows that, for any subset of n elements from X, with very high probability, a random sample of size r will be an epsilon net with $\epsilon = (c/r) \log r$ (c a constant). This means that any range "containing" at least ϵn objects of X will also contain one of the r objects in the sample. In other words and in our setting, if a trapezoid is not cut by any of the r sample lines, then it can be cut by at most $c(n/r) \log r$ of the original lines. This is exactly the result stated above.

In a subsequent paper [21] Clarkson and Shor strengthened the argument in [19] and were able to show that if we draw a random sample of r among the n lines and trapezoidalize their arrangement, then the expected number of lines cutting any particular trapezoid will be $O(n/r)$. We can focus on a particular trapezoid by always considering the trapezoid in the arrangement containing some specific point p in the plane. This improved argument removes the $\log r$ factor and gives us a result which (allowing randomization) is as good as Matoušek's for most applications. Of course all of the above [19], [41], [21] develop such results in a very general setting that will work for our use in partitioning arbitrary arrangements of algebraic manifolds in any dimension. The probabilistic argument involved is given in an elementary form and exactly for the case discussed in this section in section 5.2 of [20]. The reader is referred to that paper for the details.

5. Arrangements in Higher Dimensions

As the reader can conclude by now, complexity analysis of arrangements in the plane has been quite successful, and has yielded sharp and often tight bounds for most of the relevant problems. It is natural therefore to want to extend the results reviewed so far to arrangements of surfaces in higher dimensions. Unfortunately, the analogous problems become much harder to analyze, and most of them are still largely unsettled, except for a few special cases that will be reviewed shortly.

In this section we will review some of the problems that arise in higher dimensions. The next chapter will cover additional issues that are related to such arrangements.

5.1 Lower Envelopes of Bivariate Functions and Their Applications

While the "one-dimensional" Davenport-Schinzel problem is by now relatively well understood, its two-dimensional generalization has been largely uninvestigated (with the exception of the simple case where each f_i is a plane [54], and a few recent initial studies of more complex cases [42], [58]; see also [6] for the case where each f_i is a sphere), and appears to be much harder. In this generalization one considers a collection $\mathbf{F} = \{f_1(x, y), \ldots, f_n(x, y)\}$ of n continuous bivariate functions such that each triple of functions intersect in at most some fixed number s of points, and such that each pair of functions intersect in a curve that has at most some fixed number t of singularities, and aims to obtain sharp upper and lower bounds on the maximum complexity $\kappa(\mathbf{F})$ (i.e. number of faces, edges, and vertices) of the planar map M, which can be called the *minimization diagram* of $\mathbf{F}$, obtained by projecting the pointwise minimum of these functions onto the x-y plane. (Each region of this map consists of a maximal connected set of points at which M is attained by a particular function f_i, and the edges (resp. vertices) of this map consist of points at which M is attained simultaneously by two (resp. three) functions. We assume here that the functions in $\mathbf{F}$ are in "general position", thereby excluding degeneracies at which two functions coincide on a two-dimensional region, or three functions coincide on a one-dimensional set, or four functions coincide at all, etc.)

The two-dimensional Davenport-Schinzel problem arises as a central subproblem in many applications in computational and combinatorial geometry. As an illustration, note that almost any conceivable generalization of the (nearest neighbor) Voronoi diagram in the plane can be regarded as the minimization diagram of a certain collection of functions, each measuring the distance from a test point to one of the objects defining the diagram. Similarly, the boundary of the configuration space of free positions of a moving system with three degrees of freedom can often be defined as the envelope of a certain collection of 2-d surfaces, each representing contact positions with some obstacle. In visibility problems, one often considers some space of rays, and for each ray the object it "sees" is the one whose intersection with the ray is nearest to the view-point. Thus if the ray space is two dimensional (as in the case of rays emanating in the plane from a fixed segment, or rays emanating in 3-space from a fixed point), analysis of the visibility along these rays reduces to the calculation of the lower envelope of an appropriate collection of "distance functions" along the rays. These few examples (more applications are given below) illustrate the significance of the 2-d Davenport-Schinzel problem.

We will review two partial solutions to, and some generalizations of, this two-dimensional Davenport-Schinzel problem. Clearly, a trivial upper bound for $\kappa(\mathbf{F})$ is $O(n^3)$ (with a constant depending on s and t), and our main goal is to improve that bound, and make it as nearly quadratic as possible. Our main conjecture is that the complexity $\kappa(\mathbf{F})$ is at most $O(n\lambda_{s'}(n))$, for some s' depending on $\underline{s}$ (and possibly also on t). That $\kappa(\mathbf{F})$ can indeed be superquadratic was shown in an example given in [58]; a similar lower bound, but involving only piecewise linear functions, will be noted below.

The first result proves this conjecture for the special case in which the intersection curve of any pair of the functions in $\mathbf{F}$ intersect every plane of the form $x = c$, for some constant c, in at most two points. It is shown in [55] that in this case $\kappa(\mathbf{F}) = O(n\lambda_{s+2}(n))$, where the constant of proportionality depends on s and t, and an $O(n\lambda_{s+2}(n)\log n)$ time algorithm for the calculation of the minimization diagram of $\mathbf{F}$ is presented. The analysis of [55] proceeds by reducing this restricted case of the 2-d problem to a collection of 1-d problems involving lower or upper envelopes of certain subsets of the intersection curves of pairs of the functions f_i; as noted above, the approach is quite reminiscent of the analysis in [44], but turns out to be considerably simpler.

The extra condition assumed above is somewhat artificial, but nevertheless covers certain applications in which the functions $f_i(x_0, y)$ have relatively simple form (as functions of y) for each fixed x_0, e.g. are linear or quadratic in y. For example, such a situation arises in analysis of the pattern of changes in the convex hull of n moving points. The above result then implies that the number of combinatorial changes in the convex hull is $O(n\lambda_s(n))$, for some constant s depending on the kind of motion of the points, and that these changes can all be found in time $O(n\lambda_s(n)\log n)$. (This particular convex hull problem was also studied by Atallah [5] using a different technique.)

The second result proves a strong form of the above conjecture for the case in which each function in $\mathbf{F}$ is piecewise linear, so that all the graphs of these functions have altogether n faces. In this case we can replace $\mathbf{F}$ by a collection of $O(n)$ "triangle functions", each of which has a graph that consists of a triangle (in arbitrary position in 3-space) with three adjacent steeply rising unbounded faces. Unfortunately, the intersection curve of a pair of such triangle functions can have cross sections of the form $x = c$, for some constant c, consisting of three points, so that the preceding result does not apply here. Nevertheless Pach and Sharir [50] have shown that the complexity of the lower envelope of n such functions is at most $O(n\lambda_3(n)) = O(n^2\alpha(n))$. Moreover, using the recent constructions of [64], [59], it can also be shown that this bound is tight in the worst case. The analysis of [50] and the companion paper [31], which uses a divide and conquer approach, also yields an algorithm, based on the same technique, for calculating this envelope in time $O(n^2\alpha(n))$. Because of the particular simple form of piecewise linear functions, the problem of estimating the complexity of their envelope has been one of the major open problems in the 2-d Davenport-Schinzel theory, and, as shown below, has many applications. The results of [50], [31], and a third companion paper [27], also provide generalizations of this tight bound to the envelope of multi-variate piecewise linear functions. Specifically, it is shown that the overall complexity of the envelope of n $(d-1)$-dimensional simplices in d-space, namely the total number of faces (of any dimensions) of which the envelope is composed, is $\Theta(n^{d-1}\alpha(n))$.

Several applications of these results are given in [31], including applications to motion planning, plane transversals of polyhedra, hidden surface removal, and "cluster" Voronoi diagrams.

5.2 The Complexity of a Single Cell and Related Problems

As in the two-dimensional case, the next goal is to obtain improved bounds for the complexity of (and corresponding efficient algorithms for constructing) a single cell in an arrangement of n surfaces (or surface patches) in three or higher dimensions. Consider the three-dimensional case first. We would like to conjecture that the complexity of a single cell in an arrangement of n "well-behaving" surfaces in 3-space is only $O(n\lambda_s(n))$, where s is some constant parameter depending on the patterns of double and triple intersections of the surfaces. This conjecture covers in particular the case of algebraic surfaces or surface patches whose degrees (as well as the degrees of their bounding curves, if any) are bounded by some fixed constant. A similar conjecture has been formulated above for the complexity of the lower envelope of such surfaces, so we are seeking an appropriate extension to 3-space of the phenomenon we have seen in two dimensions, namely that the maximum complexity of a single cell and of the envelope in an arrangement are asymptotically the same. Unfortunately, this conjecture appears to be very difficult in three dimensions, and is not known to hold except in a few trivial cases. (In passing we note that if the surfaces are *planes* then both the envelope and any single cell in their arrangement have only *linear* complexity; however, this case is too simple to shed any light on the general problem.)

The conjecture is still open even in the case of triangles, where the envelope is known to have near-quadratic complexity. However, we can obtain in this case the following alternative bound. Note that each cell in the arrangement of n triangles in 3-space is either convex or non-convex. In the former case the complexity of the cell is only $O(n)$, so these cells are not "interesting". Thus as an alternative goal, we may wish to bound the overall complexity of all intereseting, that is non-convex, cells in such an arrangement. It is shown in [1,4] that this complexity is at most $O(n^{7/3})$ and that this bound is tight in the worst case. The proof in [4] relies on the results, mentioned above, on the complexity of many faces in arrangements of lines segments in the plane. In particular this implies that the complexity of any single cell in an arrangement of triangles is also at most $O(n^{7/3})$, although this still leaves a considerable gap between this bound and the one conjectured above. This result has been recently extended in [38] to other types of surface patches that arise in several motion planning problems with three degrees of freedom.

There is one interesting case where the conjecture is known to hold, when the given surfaces are spheres. Using a standard "lifting" transformation (see e.g. Aurenhammer [6]), one can transform this problem to that of analyzing the complexity of a single cell in an arrangement of hyperplanes in 4-space. This implies an $O(n^2)$ bound on the complexity of a single cell, in support of the above conjecture.

In $d > 3$ dimensions the general situation is even worse. In the case of hyperplanes, the Upper Bound Theorem implies that the complexity of a single cell is $O(n^{\lfloor d/2 \rfloor})$. In the case of spheres, the same transormation just mentioned enables us to obtain a bound of $O(n^{\lfloor (d+1)/2 \rfloor})$. In the general case, we venture the conjecture that the complexity of a single cell is at most close to $O(n^{d-1})$, perhaps $O(n^{d-2}\lambda_s(n))$ for some constant s. This appears to be completely open.

Remark. In comparing the conjectured bounds on the complexity of a single cell or of the envelope with the overall complexity of the arrangement, we note that we aim to gain roughly a factor of $O(n)$. This saving is very substantial in low dimensions, but loses some of its appeal as the dimension d increases.

Finally we consider the problem of bounding the complexity of m distinct cells in an arrangement of n surfaces in 3 or more dimensions. In the relatively simple case of planes in 3-space, it was shown by Edelsbrunner, Guibas and Sharir [32] that the complexity in question is $O(m^{2/3}n \log n + n^2)$. This has been subsequently improved by Aronov and Agarwal [1] to $O(m^{2/3}n + n^2)$, which is tight in the worst case. Similar results have been obtained for arrangements of hyperplanes in higher dimensions [32]. For other surfaces much less is known. The technique of Aronov and Sharir [4] yields a bound on this complexity, but only if the m desired cells include all non-convex cells. The simpler problem involving incidences between m points and n surfaces has been studied in [20] for the case of spheres in 3-space, where the bound $O(m^{3/4}n^{3/4}\beta(m, n) + m + n)$ has been derived, where β is a slowly growing function related to the inverse Ackermann function. This has led to an improved bound of $O(n^{3/2}\beta(n, n))$ on the maximum number of unit distances among n points in 3-space [20]. See also [32] for other results involving incidences and related problems in higher dimensions.

6. Summary

In this review paper we have discussed many combinatorial and algorithmic questions about arrangements of low-degree algebraic surfaces in low-dimensional Euclidean spaces. We hope that the reader has obtained a feel for some of the beautiful mathematical tools used, such as Davenport-Schinzel sequences, random sampling and ϵ-nets, etc., as well as for some of the most important applications of this theory to problems in robotics and other areas of geometry.

Acknowledgements. Work on this chapter has been supported by a grant from the U.S.–Israel Binational Science Foundation. Work by the second author has also been supported by National Science Foundation Grants CCR-89-01484 and CCR-91-22103, and by grants from the Basic Research Fund of the Israeli National Academy of Sciences, and the G.I.F. – the German-Israeli Foundation for Scientific Research and Development.

References

1. P. Agarwal and B. Aronov: Counting facets and incidences. Discrete Comput. Geom. **7** (1992) 359–369
2. P. Agarwal, M. Sharir and P. Shor: Sharp upper and lower bounds for the length of general Davenport-Schinzel sequences. J. Combin. Theory, Ser. A **52** (1989) 228–274
3. B. Aronov, H. Edelsbrunner, L. Guibas and M. Sharir: Improved bounds on the complexity of many faces in arrangements of segments. Combinatorica (in press)
4. B. Aronov and M. Sharir: Triangles in space, or building (and analyzing) castles in the air. Combinatorica **10** (1990) 137–173
5. M. Atallah: Some dynamic computational geometry problems. Comp. and Math. with Appls. **11** (1985) 1171–1181
6. F. Aurenhammer: Power diagrams: Properties, algorithms, and applications. SIAM J. Comput. **16** (1987) 78–96
7. C. Bajaj: Geometric modeling with algebraic surfaces. The Mathematics of Surfaces, III (D. Handscomb, ed.). Oxford Univ. Press, 1989, pp. 3–48
8. A. Baltsan and M. Sharir: On shortest paths between two convex polyhedra. J. ACM **35** (1988) 267–287
9. J. Bentley and T. Ottmann: Algorithms for reporting and counting geometric intersections. IEEE Trans. Computers **C-28** (1979) 643–647
10. R. Canham: A theorem on arrangements of lines in the plane. Israel J. Math. **7** (1969) 393–397
11. B. Chazelle: The polygon containment problem. In: Advances in Computing Research, Vol. I: Computational Geometry (F.P. Preparata, Ed.), JAI Press, Greenwich, Connecticut (1983), pp. 1–33
12. B. Chazelle and J. Friedman: A deterministic view of random sampling and its use in geometry. Combinatorica **10** (1990) 229–249
13. B. Chazelle and H. Edelsbrunner: An optimal algorithm for intersecting line segments in the plane. J. ACM **39** (1992) 1–54
14. B. Chazelle, H. Edelsbrunner, L. Guibas, and M. Sharir: A singly exponential stratification scheme for real semi-algebraic varieties and its applications. Proc. 16th Int. Colloq. on Automata, Languages, and Programming 1989, pp. 179–193. Also in: Theoretical Computer Science **84** (1991) 77–105
15. B. Chazelle, H. Edelsbrunner, L. Guibas, and M. Sharir: Lines in space — combinatorics, algorithms, and applications. Proc. 21st ACM Symp. on Theory of Computing 1989, pp. 568–579
16. B. Chazelle, L. Guibas and D.T. Lee: The power of geometric duality. BIT **25** (1985) 76–90
17. B. Chazelle and D.T. Lee: On a circle placement problem. Computing **36** (1986) 1–16
18. L.P. Chew and K. Kedem: Placing the largest similar copy of a convex polygon among polygonal obstacles. Proc. 5th ACM Symp. on Computational Geometry, 1989, pp. 167-174

19. K.L. Clarkson: New applications of random sampling in computational geometry. Discrete Comput. Geom. **2** (1987) 195–222

20. K. Clarkson, H. Edelsbrunner, L. Guibas, M. Sharir and E. Welzl: Combinatorial complexity bounds for arrangements of curves and spheres. Discrete Comput. Geom. **5** (1990) 99–160

21. K. Clarkson and P. Shor: Applications of random sampling in computational geometry II. Discrete Comput. Geom. **4** (1989) 387–421

22. R. Cole and M. Sharir: Visibility problems for polyhedral terrains. J. Symbolic Computation **7** (1989) 11–30

23. G.E. Collins: Qunatifier elimination for real closed fields by cylindric algebraic decomposition. 2nd GI Conf. Aut. Theory and Formal Lang., Springer-Verlag, LNCS **33**, Berlin (1975) 134–183

24. H. Davenport: A combinatorial problem connected with differential equations, II. Acta Arithmetica **17** (1971) 363–372

25. H. Davenport and A. Schinzel: A combinatorial problem connected with differential equations. Amer. J. Math. **87** (1965) 684–694

26. L. De Floriani, B. Falcidieno, C. Pienovi, D. Allen and G. Nagy: A visibility-based model for terrain features. Proc. Int. Symp. on Spatial Data Handling, Seattle, July 1986

27. H. Edelsbrunner: The upper envelope of piecewise linear functions: Tight bounds on the number of faces. Discrete Comput. Geom. **4** (1989) 337–343

28. H. Edelsbrunner and L. Guibas: Topologically sweeping an arrangement. J. Comp. and System Sciences, **38** (1989) 165–194

29. H. Edelsbrunner, L. Guibas, J. Pach, R. Pollack, R. Seidel and M. Sharir: Arrangements of curves in the plane: Topology, combinatorics and algorithms. Theoretical Computer Science **92** (1992) 319–336

30. H. Edelsbrunner, L. Guibas and M. Sharir: The complexity and construction of many faces in arrangements of lines and of segments. Discrete Comput. Geom. **5** (1990) 161–196

31. H. Edelsbrunner, L. Guibas and M. Sharir: The upper envelope of piecewise linear functions: Algorithms and applications. Discrete Comput. Geom. **4** (1989) 311–336

32. H. Edelsbrunner, L. Guibas and M. Sharir: The complexity of many cells in arrangements of planes and related problems. Discrete Comput. Geom. **5** (1990) 197–216

33. H. Edelsbrunner, J. O'Rourke and R. Seidel: Constructing arrangements of lines and hyperplanes with applications. SIAM J. Computing **15** (1986) 341–363

34. P. Erdős: On sets of distances of n points. Amer. Math. Monthly **53** (1946) 248–250

35. P. Erdős: On sets of distances of n points in euclidean space. Publ. Math. Inst. Hung. Acad. Sci. **5** (1960) 165–168

36. L.J. Guibas and J. Stolfi: Primitives for the manipulation of general subdivisions and the computation of Voronoi diagrams. ACM Trans. Graphics 4 (1985) 74–123

37. L. Guibas, M. Sharir and S. Sifrony: On the general motion planning problem with two degrees of freedom. Discrete Comput. Geom. **4** (1989) 491–521

38. D. Halperin and M. Sharir: Improved combinatorial bounds and efficient techniques for certain motion planning problems with three degrees of freedom. Computational Geometry: Theory and Applications **1** (1992) 269–303

39. S. Hart and M. Sharir: Nonlinearity of Davenport–Schinzel sequences and of generalized path compression schemes. Combinatorica **6** (1986) 151–177

40. J. Hershberger: Finding the upper envelope of n line segments in $O(n \log n)$ time. Inf. Proc. Letters **33** (1989) 169–174

41. D. Haussler and E. Welzl: ϵ-nets and simplex range queries. Discrete Comput. Geom. **2** (1987) 127–151

42. K. Kedem, R. Livne, J. Pach and M. Sharir: On the union of Jordan regions and collision–free translational motion amidst polygonal obstacles. Discrete Comput. Geom. **1** (1986) 59–71

43. K. Kedem and M. Sharir: An efficient motion planning algorithm for a convex polygonal object in 2-dimensional polygonal space. Discrete Comput. Geom. **5** (1990) 43–75

44. D. Leven and M. Sharir: On the number of critical free contacts of a convex polygonal object moving in 2-dimensional polygonal space. Discrete Comput. Geom. **2** (1987) 255–270

45. D. Leven and M. Sharir: An efficient and simple motion planning algorithm for a ladder moving in two-dimensional space amidst polygonal barriers. J. Algorithms **8** (1987) 192–215

46. J. Matoušek: Construction of epsilon nets. Discrete Comput. Geom. **5** (1990) 427–448

47. J. Matoušek: Cutting hyperplane arrangements. Discrete Comput. Geom. **6** (1991) 385–406

48. M. McKenna: Worst-case optimal hidden-surface removal. ACM Trans. Graphics **6** (1987) 19–28

49. C. Ó'Dúnlaing, M. Sharir and C.K. Yap: Generalized Voronoi diagrams for a ladder: II. Efficient construction of the diagram. Algorithmica **2** (1987) 27–59

50. J. Pach and M. Sharir: The upper envelope of piecewise linear functions and the boundary of a region enclosed by convex plates: Combinatorial analysis. Discrete Comput. Geometry **4** (1989) 291–309

51. J. Pach and M. Sharir: On vertical visibility in arrangements of segments and the queue size in the Bentley-Ottmann line sweeping algorithm. SIAM J. Computing **20** (1991) 460–470

52. M. Pellegrini: Stabbing and ray-shooting in 3-dimensional space. Proc. 6th ACM Symp. on Computational Geometry, 1990, pp. 177–186

53. R. Pollack, M. Sharir and S. Sifrony: Separating two simple polygons by a sequence of translations. Discrete Comput. Geom. **3** (1988) 123–136

54. F.P. Preparata and D.E. Muller: Finding the intersection of n half spaces in time $O(n \log n)$. Theoretical Computer Science **8** (1979) 44–55

55. J.T. Schwartz and M. Sharir: On the two-dimensional Davenport-Schinzel problem. J. Symbolic Computation **10** (1990) 371–393

56. M. Sharir: Almost linear upper bounds on the length of general Davenport-Schinzel sequences. Combinatorica **7** (1987) 131–143

57. M. Sharir: Improved lower bounds on the length of Davenport-Schinzel sequences. Combinatorica **8** (1988) 117–124

58. M. Sharir and R. Livne: On Minima of Functions, Intersection Patterns of Curves, and Davenport-Schinzel Sequences. Proc. 26th IEEE Symp. on Foundations of Computer Science, 1985, pp. 312–320

59. P. Shor: Simplified geometric realizations of superlinear Davenport-Schinzel sequences. Manuscript 1990

60. S. Sifrony and M. Sharir: An efficient motion planning algorithm for a rod moving in two-dimensional polygonal space. Algorithmica **2** (1987) 367–402
61. J. Spencer, E. Szemerédi and W. Trotter: Unit distances in the Euclidean plane. In: Graph Theory and Combinatorics (Proc. Cambridge Conf. on Combinatorics, B. Bollobas, ed.), 293–308, Academic Press, 1984
62. E. Szemerédi: On a problem by Davenport and Schinzel. Acta Arithmetica **25** (1974) 213–224
63. E. Szemerédi and W. Trotter: Extremal problems in discrete geometry. Combinatorica **3** (1983) 381–392
64. A. Wiernik and M. Sharir: Planar realization of non-linear Davenport-Schinzel sequences by segments. Discrete Comput. Geom. **3** (1988) 15–47

Chapter II. Backwards Analysis of Randomized Geometric Algorithms

Raimund Seidel

Abstract. The theme of this chapter is a rather simple method that has proved very potent in the analysis of the expected performance of various randomized algorithms and data structures in computational geometry. The method can be described as "analyze a randomized algorithm as if it were running backwards in time, from output to input." We apply this type of analysis to a variety of algorithms, old and new, and obtain solutions with optimal or near optimal expected performance for a plethora of problems in computational geometry, such as computing Delaunay triangulations of convex polygons, computing convex hulls of point sets in the plane or in higher dimensions, sorting, intersecting line segments, linear programming with a fixed number of variables, and others.

1. Introduction

The curious phenomenon that randomness can be used profitably in the solution of computational tasks has attracted a lot of attention from researchers in recent years. The approach has proved useful in such diverse areas as number theory, distributed computing, combinatorial algorithms, complexity theory, and others. For surveys see [34,47,55]. It is interesting that in Rabin's seminal 1976 paper [46] which initiated the study of algorithmic uses of randomness one of the two example problems considered was a computational geometry problem, namely the Euclidean closest pair problem. Of course computational geometry as a field came about a number of years later. Shamos's thesis [52] appeared in 1978. Talking about possible future research directions in the epilogue Shamos mentions "probabilistic algorithms" and writes *"This approach seems to be able to yield geometric algorithms of startling efficiency."* However, it was to take almost another decade until randomized or probabilistic methods were investigated in computational geometry in more detail.

In the mid 80's Ken Clarkson started to create and apply his random sampling technique, which in the mean time has developed into a surprising general framework with numerous applications [15,16,17,18]. Around the same time Haussler and Welzl published their important paper [32] that introduced ε-nets and the VC-dimension, which have become very useful and versatile tools in the design and analysis of randomized algorithms. In a series of papers [42,43,44] Mulmuley introduced and studied a number of probabilistic games that allow a rather tight analysis of the expected behaviour of a number of geometric algorithms. At the same time a steady stream of papers of a more specialized nature started to appear, dealing with randomized solutions for a wide range of particular problems ([7,8,9,30,1,50,12] is a non-exhaustive, random(?) list of references).

The purpose of this chapter is to popularize a rather simple trick for analyzing the expected performance of certain randomized algorithms:

Analyze an algorithm as if it were running backwards in time, from output to input.

This is based on the observation that often the cost of the "last" step of an algorithm can be expressed as a function of the complexity of the "final product" output by the algorithm.

In this chapter we apply this "backwards analysis" to a number of problems and algorithms. We start with demonstrating the idea of backwards analysis on a simple algorithm due to Paul Chew [14] for constructing the Delaunay triangulation of the vertices of a convex polygon and show that it has linear expected running time. As far as we know Chew was the first to apply this type of analysis to a computational geometry algorithm.

Next we apply backwards analysis to an algorithm due to Mulmuley for determining all intersection pairs among a set of line segments in the plane [42, 43]. Mulmuley's original analysis of the algorithm was based on probabilistic games and was rather involved. The "backwards" view leads to a considerable simplification, and also applies equally well to a more general version of the problem, where the segments need not be straight and can intersect each other more than once.

Some claim — tongue in cheek — that any method of value in computational geometry must be also applicable to the planar convex hull problem. Thus we present a planar convex hull algorithm along with the analysis of its $O(n \log n)$ expected running time. The algorithm is somewhat reminiscent of QUICKSORT. Thus we apply the principle of backwards analysis to QUICKSORT and with little effort we derive the *exact* value for the expected number of comparisons made by QUICKSORT. Moreover, backwards analysis turns out to provide a particularly easy approach for bounding the probability that the running time of QUICKSORT significantly exceeds its expectation.

Next we give a negative example. We consider a triangulation problem along with an algorithm that is a straightforward generalization of QUICKSORT. Interestingly enough, backwards analysis can apparently not be applied to this algorithm.

Following this we turn our attention to linear programming when the dimension is small. We present a very simple algorithm for solving linear programs with m constraints in d variables that has expected running time

$O(d!m)$. Again the analysis via the backwards view is very straightforward. We then describe an adaption of this method due to Welzl [56] for the problem of finding smallest enclosing balls for a finite set of points in $\mathbb{R}^d$.

Finally we turn our attention to the problem of constructing the convex hull of n points in $\mathbb{R}^d$. We consider a randomized incremental algorithm and show that for $d > 3$ there is a simple variant that via backwards analysis can easily be shown to have optimal $O(n^{\lfloor d/2 \rfloor})$ expected running time. Then we consider the "conflict graph" based algorithm due to Clarkson and Shor [18] and present a new backwards analysis due to Clarkson [21] that shows that this algorithm has "optimal" expected running time for all dimensions d.

2. Delaunay Triangulations of Convex Polygons

Let S be a set of n points in the plane. The *Delaunay triangulation of S*, for short $DT(S)$, is a plane graph whose vertices are the points in S and that connects two points $p, q \in S$ by a straight edge iff there is an open disk that has p and q on its boundary but that contains no points of S. When S is non-degenerate in the sense that no four points of S are co-circular and not all of S lies on one straight line, $DT(S)$ is always a triangulation. The bounded faces of $DT(S)$ are then exactly those triangles of points in S for which the smallest circumscribed disk contains no point of S in its interior.

Delaunay triangulations along with their dual structures, which are known as Voronoi diagrams, have been studied intensively in computational geometry. Their efficient construction and the recognition of their multifarious useful properties by Shamos and Hoey [53] were instrumental in getting the field of computational geometry started. By now these structures along with numerous generalizations are standard fare in the field (see [45,25,3,36]).

Shamos's and Hoey's big feat was an $O(n \log n)$ algorithm for the construction of Delaunay triangulations. It was also very soon realized that the $O(n \log n)$ bound was asymptotically worst case optimal for reasonable models of computation. However, for a long time the question remained whether $DT(S)$ could be computed in $o(n \log n)$ time when the point set S has some special structure. In particular one was interested in the question whether an $O(n)$ time bound was possible when S consists of the vertices of a convex polygon, given in order around the polygon. An affirmative answer was eventually given in 1986 by Aggarwal, Guibas, Saxe, and Shor [2] using an ingenious but rather involved algorithm.

In the mean time, almost unnoticed, Paul Chew had discovered a very simple *randomized* algorithm for this problem with linear expected running time [14]. Chew employed backwards anlysis, and as far as we know this was the first time that this trick was used in computational geometry. His algorithm and analysis shall serve as a first example for the concept of backwards analysis.

So let S be the n vertices of a convex polygon P given in order around P. Assume that no four points in S are co-circular, a condition that could easily be simulated using standard perturbation techniques [25, pp. 185]. Chew's algorithm proceeds as follows:

> If S consists of only three points, then the triangle spanned by them forms $DT(S)$.
>
> If S contains more than three points, then choose a random point $q \in S$, let p and r be its two neighboring vertices around P, and let $S' = S \setminus \{q\}$. Recursively compute $DT(S')$, attach the triangle p, q, r to this triangulation, and then update this triangulation D of S as follows to obtain $DT(S)$:
>
> First identify all "bad" triangles of D, namely p, q, r and all triangles of $DT(S')$ whose circumscribed disk contains q. This is done by performing a depth-first search in the dual graph G of D whose nodes are the triangles of D and that has two triangles adjacent iff they share a common edge of D. This depth-first search is to start at the triangle p, q, r. Since the set of bad triangles is known to form a connected subgraph of G (a fact not proven here) and since G has maximum degree 3, all bad triangles can thus be identified in time proportional to their number.
>
> Finally remove from D all edges that have bad triangles on both sides, and retriangulate the resulting face by introducing all diagonals that have q as an endpoint. (See Figure 1 for an illustration.)

What is the expected running time of this algorithm? The question really is, what is the expected running time of this procedure with the recursive call excluded. It is not hard to see that this cost is proportional to the number of bad triangles. So what is the expected number of bad triangles?

This question seems difficult to answer, especially if we fix our attention on one particular point q. The trick now is to express this cost not in terms of $DT(S')$ and q, but *in terms of the resulting structure* $DT(S)$. It is not hard to· see that the number of bad triangles is exactly one more than the number of

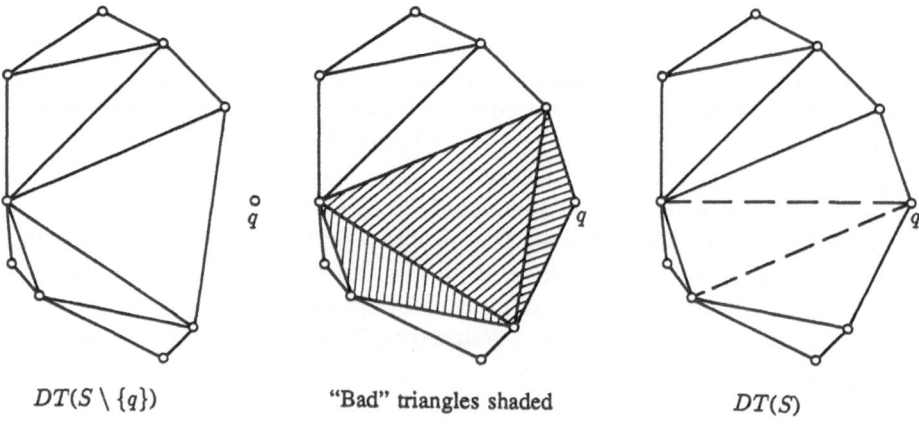

$DT(S \setminus \{q\})$ "Bad" triangles shaded $DT(S)$

Fig. 1

diagonals with endpoint q that are introduced in the last step of the algorithm. Or in other words, the number of bad triangles is proportional to the degree of q in $DT(S)$. But of course, if q is chosen from S uniformly at random, then the expected degree of q in $DT(S)$ is the sum of the degrees of all vertices in S divided by n, which is twice the number of edges of $DT(S)$ divided by n, which, since $DT(S)$ is an outer-planar graph, is $2 * (2n - 3)/n = 4 - 6/n$.

Thus the expected time necessary to perform the body of the procedure outlined above without the recursive call is constant. From this it follows immediately that the overall expected time necessary is $O(n)$.

Let us point out once more that the decisive part in the analysis of the algorithm is to express the cost of the "last step" (which we abstracted out as the body of a head-recursive procedure) as a function of the produced output. If we were to run the algorithm backwards starting with $DT(S)$ and repeatedly deleting random points of S at cost proportional to their degree, the expected cost of a deletion could be expressed as a function of the "input" (namely the average degree of the current triangulation) and it would be clear that this is constant.

3. Intersecting Line Segments

Let S be a set of n straight line segments in the plane. We are interested in finding all intersecting pairs of segments in S.

The first non-trivial algorithm for solving this problem was given by Bentley and Ottmann in 1979 [5]. It was based on the sweep paradigm and achieved a worst case running time of $O((K + n) \log n)$, where K is the output size, namely the number of intersecting pairs of segments in S. Since it is possible that all segments of S intersect each other, this algorithm can have an $O(n^2 \log n)$ running time, which is inferior to the $O(n^2)$ time of the trivial method of checking every pair in S. Thus the question arose, whether a bound of the form $O(K + n \log n)$ was possible. Since it is easy to show by reduction from element uniqueness that $\Omega(n \log n)$ is a lower bound to the segment intersection problem, and since $\Omega(K)$ is also a lower bound as at least this much time has to be spent on output, one could not hope for anything better than $O(K + n \log n)$.

In 1983 Chazelle [10] came close to this goal with a rather complicated algorithm whose worst case running time was $O(K + n \log^2 n / \log \log n)$. Finally, five years later he and Edelsbrunner [11] designed an even more complicated deterministic algorithm that did achieve the $O(K + n \log n)$ worst case running time. Around the same time independently Mulmuley [42] as well as Clarkson and Shor [18] developed rather simple randomized algorithms with $O(K + n \log n)$ expected running time. Clarkson and Shor based the analysis of the running time of their algorithm on the general theory of random sampling. They even managed to come up with a version of the algorithm that required only $O(n)$ space. Mulmuley analyzed the performance of his algorithm via probabilistic games that he developed for this purpose. His analysis is reasonably complex, however it yields rather tight constants.

In this section we present Mulmuley's algorithm and give a very simple analysis of its expected performance that is based on our backwards view.

For the sake of ease of presentation let us assume we are dealing with a set S of n segments that is non-degenerate in the sense that no two segments of S have the same endpoint, no three segments intersect in a common point, no two segment endpoints have the same x-coordinate, and that no segment endpoint lies in the relative interior of some other segment. As usual such non-degeneracy could be simulated by standard perturbation methods [25, pp. 185], or also the algorithm could easily be modified so that none of these assumptions are necessary.

Mulmuley's algorithm does more than just determine which pairs of segments in S intersect. It constructs what we call the *trapezoidal decomposition* induced by S. This decomposition $\mathcal{T}(S)$ can be intuitively defined as follows: First draw a sufficiently large axis-parallel rectangle frame F that contains in its interior all segments of S. Next draw all segments of S in the rectangle F. Finally, from each intersection point and from each segment endpoint draw its *vertical extensions*, i.e. start drawing two vertical rays, one going up, the other going down, that extend until they hit a segment of S or the boundary of F (see Figure 2). Thus F is decomposed into trapezoids that each have two vertical sides (one of which can have length 0). Using a sweep argument it is not hard to prove that $\mathcal{T}(S)$ contains exactly $3(n + K) + 1$ trapezoids; thus, when viewed as a planar graph, $\mathcal{T}(S)$ has $O(K + n)$ faces, edges, and vertices.

Let us define for any subset $R \subset S$ the trapezoidal decomposition $\mathcal{T}_S(R)$ in a similar way as follows: Draw all segments of R in the rectangle frame F, and for each intersection point of segments in R as well as for each endpoint of a segment in S (note: *in S*) draw its vertical extensions. But now the rays that form the vertical extensions extend only until they hit a segment of R or the boundary of the rectangle F (see Figure 3). The decomposition $\mathcal{T}_S(R)$ partitions F into $2n + r + 3K_R + 1$ trapezoids, where $r = |R|$ and K_R is the number of pairs of intersecting segments in R, and thus $\mathcal{T}_S(R)$ as a planar graph has $O(n + K_R)$ faces, edges, and vertices.

Mulmuley's algorithm for computing $\mathcal{T}(S) = \mathcal{T}_S(S)$ is very simple: First compute $\mathcal{T}_S(\emptyset)$ and then insert the segments of S into the trapezoidation in random order to compute $\mathcal{T}_S(R)$ for an ever increasing $R \subset S$.

We will express his algorithm for computing $\mathcal{T}_S(R)$ recursively:

If $R = \emptyset$ then compute $\mathcal{T}_S(R)$ directly by sorting the endpoints of S by their x-coordinates.

Otherwise randomly pick a segment $s \in R$, recursively compute $\mathcal{T}_S(R \setminus \{s\})$, and introduce s into this trapezoidation to obtain $\mathcal{T}_S(R)$.

We still have to specify how a segment s is introduced into the current trapezoidation. In order to do this we need to give more detail about the representation of trapezoidations. Mulmuley chooses a somewhat idiosyncratic

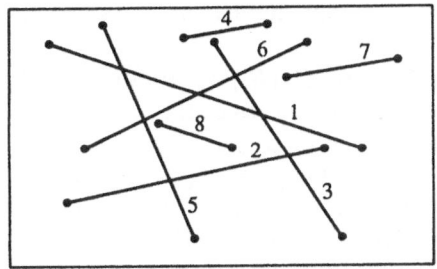

Fig. 2a. Set S of 8 segments in a frame

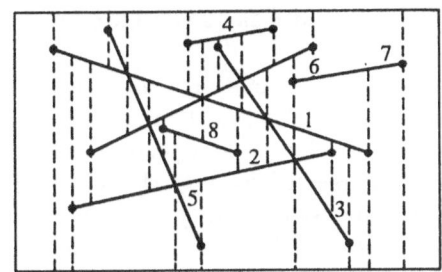

Fig. 2b. Trapezoidation $\mathcal{T}(S)$

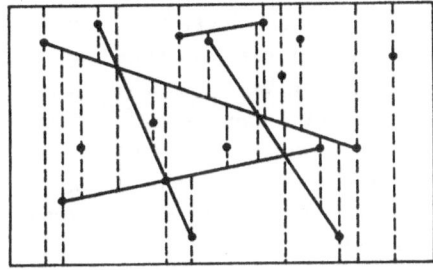

Fig. 3. $\mathcal{T}_S(R)$ for $R = \{1,2,3,4,5\}$

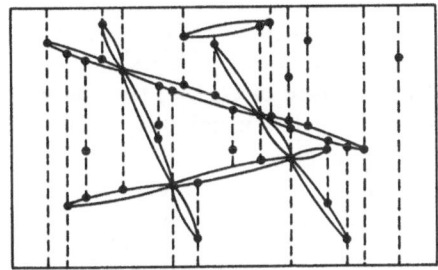

Fig. 4. Introduction of zero-width faces for pieces yields $G_S(R)$

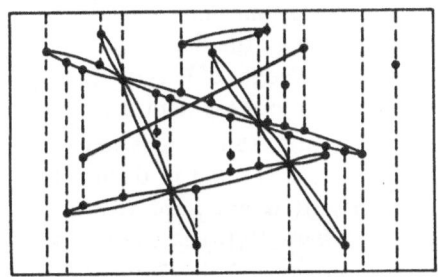

Fig. 5a. Inserting segment 6

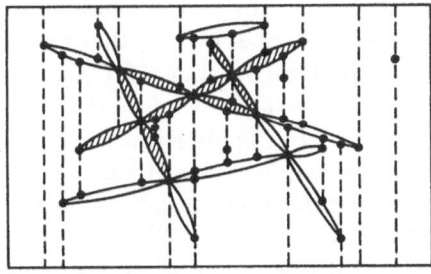

Fig. 5b. $\mathcal{P}_R(6)$ shaded

representation that essentially works as follows: Consider $\mathcal{T}_S(R)$, and consider some segment $s \in R$. Assume s is intersected by i other segments in R. Thus s is partitioned into $i+1$ *pieces*. Conceptually Mulmuley makes each piece of each segment in R into a narrow zero-width face, and obtains this way from $\mathcal{T}_S(R)$ a plane graph that we denote by $G_S(R)$. Figure 4 should make the idea clear. Note that $G_S(R)$ has the same number of vertices as $\mathcal{T}_S(R)$ and thus it also has complexity $O(n + K_R)$. Also note that in the graph $G_S(R)$ faces that correspond to trapezoids of $\mathcal{T}_S(R)$ have at most six edges around them, only the zero-width faces can have more than a constant number of edges around them.

How does one now introduce segment s into $G_S(R')$ to obtain $G_S(R)$, where $R' = R \setminus \{s\}$? It works in two phases (that of course could be combined into one). In the first phase one determines which faces and edges of $G_S(R')$ are intersected by s. In the second phase the new faces of $G_S(R)$ are created. This involves splitting the faces of $G_S(R')$ that are intersected by s, creating the zero-width faces for the pieces of s, introducing the vertical extensions from the intersection points of s with the other segments of R, and merging faces that are separated by vertical edges that are not part of vertical extensions any more because of the introduction of s. We leave the details of the second phase to the reader.

The first phase can simply be done as follows: Since the two endpoints of s are vertices of $G_S(R')$ already, we can determine in constant time, say, the leftmost face of $G_S(R')$ that is intersected by s. Now we "thread" s through $G_S(R')$ in the usual way, similar to the incremental line arrangement construction algorithm [27,13]: We walk along the segment s; assume we just entered a face f through some edge e; we determine through which edge the segment s leaves f and which new face the segment enters by simply testing all edges of f (say, in clockwise order starting after e). Now we have reached a new face, and we repeat. Obviously, this procedure can also tell when we have reached the right endpoint of s.

What is the time necessary for introducing segment s? It is not hard to see that the cost of phase one dominates the cost of phase two. Thus it suffices to consider just the cost of phase one. This cost is clearly the sum of the degrees of all the faces of $G_S(R')$ that are intersected by s (here the degree of a face f of $G_S(R')$, for short $deg(f, G_S(R'))$, is the number of edges of $G_S(R')$ incident to f). However, what does this evaluate to for a random segment $s \in R$?

Now let us apply backwards analysis. The first important step is to express the cost of phase one of introducing a segment s into $G_S(R')$ in terms of the result graph $G_S(R)$ and not in terms of $G_S(R')$. It is not hard to see that this cost is given by $\sum_{f \in \mathcal{P}_R(s)} deg(f, G_S(R))$, where $\mathcal{P}_R(s)$ is the set of all zero-width faces of $G_S(R)$ that either derive from pieces of s in $G_S(R)$ or from those pieces of other segments in R that are incident to intersection points with s (see Figure 5).

It follows that if s is randomly chosen from the r segments in R, then the expected cost of adding s to $G_S(R \setminus \{s\})$ is proportional to

$$\frac{1}{r} \sum_{s \in R} \sum_{f \in \mathcal{P}_R(s)} deg(f, G_S(R)) \ .$$

But in this double sum every piece of a segment in R contributes at most three times. Thus if $\mathcal{P}(R)$ denotes the set of all faces that derive from pieces of segments in R, then this double sum is at most

$$\frac{3}{r} \sum_{f \in \mathcal{P}(R)} deg(f, G_S(R)) .$$

Since $G_S(R)$ is a planar graph, this sum is clearly proportional to the complexity of the graph, and thus it is $O(n + K_R)$. It follows that the expected cost of introducing the last segment of R is $O(\frac{n}{r} + \frac{K_R}{r})$.

But what is the expected value of K_R? By the way the algorithm proceeds it is clear that R is a random subset of S of size r. Now if $\{s, t\}$ is one of the K pairs of intersecting segments in S, what is the probability that they are both in R? Clearly $\frac{r(r-1)}{n(n-1)}$. It follows that the expected number of pairs of intersecting segments in R, i.e. the expectation K_r of K_R is $\frac{r(r-1)}{n(n-1)}K$.

Thus the expected cost of introducing the last segment into $G_S(R)$ is $O(\frac{n}{r} + \frac{r-1}{n(n-1)}K)$. To obtain the expected cost for all recursive calls of the entire algorithm one clearly only needs to sum this expression for $1 \le r \le n$, which yields

$$O(nH_n + K),$$

where $H_n = 1 + 1/2 + \ldots + 1/n \approx \log n$. Since computing $G_S(\emptyset)$ just amounts to sorting $2n$ numbers it follows that the expected running time of the entire algorithm is $O(K + n \log n)$.

Remarks. As an exercise the reader may want to try this type of analysis on a version of this algorithm that does not use zero-width faces and uses instead of $G_S(R)$ simply a planar graph representation of $T_S(R)$. Thus the trapezoids can have arbitrarily many edges around them.

In the presentation of Mulmuley's algorithm we assumed non-degeneracy. This is not too much of an issue, except in cases where many segments intersect in one point. In such a situation it would be desirable to obtain an expected running time of $O(I + n \log n)$, where is I is the number of intersection points between segments. (Note that if all segments intersect in one point, then $I = 1$ but $K = \binom{n}{2}$.) By fairly obvious modifications of the algorithm outlined above it is possible to achieve this $O(I + n \log n)$ bound. The only complications arise in the analysis, since I_r, the analogue of K_r, seems to be difficult to express in a nice closed form. However, the following can be shown and saves the analysis: If X is the set of intersection points between segments of S and if for $p \in X$ the number of segments of S that intersect in p is denoted by $d(p)$, then[1]

$$\sum_{1 \le r \le n} \frac{I_r}{r} = \sum_{p \in X} (H_{d(p)} - 1) .$$

But since $\sum_{p \in X} d(p) = O(I + n)$, clearly $\sum_{p \in X} (H_{d(p)} - 1)$ and therefore also $\sum_{1 \le r \le n} I_r/r$ is $O(I + n)$.

Finally we should point out that the algorithm described here does not exploit at all the straightness of the segments in S or the fact that any two segments intersect at most once. With very straightforward modifications the algorithm can be adapted to construct the trapezoidal decomposition induced by a set S of "segments," where each member $s \in S$ is a bounded x-monotone curve (i.e. every vertical line intersects s at most once), where every pair $s, s' \in S$ intersect in a finite number of points, and where for any vertical line ℓ one can determine in $O(1)$ time the "first" intersection point between s and s' to the right of ℓ. No changes in the analysis are necessary at all. It is still $O(I + n \log n)$, where I now stands for the number of intersection points and can be arbitrarily large.

4. Constructing Planar Convex Hulls

The convex hull $conv\, S$ of a set S of n points in the plane is the smallest convex polygon that contains S. Computing such a convex hull amounts to determining the circular sequence of points in S that constitute the corners around the polygon $conv\, S$.

The planar convex hull problem has received an extraordinary amount of attention in the computational geometry literature. We will not attempt to give a complete history here, but just list three mile stones: in 1972 Graham gave the first $O(n \log n)$ algorithm [29]; in 1978 Bentley and Shamos showed that for a large class of geometric distributions the convex hull of a set of n points drawn according to such a distribution could be computed in $O(n)$ expected time [6] (here the expectation is with respect to the input distribution); in 1983 Kirkpatrick and Seidel gave an algorithm whose worst case running time also depends on the output-size and comes to $O(n \log H)$, where H is the number of corners of the output polygon [35].

The attractiveness of the planar convex hull problem to computational geometers stems partly from the fact that most computational paradigms can

---

[1] The derivation of this formula — at least as done by the author — is not completely straightforward and requires some massaging of sums involving quotients of binomial coefficients. In particular one needs to show that

$$\sum_{1 \le r \le n} \frac{1}{r} \left[1 - \frac{\binom{n-d}{r}}{\binom{n}{r}} - d \frac{\binom{n-d}{r-1}}{\binom{n}{r}} \right] = H_d - 1 ,$$

where the quantity in the square brackets is the probability that an intersection point among d segments of S exists in a random sample of r segments.

be successfully applied to this problem. This is also the case with our paradigm of backwards analysis. In this section we present a randomized algorithm and anlyze its expected running time exploiting this backwards view. I am not sure whom to attribute this algorithm to. It certainly owes a lot to the conflict graph based algorithm of Clarkson and Shor [18], and it seems to have grown out of discussions among several researchers at a DIMACS workshop in the fall of 1989.

For the sake of ease of presentation we will again assume that we are dealing with a set S of n points in non-degenerate position, which in this case is to mean that no three points in S are colinear. Moreover, we will assume that $n > 2$. Our algorithm again works incrementally. It puts the points of S in a random order and then computes the convex hulls of ever increasing subsets of S. We will again describe the algorithm recursively. In this description we will assume the existence of the following kind of oracle: If for $T \subset S$ we know $P_T = conv\, T$, then the oracle can tell for any point $q \in S \setminus T$ whether q is contained in P_T, and if q is not contained, the oracle can tell one edge of P_T that is "visible" from q, i.e. the straight line through that edge separates P_T and q.

Here is the algorithm for computing $P_R = conv\, R$, where $R \subset S$.

If $|R| = 3$ then P_R is the triangle formed by the three points in R.
If $|R| > 3$ then randomly choose a point q from R, and let $R' = R \setminus \{q\}$.
Recursively compute $P_{R'} = conv\, R'$, and then "insert" point q as follows to obtain $P_R = conv\, R$:
Query the oracle about q and $P_{R'}$. If q is contained in $P_{R'}$, then $P_R = P_{R'}$ and nothing needs to be done.
Otherwise, the oracle returns an edge e of $P_{R'}$ that is visible form q. Starting at e perform a search to determine all edges of $P_{R'}$ that are visible from q. These edges form a chain. To obtain P_R replace that chain by a new chain of two edges that have q as common endpoint (see Figure 6).

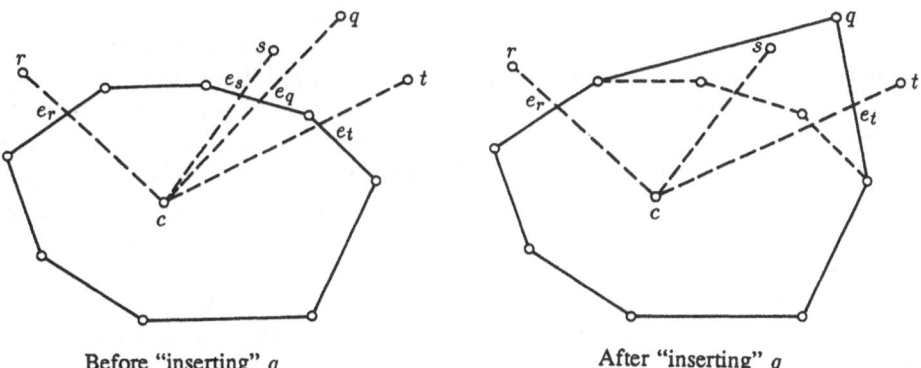

Before "inserting" q After "inserting" q

Fig. 6

If one is allowed to disregard the costs incurred by the oracle, then by the following amortization argument the running time of this algorithm is easily seen to be $O(n)$. The cost of computing P_R from $P_{R'}$, provided they are different, is clearly proportional to the number of edges of $P_{R'}$ that are found to be visible from q. There can be a large number of such edges. However, as they are all deleted and never re-appear, one can charge this cost to the creation of those edges. But whenever a point of S is added at most two new edges are created, and thus the overall cost for all creations and deletions and hence for the entire algorithm (without the cost of the oracle calls) is $O(n)$.

How can one implement the oracle? The idea is to maintain for each point p in $S \setminus R$ a canonical edge e_p of P_R that is visible from p (or to note that no visible edge exists for p). This canonical edge is defined as follows: Let c be a fixed point (not necessarily in S) in the interior of P_R. For each point $p \in S \setminus R$ that lies outside of P_R, its visible "conflict" edge e_p is the unique edge of P_R that is intersected by the straight line segment that connects c and p (uniqueness can be enforced by considering each corner of P_R as part of exactly one of its two incident edges). With this information clearly each oracle call can be answered in constant time. It remains to specify how this information is maintained.

So assume that in the procedure outlined above the recursive call has produced for each point $p \in S \setminus R'$ its canonical conflict edge e_p of $P_{R'}$. If for $p \in S \setminus R$ the conflict edge e_p is not visible from q, then this edge is also an edge of P_R and remains to be the conflict edge for p. (Note that if c is in $P_{R'}$, then it is also in P_R.) If, on the other hand, e_p is visible from q and gets deleted, then the only candidates for being the new conflict edge for p are the two new edges of P_R that are incident to the new point p. Assuming that for each edge e of the current convex hull we maintain the set of all p for which $e_p = e$, it should be clear that the cost of producing the conflict information with respect to P_R given the conflict information for $P_{R'}$ is proportional to the number points in $S \setminus R$ for which the conflict edge changes.

We estimate the expected cost of maintaining the conflict information by considering each point $p \in S$ individually. Since in case of a change p's new conflict edge can be determined in constant time, it suffices to estimate the expected number of changes for p. So what is the probability that for $p \in S$ the conflict edge e_p changes when computing P_R from $P_{R'}$? Backwards analysis suggests that we should express this probability in terms of the "output" P_R. Clearly e_p is new for p iff q is one of the two endpoints of e_p. But since the algorithm chose q to be a random element of R, the probability that q happens to be one of the two endpoints of e_p is $2/r$. Thus the expected number of conflict edge changes for a point $p \in S$ when computing P_R from $P_{R'}$ is at most $2/r$ (note that p could be already in R in which case no change can occur any more). Summing over all $r \leq n$ now yields that the expected total number of conflict edge changes for a point $p \in S$ is at most $2H_n$, which is $O(\log n)$.

Observing that creating the initial conflict information in the "bottoming out" case where $|R| = 3$ takes $O(n)$ time, we can now conclude that the entire maintenance of the conflict information and also the entire algorithm takes expected time $O(n \log n)$.

Remarks. Using the techniques described in the next section it is possible to show that the probability that the running time of the convex hull algorithm presented here exceeds its expected value by a multiplicative factor of c is only $O(n^{-c(\log c - 1)})$.

5. Backwards Analysis of QUICKSORT

QUICKSORT constitutes the archetypical example of a randomized[2] algorithm. Invented by Hoare in 1960 [33], it has since been amply analyzed (see for instance Sedgewick's book [49]) and with its various versions it has become the maybe most frequently used sorting algorithm in practice.

We will consider a somewhat different version of QUICKSORT that is more amenable to backwards analysis than the usual version. However, we also show that both versions have exactly the same running time distribution. Thus the results of our analysis carry over to ordinary QUICKSORT.

Let S be a set of n distinct keys. (The presence of non-distinct keys does not increase the running time of QUICKSORT.) Our algorithm at first puts those keys into a random order $p_1, \ldots, p_n$. For $0 \le r \le n$ let $S_r = \{p_i | i \le r\}$. Our algorithm will make n iterations (or "rounds"), maintaining the following invariant I_r upon completion of each round r:

$$I_r : \begin{cases} \text{The } r \text{ keys in } S_r \text{ have been sorted correctly; say, their order is} \\ \qquad q_0 = -\infty < q_1 < q_2 < \cdots < q_r < \infty = q_{r+1} \ . \\ \text{The remaining keys in } S \setminus S_r \text{ have been partitioned into} \\ r + 1 \text{ sets } B_0, B_1, \ldots, B_r, \text{ where} \\ \qquad B_j = \{q \in S \setminus S_r | q_j < q < q_{j+1}\} \ . \end{cases}$$

At the beginning of execution obviously invariant I_0 holds with $B_0 = S$. In the end invariant I_n must hold, which implies that the set $S = S_n$ has been correctly sorted.

What needs to happen in round r, so that, assuming invariant I_{r-1}, one can establish I_r? At the beginning of the round key p_r has to be contained in some B_j. Thus p_r lies between q_j and q_{j+1}, and hence we now know the sorted order of S_r, as desired. To establish the second part of invariant I_r we only need to split the set $B_j \setminus \{p_r\}$ into two subsets comprising the keys smaller than "pivot" p_r and larger than "pivot" p_r, respectively (see Figure 7).

---

[2] Of course the original version of QUICKSORT was deterministic and it was probabilistically analyzed with respect to an assumed input distribution, namely all permutations of the output occur equally likely as input.

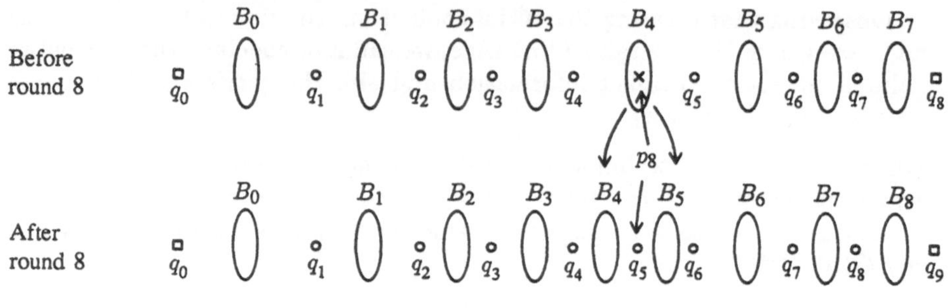

B_0 B_1 B_2 B_3 B_4 B_5 B_6 B_7

Before round 8

q_0 q_1 q_2 q_3 q_4 q_5 q_6 q_7 q_8

p_8

B_0 B_1 B_2 B_3 $B_4 \mid B_5$ B_6 B_7 B_8

After round 8

q_0 q_1 q_2 q_3 q_4 q_5 q_6 q_7 q_8 q_9

Fig. 7

This already completes our description of the algorithm. In this description we have omitted all data structuring aspects. This is justified since we don't intend this algorithm to be implemented but rather to serve as a vehicle for analyzing the number of key comparisons that happen in the usual version of QUICKSORT.

Where do key comparisons happen in this algorithm? They only happen when a set $B_j \setminus \{p_r\}$ is split into two. In that case each element of that set has to be compared with "pivot" p_r. We want to estimate the expected number of such comparisons. (It makes sense to talk about expectations since our algorithm starts by putting the elements of S into random order.)

Let us fix our attention at an arbitrary key $p \in S$ and ask how often p is involved in a comparison where it is not the "pivot." Let us just concentrate on some round r. If $p \in S_r$, then it certainly did not participate in such a comparison. So assume $p \notin S_r$, which means p must be in some B_j. In that case a comparison involving p only happened if B_j is "new," i.e. did not exist in round $r-1$. But B_j can only be new if one of q_j or q_{j+1} was p_r. Backwards analysis now says that any one of $q_1, \ldots, q_r$ has the same probability of being p_r. Thus with probability at most $2/r$ one of q_j or q_{j+1} was p_r (we say "at most" since the fictitious q_0 and q_{r+1} cannot be p_r). This means that the expected number of comparisons involving p in round r is at most $2/r$, and hence the the expected number of comparisons involving p (not as pivot) over all rounds is at most $2H_n$. Considering every key $p \in S$ in turn immediately yields a $2nH_n$ upper bound for the expected number of comparisons made by our algorithm.

Let us try to tighten this analysis. Call a comparison in our algorithm between a key p and a pivot p_r an L-comparison if $p_r < p$; call it an R-comparison otherwise. Let p be the k-th smallest key in S and let $L_k = \{q \in S \mid q \leq p\}$. Let us try to estimate the expected number of L-comparisons for p, i.e. comparisons that involve p as non-pivot and some other member of L_k as pivot, in other words, we want to estimate the number of times that a set B_j that contains p is split and p turns out to be larger than the splitting pivot key. For this purpose it suffices to consider only L-rounds of our algorithm, i.e. rounds r where the pivot p_r is in L_k. Clearly there are exactly k such

L-rounds. Let us number them from 1 to k. Consider now L-round i, where $1 \leq i \leq k$. What is the probability P_i that p is involved in an L-comparison, i.e. what is the probability that the set B_j that contains p was just split? Well, out of the i pivots chosen from L_k so far, q_i must have been the last one. By backwards analysis the probability for this event is $1/i$. However, for an L-comparison involving p to occur it must also be the case that p was not one of the i pivots chosen so far. The probability for this event is $(1 - i/k)$. It follows that $P_i = (1 - i/k) \cdot (1/i)$ and thus the expected number L-comparisons in L-round i that involve p is $1/i - 1/k$. Summing over all k L-rounds we get that the expected number of L-comparisons involving p is $H_k - 1$. Summing now over all n keys of S we find that the expected number of all L-comparisons is $\sum_{1 \leq k \leq n} H_k - n$, which is $(n+1)H_n - 2n$. By symmetry the expected number of R-comparisons is the same and thus the expected total number of key comparisons made by our QUICKSORT algorithm is $2(n+1)H_n - 4n$.

We should now convince the reader that our version of QUICKSORT has the same running time behaviour as the usual version, so that we can claim that our analysis applies to the ususal version also. The usual version sorts a set S of n keys as follows: Choose a $p \in S$ uniformly at random, and compute the sets $S_< = \{q \in S | q < p\}$ and $S_> = \{q \in S | q > p\}$. Output the result of applying the algorithm recursively to $S_<$, followed by p, followed by the the result of applying the algorithm to $S_>$.

With every run of this algorithm we can associate a binary n-node tree T. It is recursively defined as having p as its root whose left child is the tree associated with the sort of $S_<$ and whose right child is the tree associated with the sort of $S_>$. The number of key comparisons made in a particular run can now be expressed as a function $C(T)$ of the associated tree T, namely the sum of the sizes of all n rooted subtrees, minus n. Thus the expected running time of the ordinary version of QUICKSORT can be expressed as the sum over all possible such trees T of the product of $C(T)$ and the probability that T arises.

We can also associate a binary n-node tree T with every run of our version of QUICKSORT. We do this inductively by associating a tree T_r with every round r of our algorithm. T_r has r interior nodes, namely the keys in S_r, and it has $r+1$ leaves, the sets B_j. Tree T_{r+1} is obtained from T_r by replacing the leaf B_j that contains p_{r+1} by a tree whose root is p_{r+1} and whose left and right children are the newly generated B_j and B_{j+1}, respectively. The final tree T is then T_n stripped of all its leaves. Again the number of key comparisons made in a particular run of our algorithm is expressed exactly by $C(T)$. It follows that the two versions of QUICKSORT have the same running time distribution if every tree T is generated by our version with the same probability as by the usual version. But it is easy to check that indeed both versions produce a particular tree T with probability $\prod_{p \text{ node of } T} 1/(\text{size of subtree rooted at } p)$.

That the expected number of comparisons of QUICKSORT is $2(n+1)H_n - 4n$ is a well known result and has been derived before without using backwards analysis (see e.g. [28]). We will now use backwards analysis to estimate in a reasonably painless way the probability that the running time of QUICK-SORT is significantly larger than its expectation (see [47]) for a similar result). For this purpose we will artificially slow down our algorithm as follows: If in round r the new pivot p_r is in B_0, then all keys in B_r are compared with p_r also. Similarly, if p_r is in B_r, then all keys in B_0 are compared with p_r also. Conceptually we are now performing a cyclical sort of S, where after round r there are r sets B_j, for which index arithmetic is done modulo r. This "cyclical" modification of the algorithm removes all "boundary" effects and makes all keys appear symmetrically the same. We will also slow down our algorithm even further. In each round r the pivot p_r will make two extra comparisons (say, with itself).

We will again partition the comparisons made by our slowed down algorithm into L-comparisons and R-comparisons. Again, an L-comparison is a comparison between a key p and a pivot p_r, where $p_r < p$. We will also consider to be an L-comparison one of the two extra comparisons made by every pivot p_r.

For a key $p \in S$ let L_p be a random variable that counts the number of L-comparisons that p is involved in plus the one L-comparison between p and itself when p is the pivot. Define R_p analogously. Let now $X = \sum_{p \in S}(L_p + R_p)$ be a random variable counting the total number of comparisons made by our slowed down algorithm. We are interested in estimating the probability that X exceeds its expectation by a multiplicative factor of c. This can clearly only happen if at least one of the random variables L_p or R_p exceeds its expectation by a factor of c. Thus we have

$$\Pr(X > c \cdot \mathrm{E}[X]) \leq \sum_{p \in S} \Pr(\mathrm{L_p} > c \cdot \mathrm{E}[\mathrm{L_p}]) + \sum_{p \in S} \Pr(\mathrm{R_p} > c \cdot \mathrm{E}[\mathrm{R_p}]) \ .$$

Since our setup is completely symmetric all random variables L_p and R_p have the same distribution. Thus for any $p \in S$ we have

$$\Pr(X > c \cdot \mathrm{E}[X]) \leq 2\mathrm{n} \cdot \Pr(\mathrm{L_p} > c \cdot \mathrm{E}[\mathrm{L_p}]) \ . \tag{1}$$

Let us now fix our attention on some key $p \in S$ and let $Y = L_p$. For $1 \leq r \leq n$ let Y_r be a 0-1 random variable counting the contribution of *round r* to Y. Using the ideas of the previous paragraphs and observing the slow down modifications of our algorithm it is easy to see that Y_r is 1 with probability exactly $1/r$. Thus $\mathrm{E}[Y] = \sum_{1 \leq r \leq n} \mathrm{E}[Y_r] = \sum_{1 \leq r \leq n} 1/r = \mathrm{H}_n$ and therefore $\mathrm{E}[X]$ the expected number of comparisons in our slowed down algorithm is $2nH_n$.

To estimate $\Pr(Y > c \cdot \mathrm{E}[Y])$ we can now use the well-known Chernoff bound (see [47,31]), which in one form states that if a random variable Z is the sum of n independent 0-1 random variables and the expectation of Z is E, then for $c \geq 1$

$$\Pr(Z > c \cdot E) \leq e^{-E(1-c+c\log c)} \ .$$

In our case the Y_i's can easily be proven to be independent. Thus we obtain

$$\Pr(Y > c \cdot \mathrm{E}[Y]) \leq \mathrm{e}^{-\mathrm{H_n}(1-c+c\log c)} = \mathrm{O}(\mathrm{n}^{-(1+c(\log c-1))})\ .$$

From this and inequality 1 we can now conclude that the probability that the running time of our modified algorithm exceeds its expectation by a multiplicative factor of $c \geq 1$ is $O(n^{-c(\log c-1)})$. Since this algorithm always performs more comparisons than ordinary QUICKSORT we can conclude that the probability that QUICKSORT makes more than $2cnH_n$ comparisons is also $O(n^{-c(\log c-1)})$.

6. A Bad Example

Consider the following higher-dimensional triangulation problem: Suppose a set S of n points in $\mathbb{R}^d$ is contained in the interior of a d-simplex D with vertex set $Q = \{q_0, q_1, \ldots, q_d\}$. We are to *triangulate* S, i.e. construct a collection of simplices T, so that $\bigcup\{\Delta \in T\} = D$, each simplex $\Delta \in T$ has its vertex set in $S \cup Q$, every $p \in S$ is vertex of some simplex in T, and every two simplices in T intersect in a common face (which can be the empty set).

Deterministic algorithms for solving this problem in $O(n \log n)$ time have been presented in [4,26]. Here we consider the following randomized incremental algorithm that was inspired by the QUICKSORT algorithm of the previous section. For the purpose of illustration we will assume that the points in $S \cup Q$ are in non-degenerate position, i.e. no $d+1$ points lie in a common hyperplane.

Our algorithm will at first put the points in S in some random order $p_1, \ldots, p_n$. For $0 \leq r \leq n$ let now $S_r = \{p_i | i \leq r\}$. The algorithm works in n rounds. Upon completion of round r a triangulation T_r will have been computed, and for each simplex $\Delta \in T_r$ the set B_Δ will contain the points in $S \setminus S_r$ that lie in the interior of Δ. Initially we have $T_0 = \{D\}$ and $B_D = S$.

Assuming inductively that the algorithm has correctly completed round $r-1$, our algorithm only needs to do the following in round r: Let Δ be the simplex in T_{r-1} that has p_r in its interior and let B_Δ be the corresponding subset of $S \setminus S_{r-1}$. The simplex Δ is split into $d+1$ simplices $\Delta_0, \ldots, \Delta_d$, each being a pyramid with a facet of Δ as base and with p_r as apex. The set $B_\Delta \setminus \{p_r\}$ is split accordingly.

In any reasonable implentation the running time of this algorithm will be proportional to the number of times that during the execution the containing simplex changes for a point $p \in S$. It is now tempting to apply backwards analysis to obtain the expectation of this quantity. Fixing attention at a point $p \in S$ and at some round r it seems that the probability that the containing simplex of p changed in round r is at most $(d+1)/r$. After all, either p is contained in S_r, in which case no change occurs, or otherwise p must be in some simplex $\Delta \in T_r$, which is "new" iff one of its $d+1$ vertices happened to be p_r. By backwards analysis this happens with probability at most $(d+1)/r$. Thus the expected number of containing simplex changes for p in the entire algorithm is at most $(d+1)H_n$, and hence the expected running time of the algorithm is $O((d+1)nH_n)$.

However, this argument is quite wrong. The fallacy lies in the fact that the triangulation T_r of S_r is not canonical, i.e. it depends on the ordering of the points in S_r. In particular this means that the containing simplex of p is not canonical. Thus it is difficult to argue that p_r is a vertex of that simplex with probability at most $(d+1)/r$.

For dimension $d = 1$ this algorithm indeed does have $O(nH_n)$ expected running time, since any 1-dimensional point set admits exactly one triangulation, and hence T_r is always canonical. But of course for $d = 1$ the triangulation algorithm of this section is nothing but QUICKSORT. It remains to be seen whether for fixed $d > 1$ the expected running time of this triangulation algorithm is indeed $O(n \log n)$.

7. Linear Programming for Small Dimension

General linear programming has a long history. In this paper we are only interested in the case where the dimension (or number of variables) d is a small constant and m, the number of halfspaces (or constraints), can be quite large. In the last decade deterministic algorithms were developed that solve such linear programs in $O(m)$ worst case time [39,40,22,23,19]. However those algorithms are mostly of theoretical interest only since they are quite complicated and the dependence of their running times on the dimension d is exponential and has only been shown to be 3^{d^2} at best. More recently Ken Clarkson [20] proposed a randomized algorithm with a remarkable running time of $O(d^2 m) + (\log m)O(d)^{d/2+O(1)} + O(d^4 \sqrt{m} \log m)$. Here we briefly present another randomized algorithm that was first described in [50]. The main virtues of this algorithm are its simplicity and its amenability to backwards analysis (well — for the purpose of this paper this is a virtue).

Geometrically, linear programming amounts to the following: One is given a set $\mathcal{H}$ of m halfspaces and a vector a in $\mathbb{R}^d$, and one wants to find a vertex v of the polyhedron P formed by the intersection of the halfspaces, so that v maximizes the linear functional specified by a; in other words, v must be contained in the tangent hyperplane of P whose outward normal is a.

The reader might wonder about our specification of linear programming. The required optimum vertex v of P might not exist, either because P is empty or because P is unbounded. Thus we amend our specification. In case of emptiness of P we require this fact to be reported by the algorithm. For the sake of ease of presentation we will ignore for the time being the unboundedness situation and assume that our problem and all subproblems to be encoutered are very well behaved in the sense that if a problem is feasible a unique optimum vertex exists and that this vertex is the intersection of the bounding hyperplanes of exactly d of the given halfspaces. We will show later how those assumptions can be removed.

Here is our algorithm for solving such a linear program given by a set $\mathcal{H}$ of $m \geq d$ halfspaces and a direction a in $\mathbb{R}^d$.

If $d = 1$, then the problem amounts to finding the smallest (or largest) real number satisfying m inequalities. With m comparisons this number can easily be found or it can be established that a number satisfying all inequalities does not exist.

Now assume $d > 1$. If $|\mathcal{H}| = d$, then by our assumptions the solution is the intersection of the d hyperplanes that bound the halfspaces in $\mathcal{H}$. Thus this optimum vertex can be found in $O(d^3)$ time.

So assume that $|\mathcal{H}| > d$. Choose a halfspace $H \in \mathcal{H}$ uniformly at random. Recursively solve the d-dimensional linear program given by the $m - 1$ halfspaces $\mathcal{H} \setminus \{H\}$ and direction a. This yields an optimum point w. (If such an optimum does not exist, the original problem does not have a solution.)

Now, if w is contained in H, then clearly w is also the solution for the original problem and we are done. If w is not contained in H, then the optimum vertex v for $\mathcal{H}$, if it exists at all, must be contained in the hyperplane h that bounds H. As a matter of fact v must be the solution of the $(d - 1)$-dimensional linear programming problem given by the $m - 1$ constraints $\mathcal{H}' = \{I \cap h \mid I \in \mathcal{H} \setminus \{H\}\}$ and direction a', the orthogonal projection of a into h. The solution of that problem is now found recursively.

Let us now analyze the expected running time of this algorithm. The most important issue seems to be to estimate the probability that the $(d - 1)$-dimensional problem for $\mathcal{H}'$ needs to be solved. This is exactly where backwards analysis comes into play. Note that this recursive call is necessary iff the optimum vertex v for $\mathcal{H}$ is different from the optimum vertex w for $\mathcal{H} \setminus \{H\}$. But such a difference can only occur if the bounding hyperplane of H is one of the d hyperplanes that define v. But since H is chosen from the m hyperplanes in $\mathcal{H}$ at random this happens with probability d/m.

Now let $T(d, m)$ be the expected running time of our algorithm for solving a linear program with $m \geq d$ halfspaces in $\mathbb{R}^d$. Assuming that testing whether a point v lies in a halfspace H takes $O(d)$ time and that computing the intersection of a halfspace with a hyperplane takes also $O(d)$ time, $T(d, m)$ is defined recursively as follows:

$$T(d, m) = \begin{cases} O(m) & \text{if } d = 1 \\ O(d^3) & \text{if } m = d \\ T(d, m - 1) + O(d) + \frac{d}{m}O(dm) + \frac{d}{m}T(d - 1, m - 1) & \text{otherwise} \end{cases}$$

It is now easy to check that $T(d, m) = O\left(\sum_{1 \leq i \leq d} \frac{i}{(i-1)!} d! m\right)$, which is $O(d!m)$ since the sum converges even without an upper bound for i.

We still have to deal with the various assumptions we made initially. Uniqueness of the optimum vertex can be achieved by standard perturbation techniques, or by requiring the algorithm always to return the optimum vertex that has the lexicographically smallest coordinate representation. Note that in light of the previous section it is crucial for the analysis of our algorithm that the optimum vertex is defined uniquely and canonically.

The assumption that an optimum vertex must always be the intersection of the bounding hyperplanes of exactly d halfspaces can be dropped altogether. Involvement of more than d halfspaces in v makes it less likely that v is different from w and hence cannot increase the running time of our algorithm.

The boundedness assumption appears to be the most difficult to remove. One approach is to enforce it simply by stipulating that we are not interested in all of $\mathbb{R}^d$ but just some "bounding box" B (i.e. we impose explicit lower and upper bounds for the variables; see [50] for details). Another approach involves generalizing the notion of "optimum vertex:" in case of unbounded-edness define the "optimum" to be the unit vector in the direction of a ray in the feasibility region that maximizes the inner product with the objective direction a. Both approaches require slight changes in the "bottoming-out" part of our procedure. See the next section for an abstract description.

8. Welzl's Minidisk Algorithm

Here we present an algorithm due to Emo Welzl [56] for constructing the smallest enclosing ball for a finite point set $T \subset \mathbb{R}^d$. The algorithm is very similar to the linear programming algorithm of the previous section. However, the idea of recursively solving a problem of *smaller dimension* has to be viewed now as recursively solving a problem with *more equality constraints*.

Below we describe a function minidisk(T, C), which takes as input two disjoint finite sets $T, C \subset \mathbb{R}^d$ and which is to return the ball of smallest radius that contains T and has all points of C on its boundary. Of course, for arbitrary sets T and C such a ball need not exist. However, we will assume that the function will only be called with parameters T and C for which the existence of such a ball is guaranteed. In particular note that minidisk $(T, \emptyset)$ simply computes the smallest enclosing ball of T, and such a ball of course always exists (for the case $T = \emptyset$ we consider the empty set to be a degenerate ball).

Note that the smallest enclosing ball of a set T will always be determined by at most $d + 1$ points of T. For our implementation of minidisk() we will assume the existence of a function primitive_ball (D), which given any set D of at most $d + 1$ points returns in time $O(d^3)$ the smallest ball that has all of D on its boundary (again assuming the existence of such a ball).

```
minidisk(T, C)
        if T = ∅ then return primitive_ball(C);
        choose some p from T uniformly at random and let T' := T \ {p};
        B' := minidisk(T', C);
        if p ∈ B' then return B'
                else return minidisk(T', C ∪ {p});
```

The correctness of this function follows from the following two lemmas:

Lemma 8.1   Let T and C be two sets in $\mathbb{R}^d$. The smallest ball that contains T and has all points of C on its boundary is unique (provided it exists).

Proof. Assume B_1 and B_2 are two distinct smallest enclosing balls for T that have C on their boundary. Let c_1 and c_2 be the two centers and let R be the common radius. Then it is easy to check that the smaller ball with center $(c_1 + c_2)/2$ and radius r given by $r^2 = R^2 - \langle c_1 - c_2, c_1 - c_2 \rangle$ also contains T and has all points of C on its boundary. $\qquad\square$

Lemma 8.2   Let T' and C be two sets in $\mathbb{R}^d$ so that the smallest enclosing ball B' for T' that has all points of C on its boundary exists.

If some point $p \in \mathbb{R}^d$ is not contained in B', then the smallest enclosing ball B for $T \cup \{p\}$ that has all points of C on its boundary also has p on its boundary (provided it exists).

Proof. Assume B does not have p on its boundary, i.e. p lies in the interior of B. But in this case B must also be the smallest enclosing ball for T' hat has C on its boundary. By the previous lemma that ball is unique, i.e. $B = B'$, which would mean $p \in B$ and $p \notin B$, a contradiction. $\qquad\square$

For the analysis of the expected running time of $\mathtt{minidisk}(T, C)$ the most important step is to estimate the probability that a point p chosen randomly from T is not contained in the ball B'. Let B be the smallest enclosing ball of T that has all points of C on its boundary. Lemma 8. tells us that $p \notin B'$ implies that p lies on the boundary of B and that $B \neq B'$. In the non-degenerate case where no $d + 1$ points of $T \cup C$ are co-spherical it is clear that there are at most $d + 1 - |C|$ choices for p that render B and B' different. It is not too hard to see that this number cannot be larger if there are degeneracies. Thus the probability that a random point p of T is not contained in the ball B' is at most $\delta/|T|$, where $\delta = d + 1 - |C|$.

Now consider a call $\mathtt{minidisk}(T, C)$, where $T, C \subset \mathbb{R}^d$, $|T| = n$, and $|C| = d + 1 - \delta$. Let $f(n, \delta)$ denote the expected number of recursive invocations $\mathtt{minidisk}(S, D)$ for which $S \neq \emptyset$, and let $g(n, \delta)$ denote the expected number of recursive invocations for which $S = \emptyset$. Thus f and g satisfy the following recursive relationships:

$$f(n, \delta) \leq \begin{cases} 0 & \text{if } n = 0 \\ 1 + f(n - 1, \delta) + \frac{\delta}{n} f(n - 1, \delta - 1) & \text{otherwise} \end{cases}$$

$$g(n, \delta) \leq \begin{cases} 1 & \text{if } n = 0 \\ g(n - 1, \delta) + \frac{\delta}{n} g(n - 1, \delta - 1) & \text{otherwise} \end{cases}$$

It is an easy inductive exercise to check that now $f(n, \delta) \leq \sum_{1 \leq i \leq \delta} \frac{1}{i!} \delta! n$, which is $O(\delta! n)$, and that $g(n, \delta) \leq \delta! (1 + H_n)^\delta$, which is $O(\delta! \log^\delta n)$. If we assume that an invocation with $S \neq \emptyset$ takes time $O(d)$ and an invocation with $S = \emptyset$ (i.e. a call to primitive_ball) takes time $O(d^3)$ we get the following:

Theorem 8.3   *For a set T of n points $\mathbb{R}^d$ a call to minidisk$(T, \emptyset)$ computes the smallest enclosing ball of T in expected time $O(d(d+1)!n)$.*

The type of algorithm presented in this and the previous section can be successfully applied also to other problems, such as computing the smallest enclosing ellipsoid of a point set in $\mathbb{R}^d$, or computing the largest inscribed sphere of a convex polyhedron. It is possible to unify these algorithms by considering the relevant problems as special instances of a suitably axiomatized abstract optimization problem.[3]

Very recently Micha Sharir and Emo Welzl [54] proposed a new axiomatic framework along with a new randomized algorithm for the linear programming type of problems considered here and in the previous section. Their method avoids the recursion on the dimension and in the case of linear programming achieves an expected time bound of $O(2^d n)$. The complexity analysis of their new algorithm also exploits aspects of "backwards" analysis. Even more recently the same two authors together with Jiří Matoušek [38] managed to improve the analysis of that algorithm to the remarkable expected time bound of $O((nd + d^3)e^{4\sqrt{d\ln(n+1)}})$.

9. Clarkson's Backwards Analysis of the Conflict Graph Based on the Convex Hull Algorithm

In their landmark paper on applications of random sampling in computational geometry [18] Clarkson and Shor described a randomized algorithm for constructing convex hulls of point sets in $\mathbb{R}^d$ that has optimal expected running time. Their analysis of the expected running time was based on very general lemmas about random sampling. Recently Ken Clarkson [21] has discovered a new analysis that is completely self-contained and relies heavily on the idea of backwards analysis. In this section we give first a brief description of the convex hull algorithm[4] and then its new analysis.

We want to construct the convex hull of a set S of n points in $\mathbb{R}^d$, with $n > d > 1$. We will assume that S is in non-degenerate position, i.e. no $d + 1$ of its points lie in a common hyperplane. Such non-degeneracy can easily be simulated with impunity using standard perturbation techniques [25, pp. 185]. Non-degeneracy ensures that the convex hull of any subset of S is a simplicial polytope.

---

[3] Previous versions of this paper contained an attempt of such an axiomization. However, that framework turned out to be too weak, and the algorithmic results claimed in those previous versions are fallacious.

[4] We actually present a slightly different version than the one in [18] in that we do not dualize and use a slightly different notion of a conflict graph.

A few relevant basics about polytopes: Let P be a simplicial d-polytope, let V be the vertex set of P, and let $m = |V|$. It is known that P can have at most $O(m^{\lfloor d/2 \rfloor})$ faces. We call the $(d-1)$-faces of P *facets* and the $(d-2)$-faces *ridges*. Every facet is uniquely identified by the d-tuple of its vertices. Similarly every ridge can be identified by a $(d-1)$-tuple of vertices in V. Since every ridge is contained in precisely two facets one can represent the facial structure of P by its *facet graph* $\mathcal{G}(P)$, which has the facets of P as its nodes and two facets adjacent iff they share a common ridge of P. Note that for simplicial d-polytopes the facet graph is regular of degree d.

Let p be some point in $\mathbb{R}^d$ in non-degenerate position with respect to V. We call a facet F of P *visible* from p iff the hyperplane spanned by F separates P and p. We call F *obscured* otherwise. We call a face G of P visible from p iff it is only contained in facets of P that are visible from p. Obscured faces are defined analogously. We call G a horizon face with respect to x iff it is contained in some visible and some obscured facet.

This terminology allows a convenient characterization of the facial structure of the polytope $P' = conv(P \cup \{x\})$ in terms of the faces of P: No visible face of P is a face of P'; all obscured and all horizon faces of P are faces of P'; for each horizon face G of P the pyramid $conv(G \cup \{x\})$ is a face of P'; this yields all faces of P'.

This characterization justifies the following method for obtaining P' from P and x. We assume here that the polytopes are represented by their facet graphs. Thus, to be more precise, the procedure outlined below is intended to compute the facet graph $\mathcal{G}(P')$ from x and the facet graph $\mathcal{G}(P)$, and when we talk about facets or ridges we are simultaneously referring to nodes and arcs of a facet graph (which are identified by d-tuples and $(d-1)$-tuples, respectively, of the points in $S \cup \{x\}$).

(i) Determine the set $Vis(x, P)$ of all facets of P that are visible from x. (In case no visible facets exist at all, x must be contained in P already, i.e. $\mathcal{G}(P') = \mathcal{G}(P)$, and nothing further needs to be done.)

(ii) Partition the ridges contained in facets of $Vis(x, P)$ into the set of visible and the set of horizon ridges of P with respect to x, by checking for each ridge whether both containing facets are in $Vis(x, P)$. Delete all visible facets and ridges.

(iii) For each horizon ridge G of P generate the new facet $conv(G \cup \{x\})$ of P' (i.e. a new node for the facet graph).

(iv) Generate the new ridges of P' (i.e. the edges between the new nodes of the facet graph).

Step (i) of this algorithm is still rather vaguely specified. We defer the details of how the visibility set $Vis(x, P)$ can actually be obtained. Let us first analyze the cost of this algorithm, but ignoring the cost incurred by step (i).

The cost of step (ii) is clearly proportional to the number of facets in $Vis(x, P)$. But since all those facets are deleted, and every facet can be deleted at most once, we can charge the deletion cost of each facet to its creation, and thus in the amortized sense, step (ii) incurs no cost at all.

Step (iii) has cost proportional to the number of new facets created. These are exactly the facets of P' that contain x. Let us denote their number by $deg(x, P')$.

The number of new ridges created in step (iv) is proportional to the number of new facets, to be precise, their number is $(d-1)deg(x, P')/2$. How can they be found? For every new facet generated in step (iii) the $d-1$ new ridges contained by it can be determined "locally." Radix sorting the $(d-1)$-tuples of vertices (or rather vertex indices) that identify these ridges then allows to match them up and to form the new edges of the facet graph $\mathcal{G}(P')$ in time proportional to $n + deg(x, P')$. When $d < 4$ the radix sort can be avoided: In the case $d = 2$ there are only two new facets and one new ridge, namely x. In the case $d = 3$ one can exploit the planar graph nature of the facet graphs to find the new ridges in time proportional to their number. We omit here the details of how to do this.

We conclude that, ignoring step (i), the total cost of this insertion algorithm is proportional to $deg(x, P')$ in case $d = 2, 3$ and proportional to $n + deg(x, P')$ for $d > 3$.

Consider now the following algorithm for constructing the convex hull of a set S of $n > d$ points in $\mathbb{R}^d$ in non-degenerate position.

1. Put the points of S in a random order $p_1, \ldots, p_n$. For $1 \le r \le n$ let S_r denote $\{p_1, \ldots, p_r\}$, and let P_r denote $conv\, S_r$.
2. Form the facet graph $\mathcal{G}(P_{d+1})$. (Note that this graph is simply the complete graph on $d + 1$ vertices.)
3. For $d + 1 < r \le n$, using the insertion procedure outlined above, form the facet graph $\mathcal{G}(P_r)$ from $\mathcal{G}(P_{r-1})$.

We want to determine the expected running time of this algorithm. Obviously the most important question is to determine the expected cost of step 3. We know that, ignoring step (i) of the insertion algorithm, the expected cost of the insertion in iteration r is determined by the expectation of $deg(p_r, P_r)$. Now apply backwards analysis. With probability $1/r$ point p_r was the last one in the random permutation of S_r. Thus the expected value of $deg(p_r, P_r)$ is $(1/r) \cdot \sum_{p \in S_r} deg(p, S_r)$, which, since every facet contains exactly d vertices, is $(d/r) \cdot F(P_r)$, where $F(P_r)$ denotes the number of facets of P_r. But P_r has at most r vertices. Thus by the upper bound theorem for polytopes $F(P_r) = O(r^{\lfloor d/2 \rfloor})$, and the expected value of $deg(p_r, P_r)$ is therefore $O(r^{\lfloor d/2 \rfloor - 1})$. We conclude that for $d > 3$ the expected running time of the entire algorithm is $\sum_{d+1 < r \le n} O(r + r^{\lfloor d/2 \rfloor - 1})$, which is $O(n^{\lfloor d/2 \rfloor})$. For $d = 2, 3$ we get that the expected running time of the algorithm is $\sum_{d+1 < r \le n} O(1)$, which is $O(n)$.

But recall that this analysis does not take into account the cost incurred by step (i) of the insertion algorithm. Let us now turn to the details of how that step can be implemented. How can one determine the set $Vis(p_r, P_{r-1})$ of all facets of P_{r-1} that are visible from p_r?

As pointed out in [50] there is a simple solution for this problem that turns out to be reasonably efficient for the case $d > 3$. Since the facets in $Vis(p_r, P_{r-1})$ induce a connected subgraph in the facet graph $\mathcal{G}(P_{r-1})$ it suffices to find just one visible facet. The remaining ones can then be determined by a depth-first search in time proportional to their number. All the visible facets found will be deleted, never to reappear again, and thus we can charge their discovery cost to their creation. In other words, in the amortized sense this depth-first search incurs no cost at all, and we only have to worry about the time necessary to discover one facet of $Vis(p_r, P_{r-1})$. However, this problem is nothing but a linear programming problem with r constraints and in d variables and can thus, as we saw in Section 7, be solved in $O(r)$ expected time. Summing over all n insertions this yields an overall expected cost of $O(n^2)$, which for $d > 3$ is subsumed by the $O(n^{\lfloor d/2 \rfloor})$ expected running time of the remaining parts of the algorithm.

A solution to this "visibility problem" that performs satisfactorily in all dimensions, and not just for $d > 3$, was invented by Clarkson and Shor [18]. In essence, they proposed to maintain at each iteration r of the algorithm the complete visibility set[5] $Vis(p, P_r)$ for each point $p \in S \setminus S_r$. We will here describe a variant of this approach that was also already considered in [18], where for each point $p \in S \setminus S_r$ only one representative visible facet $VF(p, P_r) \in Vis(p, P_r)$ is maintained (provided such a facet exists at all).

Initially some $VF(p, P_{d+1})$ can be computed for all $p \in S \setminus S_{d+1}$, in $O(n)$ time. For $r > d + 1$ how can one compute $VF(p, P_r)$ from $VF(p, P_{r-1})$? For some point $p \in S \setminus S_r$ let $F = VF(p, P_{r-1})$. If F is undefined (because p is contained in P_{r-1}), then also $VF(p, P_r)$ is undefined. If the facet F is also a facet of P_r, then one can choose $VF(p, P_r) = F$. We only have to actually do something if the facet F of P_{r-1} is not a facet of P_r any more, i.e. $F \in Vis(p_r, P_{r-1})$ and is thus one of the facets that gets deleted in step (ii) of the insertion algorithm. In order to discover all $p \in S \setminus S_r$ for which we actually have to do something we need to maintain for each facet the set of points p for which the facet is the representative visible facet.

To find the replacement p-visible facet of P_r we now start at F a depth-first search in the facet graph $\mathcal{G}(P_{r-1})$ to discover a connected component of facets in $Vis(p_r, P_{r-1}) \cap Vis(p, P_{r-1})$ in time proportional to their number. Let D be the set of horizon ridges (with respect to p_r) contained in those facets. For each ridge $G \in D$ now check if its containing facet of P_{r-1} that is not in $Vis(p_r, P_{r-1})$ (which is therefore a facet of P_r) is visible from p, and check if the "new" facet $conv(G \cup \{p_r\})$ of P_r is visible from p. If no p-visible

---

[5] Clarkson and Shor have the notion of a "conflict graph," which in our case would be a bipartite graph whose nodes are the facets of P_r and the points in $S \setminus S_r$ and that has an arc joining facet F with point q iff F is visible from q.

facet is found this way, then we make $VF(p, P_r)$ undefined, otherwise we set $VF(p, P_r)$ to one of those p-visible facets.

The cost of finding the new representative p-visible facet on P_r is thus proportional to the size of $Vis(p_r, P_{r-1}) \cap Vis(p, P_{r-1})$, or in other words, the number of visibilities between facets and p that cease to exist with the insertion of p_r. Therefore, in order to estimate the cost of maintaining representative visible facets for all p over the entire algorithm, we need to determine the expected number of visibilities between facets and points that cease to exist in the course of the algorithm. Obviously this is the same as the expected number of visibilities that come into existence. We will now estimate the latter quantity.

Let $R = \{p_1, \ldots, p_r\}$. Since the p_i's are in a random order R is now a random subset of S of size r. What is the expected number of visibilites between facets of $conv\,R$ and points in $S \setminus R$ that came into existence when the last point of R was inserted? Let's do it backwards! Which visibilities would disappear if a random point q of R was removed? Exactly those that involved a facet that contained q. Since every facet is determined by exactly d points of R the probability that any particular facet contains q is d/r. It follows that the expected number of visibilities that disappear when a random point of R is removed (or visibilites created when q is inserted) is

$$\frac{d}{r} \sum_{q \in S \setminus R} |Vis(q, conv\,R)|\,.$$

Let $vis(q, R)$ denote $|Vis(q, conv\,R)|$. For $A \subset S$ let now $f(A)$ denote the number of facets of $conv\,A$, and for $a \in A$ let $deg(a, A)$ now denote the number of facets of $conv\,A$ that contain a. Here comes an ingenious observation due to Ken Clarkson [21]:

$$vis(q, R) = f(R) - f(R \cup \{q\}) + deg(q, R \cup \{q\})\,, \tag{2}$$

since the facets of $conv\,R$ that are not visible from q are exactly the facets of $conv\,(R \cup \{q\})$ that do not contain q.

It follows that C_r, the expected number of visibilities created when the r-th point of S is inserted, is

$$C_r = \frac{1}{\binom{n}{r}} \sum_{\substack{R \subset S \\ |R| = r}} \frac{d}{r} \sum_{q \in S \setminus R} vis(q, R)$$

$$= \frac{1}{\binom{n}{r}} \sum_{\substack{R \subset S \\ |R| = r}} \frac{d}{r} \sum_{q \in S \setminus R} \left(f(R) - f(R \cup \{q\}) + deg(q, R \cup \{q\}) \right)\,.$$

Now let $f_r = \frac{1}{\binom{n}{r}} \sum_{\substack{R \subset S \\ |R| = r}} f(R)$ denote the expected value of $f(R)$. Note that f_r actually also depends on the set S. We will estimate each of the three main summands of the sum above separately.

$$\frac{1}{\binom{n}{r}} \sum_{\substack{R \subseteq S \\ |R|=r}} \frac{d}{r} \sum_{q \in S \setminus R} f(R) = \frac{d}{r}(n-r)f_r$$

$$\frac{1}{\binom{n}{r}} \sum_{\substack{R \subseteq S \\ |R|=r}} \frac{d}{r} \sum_{q \in S \setminus R} f(R \cup \{q\}) = \frac{1}{\binom{n}{r}} \sum_{\substack{R' \subseteq S \\ |R'|=r+1}} \frac{d}{r} \sum_{q \in R'} f(R')$$

$$= \frac{1}{\binom{n}{r+1}} \sum_{\substack{R' \subseteq S \\ |R'|=r+1}} \frac{\binom{n}{r+1}}{\binom{n}{r}} \frac{d}{r}(r+1)f(R')$$

$$= \frac{1}{\binom{n}{r+1}} \sum_{\substack{R' \subseteq S \\ |R'|=r+1}} \frac{d}{r}(n-r)f(R') = \frac{d}{r}(n-r)f_{r+1}$$

$$= \frac{d}{r+1}\left(n-(r+1)\right)f_{r+1} + \frac{dn}{r(r+1)}f_{r+1}$$

$$\frac{1}{\binom{n}{r}} \sum_{\substack{R \subseteq S \\ |R|=r}} \frac{d}{r} \sum_{q \in S \setminus R} deg(q, R \cup \{q\}) = \frac{1}{\binom{n}{r+1}} \sum_{\substack{R' \subseteq S \\ |R'|=r+1}} \frac{\binom{n}{r+1}}{\binom{n}{r}} \frac{d}{r} \sum_{q \in R'} deg(q, R')$$

$$= \frac{1}{\binom{n}{r+1}} \sum_{\substack{R' \subseteq S \\ |R'|=r+1}} \frac{n-r}{r+1} \cdot \frac{d}{r}(d \cdot f(R'))$$

$$= \frac{d^2}{r(r+1)}(n-r)f_{r+1}$$

Thus the total expected number of visibilities created in the entire course of the algorithm is

$$\sum_{d+1 \leq r < n} C_r = \sum_{d+1 \leq r < n} \left(\frac{d}{r}(n-r)f_r - \frac{d}{r+1}(n-(r+1))f_{r+1} \right.$$
$$\left. - \frac{dn}{r(r+1)}f_{r+1} + \frac{d^2}{r(r+1)}(n-r)f_{r+1} \right) .$$

This is a telescoping sum, and therefore we get

$$\sum_{d+1 \leq r < n} C_r = \frac{d}{d+1}(n-d-1)f_{d+1} +$$

$$+ d(d-1)n \sum_{d+1 \leq r < n} \frac{f_r}{r(r+1)} - d^2 \sum_{d+1 \leq r < n} \frac{f_{r+1}}{r+1} .$$

But since any d-polytope with r vertices has $O(r^{\lfloor d/2 \rfloor})$ facets it is certainly the case that $f_r = O(r^{\lfloor d/2 \rfloor})$. Thus it is easy to see that for $d > 3$ the expected total number of visibilities created is $O(n^{\lfloor d/2 \rfloor})$, whereas for $d = 2, 3$ this number is $O(nH_n)$, which is $O(n \log n)$.

We can therefore conclude that the randomized incremental algorithm for constructing the convex hull of n points in $\mathbb{R}^d$ has an expected running time of $O(n \log n)$ for $d = 2, 3$ and $O(n^{\lfloor d/2 \rfloor})$ for $d > 3$, which was the best we could hope for. Note that this analysis gives amazingly tight bounds for the expected number of visibilities for the case $d = 2, 3$.

10. Odds and Ends

It should be pointed out that the type of analysis presented in the previous section is not particular to the convex hull problem but can be applied to randomized incremental construction in the formal framework of Clarkson and Shor [18]. In their terminology the generalization of the crucial insight (2) is the observation that the regions defined by R that do not conflict with object q are exactly those regions defined by $R \cup \{q\}$ that do not involve q. Note that most of the problems and algorithms presented in this paper actually fall into Clarkson and Shor's framework — maybe this paper should have been made much shorter.

There are a number of problems and algorithms that should have been included in this survey, but were not because of time constraints. Maybe the most serious gap concerns geometric searching, in particular planar subdivsion searching. A search structure is constructed using a randomized algorithm; query times are then random variables with respect to the "coin flips" made during the construction. Backwards analysis works very well for determining the expectation of those query times; see for instance [51]. In that paper a little twist is also added to this approach, which yields a rather straightforward randomized method for triangulating a simple polygon in nearly optimal $O(n \log^* n)$ expected time.

In section 5 we gave some analysis of how tightly the running time of QUICKSORT is concentrated around its expectation. What about tail estimates for the running times of the other algorithms presented in this paper? What is the probability that the actual running time on a problem of size n exceeds the expectation by a multiplicative factor of c? For the polygon triangulation algorithm of secion 2 one can use a general result of Mehlhorn [41] to show that this probability is at most $\frac{1}{e}\left(\frac{e}{c}\right)^c$. For Mulmuley's algorithm of section 3 Matoušek and Seidel [37] have recently shown a tail estimate of $O(n^{-c})$, provided K, the number of intersecting segment pairs, is not too small relative to n. For the linear programming algorithm of section 7 a bound of $O(c^{-d!})$ is given in [50], where d is the dimension. A similar bound applies to the algorithm of section 8. To my knowledge no non-trivial tail estimate is known for the convex hull algorithm of section 9.

Acknowledgements. Work on this chapter has been supported by NSF Presidential Young Investigator Award CCR-9058440. Many people have wittingly and unwittingly contributed to this paper in one form or another. I would like to give credit to all the participants of the various DIMACS workshops on computational geometry in 1989/90. In particular I would like to thank Ken Clarkson, Emo Welzl, Günter Rote, Peter Shor, Kurt Mehlhorn, Ricky Pollack, Leo Guibas, Micha Sharir, Herbert Edelsbrunner, Ketan Mulmuley, and Otfried Schwarzkopf. Finally, I am grateful to János Pach for his seemingly infinite patience.

References

1. P.K. Agarwal, H. Edelsbrunner, O. Schwarzkopf, and E. Welzl: Euclidean minimum spanning trees and bichromatic closest pairs. Proc. 6th ACM Symp. on Computational Geometry 1990, pp. 203–210
2. A. Aggarwal, L.J. Guibas, J. Saxe, and P.W. Shor: A Linear time algorithm for computing the Voronoi diagram of a convex polygon. Proc. 19th ACM Symp. on Theory of Computing 1987, pp. 39–47
3. F. Aurenhammer: Voronoi diagrams – A survey. To appear in ACM Computing Surveys
4. D. Avis and H. ElGindy: Triangulating simplicial point sets in space. Proc. 2nd ACM Symp. on Computational Geometry 1986, pp. 133–141
5. J.L. Bentley and T.A. Ottmann: Algorithms for reporting and counting geometric intersections. IEEE Transactions on Computers 28 (1979) 643–647
6. J.L. Bentley and M.I. Shamos: Divide-and-conquer for linear expected time. Information Processing Letters 7 (1978) 87–91
7. J.D. Boissonnat, O. Devillers, R. Schott, M. Teillaud, and M. Yvinec: Applications of random sampling to on-line algorithms in Computational Geometry. INRIA Tech. Report 1285 (1990)
8. J.D. Boissonnat, O. Devillers, and M. Teillaud: A randomized incremental algorithm for constructing higher order Voronoi diagrams. To appear in Algorithmica
9. J.D. Boissonnat, O. Devillers, R. Schott, M. Teillaud, and M. Yvinec: On-line algorithms with good expected behaviours. Manuscript (1991)
10. B. Chazelle: Reporting and counting segment intersections. J. Computer System Science 32 (1986) 156–182
11. B. Chazelle and H. Edelsbrunner: An optimal algorithm for intersecting line segments in the plane. Proc. 29th IEEE Symp. on Foundations of Computer Science 1988, pp. 590–600
12. B. Chazelle, H. Edelsbrunner, L.J. Guibas, and M. Sharir: Computing a face in an arrangement of line segments. Proc. 2nd ACM-SIAM Symp. on Discrete Algorithms 1991, pp. 441–448
13. B. Chazelle, L.J. Guibas, and D.T. Lee: The power of geometric duality. BIT 25 (1985) 76–90
14. P. Chew: Building Voronoi diagrams for convex polygons in linear expected time. Manuscript (1986)
15. K.L. Clarkson: A probabilistic algorithm for the post office problem. Proc. 17th ACM Symp. on Theory of Computing (1985), pp. 175–184
16. K.L. Clarkson: New applications of random sampling in computational geometry. Discrete & Computational Geometry 2 (1987) 195–222

17. K.L. Clarkson and P.W. Shor: Algorithms for diametral pairs and convex hulls that are optimal, randomized, and incremental. Proc. 4th ACM Symp. on Computational Geometry (1988), pp. 12–17

18. K.L. Clarkson and P.W. Shor: Applications of random sampling in computational geometry, II. Discrete & Computational Geometry 4 (1989) 387–421

19. K.L. Clarkson: Linear programming in $O(n3^{d^2})$ time. Inf. Proc. Letters 22 (1986) 21–24

20. K.L. Clarkson: A Las Vegas algorithm for linear and integer programming when the dimension is small. Manuscript; a preliminary version appeared in Proc. 29th IEEE Symp. on Foundations of Computer Science (1988), pp. 452–456

21. K.L. Clarkson: Personal Communication, September 10 (1990)

22. M.E. Dyer: Linear algorithms for two and three-variable linear programs. SIAM J. on Computing 13 (1984) 31–45

23. M.E. Dyer: On a multidimensional search technique and its applications to the Euclidean one-centre problem. SIAM J. on Computing 15 (1986) 725–738

24. M.E. Dyer and A.M. Friez: A randomized algorithm for fixed-dimensional linear programming. Mathematical Programming 44 (1989) 203–212

25. H. Edelsbrunner: Algorithms in combinatorial geometry. Springer, Berlin Heidelberg New York 1987

26. H. Edelsbrunner, F.P. Preparata, and D.B. West: Tetrahedrizing point sets in three dimensions. Tech. Rep. UIUCDCS-R-86-1310, Univ. of Illinois, Dept. Computer Science (1986)

27. H. Edelsbrunner, J. O'Rourke, and R. Seidel: Constructing arrangements of hyperplanes and applications. SIAM J. on Computing 15 (1986) 341–363

28. G.H. Gonnet. Handbook of algorithms and data structures. Addison-Wesley, 1984

29. R.L. Graham: An efficient algorithm for determining the convex hull of a finite planar set. Inform. Proc. Lett. 1 (1972) 132–133

30. L.J. Guibas, D.E. Knuth, and M. Sharir: Randomized incremental construction of Delaunay and Voronoi diagrams. Proc. ICALP (1990)

31. T. Hagerup and C. Rüb: A guided tour of Chernoff bounds. Inform. Proc. Letters 33 (1989/90) 305–308

32. D. Haussler and E. Welzl: Epsilon-nets and simplex range queries. Discrete & Computational Geometry 2 (1987) 127–151

33. C.A.R. Hoare: Quicksort. Computer Journal 5.1 (1962) 10–15

34. R.M. Karp: An introduction to randomized algorithms. To appear in Discr. Appl. Math.

35. D.G. Kirkpatrick, R. Seidel: The ultimate planar convex hull algorithm? SIAM J. on Comput. 15, no. 1 (1986) 287–299

36. R. Klein: Concrete and abstract Voronoi diagrams. Springer, Lecture Notes in Computer Science 400 (1989)

37. J. Matoušek and R. Seidel: On tail estimates for Mulmuley's segment intersection algorithm. In preparation

38. J. Matoušek, M. Sharir, and E. Welzl: A subexponential bound for linear programming. To appear in Proc. of 8th ACM Symp. on Computational Geometry (1992)

39. N. Megiddo: Linear-time algorithms for linear programming in $\mathbb{R}^3$ and related problems. SIAM J. on Computing 12 (1983) 759–776

40. N. Megiddo: Linear programming in linear time when the dimension is fixed. Journal of the ACM **31** (1984) 114–127

41. K. Mehlhorn: Personal Communication, October (1990)

42. K. Mulmuley: A fast planar partition algorithm: Part I. Proc. 29th IEEE Symp. on Foundations of Computer Science (1988), pp. 580–589

43. K. Mulmuley: A fast planar partition algorithm: Part II. Proc. 5th ACM Symp. on Computational Geometry (1989), pp. 33–43

44. K. Mulmuley: On obstructions in relation to a fixed viewpoint. Proc. 30th IEEE Symp. on Foundations of Computer Science (1989), pp. 592–597

45. F.P. Preparata and M.I. Shamos: Computational geometry – An introduction. Springer, Berlin Heidelberg New York 1985

46. M.O. Rabin: Probabilistic algorithms. In: J.F. Traub, ed., Algorithms and Complexity, Recent Results and New Dierections. Academic Press, New York (1976), pp. 21–39

47. P. Raghavan: Lecture notes on randomized algorithms. IBM T.J. Watson Research Center Computer Science Report RC 15430 (1990)

48. G. Rote: Personal communication. October 15 (1990)

49. R.Sedgewick: Quicksort. Garland, New York (1978)

50. R. Seidel: Linear programming and convex hulls made easy. Proc. 6th ACM Symp. on Computational Geometry (1990), pp. 211–215

51. R. Seidel: A simple and fast incremental algorithm for computing trapezoidal decompositions and for triangulating polygons. To appear in Computational Geometry: Theory and applications (1991)

52. M.I. Shamos: Computational geometry. Ph.D. thesis, Dept. of Computer Science, Yale Univ. (1978)

53. M.I. Shamos and D. Hoey: Closest point problems. Proc. 16th IEEE Symp. on Foundations of Computer Science (1975), pp. 151–162

54. M. Sharir and E. Welzl: A combinatorial bound for linear programming and related problems. Proc. of 9th Symp. on theoretical Aspects of Computer Science (STACS 1992)

55. J.S. Vitter and Ph. Flajolet: Average-case analysis of algorithms and data structures. In: J. van Leeuwen, ed, Handbook of theoretical computer science: algorithms and complexity. Elsevier (1990), pp. 431–524

56. E. Welzl: Smallest enclosing disks (balls and ellipsoids). In: H. Maurer, ed., New results and new trends in computer science. Springer Lecture Notes in Computer Science **555** (1991) 359–370

Chapter III. Epsilon-Nets
and Computational Geometry

Jiří Matoušek

Abstract. In this chapter, we present some results from the theory of range spaces of finite VC-dimension. We introduce canonical geometric range spaces and we indicate how other range spaces encountered in computational geometry can be embedded into the canonical ones. As applications of ε-nets for computational geometry problems, we mention the Cutting lemma, the Short edge lemma and the construction of an arrangement with small zones. We then indicate some improvements of these results when the application of general ε-net results is replaced by other methods, and we also address the issue of de-randomizing the algorithms based on ε-nets and related probabilistic techniques. We give main ideas of several proofs.

1. Range Spaces and ε-Nets

A *range space* is a set system, i.e. a pair $\Sigma = (X, \mathcal{R})$, where X is a set and $\mathcal{R}$ is a set of subsets of X. We will usually call the elements of X the *points* of Σ and the elements of $\mathcal{R}$ the *ranges* of Σ.

When Y is a subset of X, we naturally get a *subspace of Σ induced by Y* – namely the range space $(Y, \{R \cap Y; R \in \mathcal{R}\})$.

A subset $S \subseteq X$ is called an *ε-net for Σ* provided that $S \cap R \neq \emptyset$ for every range $R \in \mathcal{R}$ with $|R|/|X| > \varepsilon$, i.e. S intersects every range which contains a fraction bigger than ε of the points of X. If we set $\mathcal{R}_\varepsilon = \{R \in \mathcal{R}; |R|/|X| > \varepsilon\}$, we may also say that an ε-net for Σ is a transversal of the set system $\mathcal{R}_\varepsilon$.

The above definition makes sense for a finite X only. In a more general setting, we could take X as a probability space (i.e. a set X with a measure μ, satisfying $\mu(X) = 1$) and $\mathcal{R}$ as a subcollection of measurable sets in X. Then $\mathcal{R}_\varepsilon$ would mean just the collection of all ranges of measure greater than ε. The special measure we used in the above definition was given by $\mu(R) = |R|/|X|$ and it will be sufficient in our computational geometry applications.

The ε-nets interesting for us will be the small ones which, in some sense, give an "economic approximation" of the range space in question. Since we can have a collection of $1/\varepsilon$ disjoint ranges of measure ε (unless the range space is very degenerate), the best size of ε-net we can hope for is of order $1/\varepsilon$. On the other hand, we must put some requirement of the range space if we want to get something substantially smaller than n, since if e.g. the ranges are just *all* subsets of X, then any ε-net must have at least $(1 - \varepsilon)|X|$ points.

We begin by a simple probabilistic consideration, which gives the following result:

Proposition 1.1   *Let $\Sigma = (X, \mathcal{R})$ be a finite range space and let $\varepsilon \in (0, 1)$. Then there exists an ε-net S of size at most $(1/\varepsilon) \ln |\mathcal{R}|$ for Σ.*

Proof. Let $|X| = n$, and let S be a random sample drawn from X by s independent random draws (thus elements may be drawn several times). Then for any given range R of size greater than εn, the probability that $S \cap R = \emptyset$ will be at most $(1 - \varepsilon)^s < \exp(-\varepsilon s)$, and thus the probability that any of $m = |\mathcal{R}|$ ranges is missed by S is at most $m \cdot \exp(-\varepsilon s)$. Thus for $s \geq (1/\varepsilon) \ln m$, S will be an ε-net for our range space with a positive probability. $\square$

Let us now look at a typical computational geometry example of a range space. Let X be an n-point set in the plane and let a range be any subset of X which can be formed as an intersection of X with some triangle. One easily observes that the number of ranges in this space is polynomial (at most n^6 in our case). Therefore the above proposition shows that this range space admits an ε-net of size $O((1/\varepsilon) \log n)$. It turns out that this upper bound can be further strengthened to $O((1/\varepsilon) \log(1/\varepsilon))$ (which is interesting if ε^{-1} is small compared to n, e.g., for a constant ε) . The essential property of the range space needed for such an improvement is that the number of ranges is *hereditarily* polynomial, i.e. for every $A \subseteq X$, the number of ranges of the subspace induced by A is bounded by the same polynomial in the cardinality of A. The property of having hereditary polynomial number of ranges for a range space can be nicely characterized by a single type of a forbidden subspace.

Let us say that a subset $A \subseteq X$ is *shattered* (by $\mathcal{R}$), if every possible subset in the subspace induced by A is a range, i.e. if $\{A \cap R; R \in \mathcal{R}\} = 2^A$. We define the *Vapnik-Chervonenkis dimension* (or VC-dimension for shortness) of the range space $\Sigma = (X, \mathcal{R})$ as the maximum size of a shattered subset of X (if there are shattered subsets of any size, then we say that the VC-dimension is infinite).

A theorem proved independently by Vapnik and Chervonenkis [VC71], Sauer [Sau72] and Shelah says that the range spaces of finite VC-dimension are just those with hereditarily polynomial number of ranges. For a range space $\Sigma = (X, \mathcal{R})$, we define so-called *shatter function* $\pi_{\mathcal{R}}$ as follows: for a subset $A \subseteq X$, we set $\Pi_{\mathcal{R}}(A) = \{A \cap R; R \in \mathcal{R}\}$ and we put

$$\pi_{\mathcal{R}}(n) = \max\{|\Pi_{\mathcal{R}}(A)|; A \subseteq X; |A| = n\}.$$

Thus the VC-dimension of $(X, \mathcal{R})$ is just the maximum number d with $\pi_{\mathcal{R}}(d) = 2^d$. The shatter function is bounded as follows:

Theorem 1.2 [VC71,Sau72]. *For a range space $(X, \mathcal{R})$ of VC-dimension $\leq d$, it is*

$$\pi_{\mathcal{R}}(n) \leq \Phi_d(n)$$

for every n, where

$$\Phi_d(n) = \binom{n}{0} + \binom{n}{1} + \cdots + \binom{n}{d} = \Theta(n^d).$$

Thus, a shatter function is either 2^n (the case of infinite VC-dimension) or it is bounded by a polynomial. The above theorem is proved by induction: Consider a range space $\Sigma = (X, \mathcal{R})$ of VC-dimension d and fix some $x \in X$. Define range spaces $\Sigma_1 = (X \setminus \{x\}, \{R \setminus \{x\}; R \in \mathcal{R}\})$ and $\Sigma_2 = (X \setminus \{x\}, \{R \in \mathcal{R}; x \notin R, R \cup \{x\} \in \mathcal{R}\})$. The number of ranges of Σ equals to the number of ranges of Σ_1 plus the number of ranges of Σ_2. Σ_1 is an $n - 1$ point range space of VC-dimension d, and Σ_2 is easily shown to be a range space of VC-dimension $d-1$ on $n-1$ points. This gives a recurrence from which the formula is obtained. □

Haussler and Welzl [HW87] have introduced the notion of an ε-net and using a technique similar to the one of [VC71], they proved that in a range space of VC-dimension d, there exists an ε-net of size at most $(8d/\varepsilon) \log(8d/\varepsilon)$. The logarithmic factor might seem to come from the proof technique, but Pach and Wöginger [PW90] gave a lower bound example which shows that in general it has to be there. The upper bound (expressed in terms of d and ε) was improved by Blumer et al. [BEHW89] and then by Komlós; these results will appear in a joint paper [KPW92], where the following is shown:

Theorem 1.3 [KPW92]. *Let $f(d, \varepsilon)$ denote the maximum size, over all range spaces of VC-dimension $\leq d$, of a smallest ε-net for that space. Then*

$$f(d, \varepsilon) = (1 + o(1)) \frac{d}{\varepsilon} \log \frac{1}{\varepsilon}.$$

We will sketch the proof here, completely omitting all details and calculations. For the lower bound, one just takes an n-point set X, and chooses a collection of ranges by m independent choices of a random εn-element subset of X, obtaining a range space. Then, by a straightforward calculation, one bounds the probability that there exists a shattered d-point subset, and the probability that there exists a s-point transversal (ε-net) of the collection of ranges. Choosing the parameters n, m, s suitably, one gets a lower bound $f(d, \varepsilon) = \Omega(d/\varepsilon \log(1/\varepsilon))$. To get the exact value of the multiplicative constant, one uses a slightly more complicated approach (see [KPW92]).

The idea of the upper bound proof goes back to [VC71]. Similarly as in the proof of Proposition 1.1, we pick a random sample S by s independent random draws and we want to establish that it is an ε-net with a positive probability. To this end, we pick another random sample $\bar{S}$ by $\bar{s}$ more independent random draws, where $\bar{s}$ is about the same as s (or slightly bigger in the proof of

Komlós). Although $\bar{S}$ seems to have nothing to do with the ε-net property of S, it will help us to prove it. We consider both S and $\bar{S}$ as multisets, i.e. for every element we also remember how many times it was picked. First we look at the sample $\bar{S}$ and consider any subset $Z \subseteq X$ of size εn. Let m be the median of the number of points in which $\bar{S}$ hits Z, i.e. m is such that $\bar{S}$ intersects Z in at least m points with probability at least $1/2$. Obviously m does not depend on the choice of Z. Since it is a median of a binomial distribution, it will be very near to the expected size of the intersection $Z \cap S$, which is $\varepsilon\bar{s}$.

If the sample S is not an ε-net for our range space, it means that there exists a range R of size bigger than εn missed by S. Fix such a range $R(S)$ for every bad S. Since the choice of $\bar{S}$ is independent of the choice of S, $\bar{S}$ hits $R(S)$ in $\geq m$ points with probability at least $1/2$, thus

$$\text{Prob}(S \text{ is not an } \varepsilon\text{-net}) \leq 2\text{Prob}(\exists R \in \mathcal{R}; R \cap S = \emptyset \text{ and } |R \cap \bar{S}| \geq m). \quad (1)$$

Now we will look at the process of selection of S and $\bar{S}$ another way. Imagine that we first fix the multiset of elements which will belong to the union $S \cup \bar{S}$ (here the union is a union of multisets, i.e. the counts are added together for every element). For every such choice, we then consider all possible distributions of the elements of the union between S and $\bar{S}$. For every choice of the multiset $S \cup \bar{S}$, one can show that a distribution of points of the union between S and $\bar{S}$ for which $R \cap S = \emptyset$ and at the same time $|R \cap \bar{S}| \geq m$ is very improbable for any given range R (since this requires putting all elements of $R \cap (S \cup \bar{S})$ into $\bar{S}$ only). We now want to show that it is even likely that such a distribution will occur for no range at all. For our purposes, two ranges behave the same way if they intersect $S \cup \bar{S}$ in the same subset, and from the finite VC-dimension we know that the number of distinct intersections of $S \cup \bar{S}$ with a range is bounded by $\pi_{\mathcal{R}}(|S \cup \bar{S}|)$. Choosing the parameters s and $\bar{s}$ suitably and performing all the calculations, we get from this that the probability (1) will tend to zero and thus that S is an ε-net for our range space with a positive probability.   □

An important notion which historically preceded ε-nets is ε-approximation (due to [VC71]). A subset $A \subseteq X$ is an ε-approximation for a range space $(X, \mathcal{R})$, provided that for every range $R \in \mathcal{R}$ it is

$$\left| \frac{|A \cap R|}{|A|} - \frac{|X \cap R|}{|X|} \right| \leq \varepsilon,$$

i.e. the relative number of points of A in every range approximates the relative size of that range with accuracy ε. Again, one can show the existence of small ε-approximations:

Theorem 1.4 [VC71]. *Any range space of VC-dimension d admits an ε-approximation of size $O(d/\varepsilon^2 \log(d/\varepsilon))$.*

The proof of this theorem is similar in spirit to the above sketched proof of ε-net existence. An alternative proof is given in [Mat91a]. The bound in this theorem has recently been improved to an almost tight one, see [MWW91].

We will indicate how this theorem together with Proposition 1.1 imply that every range space of a bounded VC-dimension admits an ε-net of size $O((1/\varepsilon)\log(1/\varepsilon))$ (a weak form of Theorem 1.3). To see this, we need a simple observation:

Observation 1.5 *If A is an ε-approximation for a range space Σ and S is a δ-net for the range space induced by A in Σ, then S is an $(\varepsilon+\delta)$-net for Σ.*

Let Σ be a range space of bounded VC-dimension and let us choose (by Theorem 1.4) an $(\varepsilon/2)$-approximation A of size polynomial in ε^{-1} for Σ. Using the bounded VC-dimension, we know that the number of ranges of the subspace of Σ induced by A is polynomial in $|A|$ and thus also in ε^{-1}, and so by Proposition 1.1, this subspace admits an $(\varepsilon/2)$-net S of size $O((1/\varepsilon)\log(1/\varepsilon))$. This will be an ε-net for Σ.

The class of range spaces of finite VC-dimension has some pleasant closure properties. It is obviously hereditary, and it is also closed under the set-theoretic operations on ranges [Dud78]. The following lemma gives a quantitative form of this fact:

Lemma 1.6 [HW87]. *Let $(X,\mathcal{R})$ be a range space of VC-dimension d. Suppose that every range of a range space $(X,\mathcal{R}')$ is defined as $\varphi(r_1,\ldots,r_k)$, where φ is a fixed set-theoretic formula (involving unions, differences etc.) and $r_1,\ldots,r_k \in \mathcal{R}$. Then the VC-dimension of $(X,\mathcal{R}')$ is at most $\max\{n; \Phi_d(n)^k \geq 2^n\} = O(dk\log(dk))$.*

The proof follows from Theorem 1.2 and from the inequality $\pi_{\mathcal{R}'}(n) \leq \pi_{\mathcal{R}}(n)^k$.
□

Another operation preserving finite VC-dimension is the construction of *dual range space* by interchanging the role of points and ranges: the dual of a range space $\Sigma = (X,\mathcal{R})$ is the range space $\Sigma^* = (\mathcal{R}, \{\{R \in \mathcal{R}; x \in R\}; x \in X\})$ (see [Ass83,CW89]). The shatter function of Σ^* (called also the *dual shatter function* of Σ and denoted by $\pi_{\mathcal{R}}^*$), has a natural interpretation in Σ: $\pi_{\mathcal{R}}^*(m)$ is the maximum number of equivalence classes (cells the Venn diagram) into which a set of m ranges of $\mathcal{R}$ may partition the points of X.

2. Geometric Range Spaces

Let us begin by an example of a geometrically defined range space which has *infinite* VC-dimension. Such a range space is e.g.

$$(E^2, \{r \subseteq E^2; r \text{ is a convex set}\}).$$

Indeed, if we choose a subset A of d convex independent points, then this set will be shattered, since $\text{conv}(Y) \cap A = Y$ for every $Y \subseteq A$. We may even let the ranges be congruent copies of a fixed convex set, and still the VC-dimension can be infinite.

In the above example, the ranges were defined by objects of arbitrarily large complexity (description size). Now we pass to objects defining range spaces of finite VC-dimension. A basic example in the plane is the range space $(\mathrm{E}^2, \{h; h \text{ a halfplane}\})$. It is not difficult to verify by elementary geometric considerations that its VC-dimension is 3 (no 4-point set can be shattered). However, the shatter function is only quadratic, i.e. smaller than the general upper bound. More generally, the range space with point set E^d and ranges defined by d-dimensional halfspaces can be shown to have VC-dimension $d+1$. Still more generally, let us call any intersection of at most k d-dimensional halfspaces a H_d^k-range, and let us define range spaces

$$H_d^k = (\mathrm{E}^d, \{h; h \text{ is a } H_d^k\text{-range}\})$$

(when $k = 1$ we omit the superscript and write just H_d). By Lemma 1.6, this range space has a finite VC-dimension depending on d, k. If X is a finite subset of E^d, we will denote by $H_d^k(X)$ the subspace of H_d^k induced by X. As we will now indicate, the range spaces of the form $H_d^k(X)$ can serve as canonical range spaces for computational geometry.

Let $(X, \mathcal{R})$ and $(Y, \mathcal{S})$ be range spaces and let $\varphi : X \to Y$ be an injective mapping. Let us call φ an *embedding* of range spaces, if for every $R \in \mathcal{R}$, $\varphi(R)$ is a range in the subspace induced by $\varphi(X)$ in $(Y, \mathcal{S})$, i.e. it can be expressed as $R = S \cap \varphi(X)$ for some $S \in \mathcal{S}$. Obviously if we can find e.g. an ε-net for the subspace induced by $\varphi(X)$, its inverse image under φ will be an ε-net for $(X, \mathcal{R})$.

As a first example, let us consider a range space Σ with point set E^2 and with ranges determined by general conic curves, i.e. of the form

$$R(a_{00}, a_{01}, a_{02}, a_{10}, a_{20}, a_{11}) = \{[x, y]; \ a_{00} + a_{01}x + a_{02}x^2 \\ + a_{10}y + a_{20}y^2 + a_{11}xy \geq 0\}.$$

Let us define a mapping $\varphi : \mathrm{E}^2 \to \mathrm{E}^5$ by $\varphi(x, y) = [x, x^2, y, y^2, xy]$. This is an embedding of Σ into H_5: Indeed, $\varphi(R(a_{00}, a_{01}, a_{02}, a_{10}, a_{20}, a_{11}))$ can be expressed as the intersection of the hypersurface $\text{Im}\,\varphi$ with the halfspace $a_{10}x_1 + a_{20}x_2 + a_{10}x_3 + a_{20}x_4 + a_{11}x_5 \geq -a_{00}$. Obviously, this "lifting" works for ranges determined by any single polynomial inequality of a bounded degree – every monomial in the defining polynomial is assigned one coordinate in the image space. This general method is sometimes quite wasteful in the dimension of the image space; e.g., for the range space defined by balls in E^3, it gives en embedding into H_6, while the well-known lifting to the paraboloid gives an embedding into H_4.

If the ranges of our range space are defined by a *system* (conjunction) of k polynomial inequalities of fixed degrees, we can extend the above method and embed this range space into the range space $H_{d'}^k$ for some d'.

What about disjunction of polynomial inequalities? Range spaces whose ranges can be represented as a union of a fixed number k of simpler ranges (e.g. a convex $(k+2)$-gon in the plane is a union of k triangles) present no problem for finding ε-nets (unless we are interested in exact values of constants of proportionality). One can proceed as follows: find an (ε/k)-net for each type of the simpler ranges, and a union of these (ε/k)-nets will be an ε-net for the combined ranges (e.g., a (ε/k)-net for triangles is an ε-net for convex $(k+2)$-gons).

This shows that if we are able to solve problems about ε-nets in range spaces of the form $H_d^k(X)$, we can handle quite a wide variety of geometric range spaces – at least those with ranges definable by a quantifier-free formula of bounded size in the theory of real closed fields (quantified formulas can be handled by known quantifier elimination techniques, but we do not want to go into this business here). Hence for most of questions, we consider $H_d^k(X)$ as *the* geometric range spaces.

The observation about "lifting" geometric range spaces is due to Yao and Yao [YY85] (there it is formulated as applying an algorithm for halfspace range queries to queries of a more general form). The applicability to ε-nets was noted by Emo Welzl.

Now one might suspect that any range space of VC-dimension d can be embedded in a space $H_{d'}^k$ for some k, d' depending on d only. It turns out that this is not possible. One may show this by a counting argument: The number of n-point range spaces of the form $H_d^k(X)$ can be bounded using the Milnor theorem on the number of connected components of algebraic varieties (or directly by the result of Goodman and Pollack [GP86] on the number of order types), while one can establish the existence of a much larger number of range spaces of a bounded VC-dimension by a probabilistic argument similar to the one of [PW90]. Alon et al. [AHWW87] gave a specific example of range spaces of VC-dimension 2 (the finite projective planes, where the ranges are the lines), whose embedding into some $H_{d'}^k$ requires arbitrarily large values of d'. Their result is stronger than what one gets by the counting argument, since it does not use the "straightness" of the halfspaces determining the ranges in $H_{d'}^k$, and so it holds also for embedding into a range space determined by "pseudohalfspaces" (such a range space corresponds to combinatorial order type, which is not necessarily realizable by a point set in E^d).

Let us remark that the structure of range spaces of VC-dimension 1 was more or less described in [PW90] (and it turns out that each such range space has an ε-net of size $O(1/\varepsilon)$), but very little is known about the structure of range spaces of higher VC-dimensions.

3. A Sample of Applications

In this section we will discuss some basic applications of ε-nets in computational geometry (we do not consider other applications as in learning theory – see e.g. [BEHW89]).

Let H be a collection of n lines in the plane. Let us consider a range space with a point set H and with ranges determined by triangles – i.e. every range is a collection of lines of H intersecting some triangle. There are several ways how to show that this range space has a finite VC-dimension (using the embedding into some H_d^k or by a direct argument). Let N be an ε-net for this range space, i.e. a subset of lines of H such that any triangle intersected by more than εn lines of H is also intersected by a line of N. When we triangulate the arrangement of N, we get a collection of triangles whose interiors are not intersected by lines of N, and hence the interior of each of these triangles is intersected by at most εn lines of H. Since the triangulation of an arrangement of m lines has $O(m^2)$ triangles and we know from Theorem 1.3 that we can assume that $|N| = O((1/\varepsilon) \log(1/\varepsilon))$, we get the following:

Given a collection H of n lines in the plane and a number $\varepsilon > 0$, the plane can be covered by $O((1/\varepsilon^2)(\log(1/\varepsilon))^2)$ triangles with disjoint interiors, such that the interior of no triangle is intersected by more than εn lines of H.

A careful reader of the previous sentence may object that it is an obvious nonsense, since the plane cannot be covered by a finite number of triangles. This is not a serious issue, since we can allow also "unbounded triangles", or restrict ourselves e.g. to a big rectangle covering all the vertices of the arrangement of H.

The reader might have observed or known from the literature that the above argument can be performed with an ε-net for a range space with ranges determined by segments rather than triangles. Indeed, if we know that the triangles of the triangulation of N contain no segment intersected by more than εn lines of H, we can conclude that their interior cannot be intersected by more than $2\varepsilon n$ lines of H, since every line intersecting a triangle must intersect one of given two its sides. However, this works for lines (and hyperplanes), but not necessarily for curved objects like e.g. circles.

To state the result in a compact form, let us define an ε-*cutting for H*, where H is a collection of n hyperplanes in $\mathbf{E}^d$, to be a collection of simplices with disjoint interiors and covering $\mathbf{E}^d$, such that the interior of each simplex is intersected by at most εn hyperplanes of H. Changing our parameterization by ε to a more suitable one by $r = 1/\varepsilon$, we can state the result as follows:

Theorem 3.1 (Cutting lemma). *Let H be a collection of n hyperplanes in $\mathbf{E}^d$ and $r \leq n$ a parameter. There exists a $(1/r)$-cutting for H, consisting of $O(r^d(\log r)^d)$ simplices.*

This statement can be further generalized to more complicated hypersurfaces, as long as a "triangulation result" for these hypersurfaces is available, which means that one can subdivide the cells in an arrangement of hypersurfaces of the considered kind into constant complexity cells (and keep the number of these new cells reasonably small). This is usually no problem in the plane, but in higher dimensions one can get into difficulties (see [CEGS89] for results in this direction).

Considerations leading to the proof of Cutting lemma appear in [HW87]. Independently, Clarkson developed a framework (different from the ε-net theory) also leading to the proof of this result and many others (see e.g. [Cla88] or the journal version [CS89]). This framework gives results equivalent to the ε-net theory in many instances and we will not attempt to survey it in the present paper.

The above Cutting lemma and its generalizations and modifications became one of basic paradigms in the design of efficient geometric algorithms (and it also has purely combinatorial applications). Roughly speaking, it gives an efficient divide and conquer strategy for a problem involving the hyperplanes (or hypersurfaces). By this phrase saying nothing to an uninitiated reader we will quit the subject of applications of Cutting lemma before we really started and refer e.g. to the paper [Aga90b].

Another significant application of ε-nets, which originated in a paper of Welzl [Wel88] and was refined in Chazelle and Welzl [CW89], is for *range searching*. Range searching problems in computational geometry have a fundamental importance and a rich history (see e.g. [YY85,HW87,Cha89,CSW90, Mat91c] for an account and contributions to the subject). In this paper, we will touch mainly so-called *halfspace* and *simplex* range searching problem. In a traditional setting, the problem is the following: Given a point set P in E^d, construct a data structure allowing to compute the number of points of P in a given query range (halfspace or simplex) quickly. The main issue is a reasonable tradeoff between the worst-case query time and the space used for the data structure. A variant of this problem asks for reporting all the points lying in the query range. In a general setting, each point $p \in P$ is associated with some value $v(p) \in S$, where S is some semigroup (let us denote its operation as addition) and we want to evaluate sums of the form $\sum_{p \in Q} v(p)$, where Q is a given query range.

In order to establish lower bounds for range searching problems, so-called *arithmetic model* was proposed by Fredman [Fre81]. In this model, one precomputes and stores some partial sums of the values assigned to points, and an answer to a query is then expressed using these precomputed sums. What determines the complexity of the query is just the number of semigroup operations needed to compute the answer from the precomputed sums, and the storage used is assumed to be proportional to the number of precomputed sums. This model ignores many issues, as e.g. how to detect which precomputed sums should be used to form an answer. Thus while a lower bound established in the arithmetic model is quite strong, an efficient solution to a range search problem in this model does not automatically lead to an efficient algorithm in the usual sense.

Chazelle [Cha89] gave lower bounds for the simplex range searching problem, which say that given m units of storage and a n point set in E^d, we cannot in general achieve a better query time than $\Omega((n/\log n)/m^{1/d})$. For $d = 2$ the $\log n$ factor in the denominator disappears, so e.g., in the plane and with a linear storage we cannot achieve a better query time than $\Omega(\sqrt{n})$.

Numerous algorithms were proposed for the simplex range searching problem; the lower bound was first attained (upto a polylogarithmic factor) in the plane ([Wel88]), then in a general dimension but in the arithmetic model only ([CW89]) and finally by efficient randomized algorithms of Chazelle et al. [CSW90] in all dimensions. Improved results are given in [Mat91c] (deterministic algorithms optimal upto polylogarithmic factors, with particularly simple query-answering algorithm based on efficient partition trees).

The method of [Wel88] and [CW89] can be formulated in terms of a general range space of a finite VC-dimension (in the arithmetic model, generalization of the semigroup range query problem for an arbitrary range space is straightforward). In fact, Chazelle and Welzl [CW89] reveal the following connection between finite VC-dimension and the existence of efficient data structures for query answering: A range space allows to build a linear-siz data structures (for its finite subspaces) for answering semigroup range queries efficiently iff it has a finite VC-dimension. Here "efficiently" means that for a n-point subspace, any query can be answered in time $O(n^\beta)$ for some constant $\beta < 1$ (in the arithmetic model).

The approach of [CW89] uses *spanning trees with low crossing number.* Let $\Sigma = (X, \mathcal{R})$ be a range space. If $x, y \in X$, we say that a range R *crosses* $\{x, y\}$ if either $x \in R$ and $y \notin R$, or $y \in R$ and $x \notin R$. Let T be a spanning tree on the set X. We define the *crossing number of T relative to a range R* as the number of edges of T (considered as two-point subsets of X) crossed by R. We let the crossing number of T be the maximum of crossing numbers of T relative to all ranges of $\mathcal{R}$. Welzl [Wel88] shows that a range space of a finite VC-dimension admits a spanning tree (even a spanning path) with a sublinear crossing number:

Theorem 3.2   [Wel88]. *Let $(X, \mathcal{R})$ be a finite range space, $|X| = n$, and let its dual shatter function satisfy $\pi^*_{\mathcal{R}}(m) = O(m^d)$ for some constant d. Then there exists (and can be constructed in polynomial time) a spanning path on X with crossing number $O(n^{1-1/d} \log n)$.*

In order to build an efficient data structure for semigroup range queries, one can try to find a collection of m distinguished subsets of the given set X, such that any intersection of X with the query range can be expressed as a disjoint union of at most q distinguished subsets. This gives storage m and query time q in the arithmetic model.

Using the spanning path guaranteed by Theorem 3.2, the collection of distinguished subsets can easily be constructed. In a simplest approach, the distinguished subsets will be the canonical intervals along the spanning path. The total number of canonical intervals is $O(n \log n)$. Any range cuts the spanning path into $O(n^{1-1/d} \log n)$ connected pieces, and every piece is expressed as a disjoint union of $O(\log n)$ canonical intervals. A more sophisticated approach allows to use a linear number of distinguished subsets and express every connected piece of the path as a disjoint union of $O(\alpha(n))$ subsets – see [CW89].

For simplex range searching in dimensions 2 and 3, one can extend this approach to get efficient range search algorithms. In higher dimensions, there seem to be substantial difficulties in passing from the arithmetic model to a realistic algorithm.

We will not give a complete proof of Theorem 3.2 here, but we concentrate on an intermediate result used in this proof:

Lemma 3.3   (Short Edge Lemma) *Let $\Sigma = (X, \mathcal{R})$ be a finite range space, $|X| = n$, and let its dual shatter function satisfy $\pi_{\mathcal{R}}^*(m) = O(m^d)$ for some constant d. Then for any set (or multiset) $Q \subseteq \mathcal{R}$, there exist points $x, y \in X$, such that the set $\{x, y\}$ is crossed by at most $O(|Q| \log n / n^{1/d})$ ranges of Q.*

Short edge lemma is proved as follows: Let us form a range space

$$\Sigma_1 = (Q, \{\{R \in Q; \ R \text{ crosses } \{x, y\}\}; \ x, y \in X\}).$$

Every range of Σ_1 can be expressed as a difference of two ranges of the dual range space Σ^*, and so from Lemma 1.6 we get that Σ_1 has a VC-dimension bounded by a constant. Let us find an ε-net $N \subseteq Q$ for Σ_1 with $\pi_{\mathcal{R}}^*(|N|) < n$, at the same time choosing ε as small as possible. By Theorem 1.3, we can afford to pick ε of order $O(\log n / n^{1/d})$. The ranges of N partition the points of X into less than n equivalence classes (by the choice of $|N|$), and thus there exist two distinct equivalent points x, y, i.e. such that $\{x, y\}$ is crossed by no range of N. Translating what it means that N is an ε-net for the range space Σ_1, we get that $\{x, y\}$ cannot be crossed by more than $\varepsilon |Q| = O(|Q| \log n / n^{1/d})$ ranges of Q.   □

The original application of ε-nets in [HW87] was a solution to the simplex range searching problem, a construction of a good partition tree. The subject of partition trees would require another survey paper. We will mention just the problem for which ε-nets were applied in [HW87], which seems to be of a basic importance and far from being completely solved:

Problem 3.4   *Given a set X of n points in the plane and a parameter r, find a collection H of lines, which is as small as possible and has the following property: For any line h, the total number of points contained in the cells of the arrangement of H intersected by h (these cells are called the* zone *of h) is at most n/r.*

This problem can be immediately generalized into higher dimensions. Let $f(r)$ denote the minimum necessary size of H (we will show that we can make the size of H independent of n). It is easily seen that we must have $f(r) \geq r$. If one draws a "random" cloud of dots representing points, one might get the impression that the problem could be solved as follows: Construct r lines so that no region of their arrangement contains more than n/r^2 points. However, this is impossible to do in general (it suffices to consider the n points in a convex position), the best we can hope for is an arrangement of r lines in which no cell contains more than n/r points, and hence we must use some more global argument about zones.

The solution of [HW87] gives

Lemma 3.5 $f(r) = O(r^2(\log r)^2)$.

With Improved cutting lemma (see Theorem 4.1 below) this can be immediately improved to $f(r) = O(r^2)$.

The proof uses an observation, whose proof is an exercise on the properties of the duality transform:

Observation 3.6   *Let P be a point lying inside a triangle $G_1 G_2 G_3$. Then if a point q lies in the zone of the line $D(P)$ in the arrangement of the three lines $D(G_1), D(G_2), D(G_3)$, then its dual line $D(q)$ intersects the triangle $G_1 G_2 G_3$.*

Lemma 3.5 is now proved as follows: Consider the set $D(X)$ of n lines dual to the given point set. Apply the Cutting lemma 3.1 to cover the dual plane by $O(r^2(\log r)^2)$ triangles in such a way that no triangle is intersected by more than n/r lines of $D(X)$. Now let G be a set of all vertices of these triangles; we claim that the set $H = D(G)$ of lines dual to G has the desired property. Let h be any line in the primal plane; we want to show that its zone in the arrangement of H does not contain more than n/r points of X. Consider the triangle $G_1 G_2 G_3$ in the dual plane which contains the point $D(h)$. By Observation 3.6, the zone of h in the arrangement of G_1, G_2, G_3 cannot contain more than n/r points of X, since such points dualize to lines of $D(X)$ intersecting the triangle $G_1 G_2 G_3$, and we know that there are no more than n/r of such lines. But the zone of h in the arrangement of the whole $D(G)$ is contained in the zone of h in the arrangement of $D(G_1), D(G_2), D(G_3)$. □

A new progress related to the problem was made by Chazelle et al. [CSW90] who, instead of bounding $f(r)$ directly, proved a weaker result which suffices for their range search application (we again formulate it in the plane, although it holds in any dimension):

Lemma 3.7   [CSW90]. *Given a set X of n points in the plane and a parameter r, one can construct a collection of $k = O(\log r)$ triangulated arrangements $H_1, \ldots, H_k$ of lines, consisting of $O(r \log r)$ lines each, such that for every line h there exists i such that the zone of h in H_i (i.e. the triangles of H_i intersected by h) contains at most n/r points of X.*

The word "triangulated" in the above statement is important. If one could prove a similar statement for arrangements which are not triangulated, one could straightforwardly merge the collections of lines defining $H_1, \ldots, H_k$ together, obtaining a single collection of $O(r(\log r)^2)$ lines for which the zone of any line contains at most n/r points, and thus prove an almost optimal result $f(r) = O(r(\log r)^2)$. The problem with triangulated arrangements is that if we superimpose several of them, we may get a space subdivision of much larger complexity than the original triangulated arrangements. For dimensions 2 and 3, Chazelle et al. [CSW90] obtained results which say that if we use a special type of triangulation, the triangulated arrangements *can* be overlayed without a drastic increase in the complexity of the resulting subdivision (so one gets a theorem saying that a triangulation of the plane consisting of $O(r^2\text{polylog}\,r)$ triangles can be constructed in such a way that the zone of every line contains at most n/r points, and similarly in E^3).

In the proof of Lemma 3.7, we begin similarly as in the proof of the previous Lemma 3.5, namely by covering the dual plane by triangles and constructing the set $H = D(G)$. If h is a line in the primal plane, let $G_1 G_2 G_3$ be the triangle in the dual plane containing the point $D(H)$; we will call the three lines $D(G_1), D(G_2), D(G_3)$ the *guarding triple* of h. Let S be a subdivision of the plane; we observe that for any line h and its guarding triple $\{g_1, g_2, g_3\}$ (actually for any set of lines), the zone of any line h in S is contained in the union of the zone of h in the arrangement of g_1, g_2, g_3 and of the zones of g_1, g_2 and g_3 in S. Since we know from the proof of Lemma 3.5 that the zone of h in the arrangement of its guarding triple is small, we get that it is enough to guarantee small zones for all guarding triples, i.e. to prove the following:

Given a set X of n points and a set T of m triples of lines, one can find $k = O(\log m)$ triangulated arrangements $H_1, \ldots, H_k$ of $O(r \log r)$ lines each, such that for every triple $\{g_1, g_2, g_3\} \in T$, there exists i such that H_i is good for this triple, i.e. the zone of each of g_1, g_2, g_3 in H_i contains at most n/r points of X.

We will show that under the assumptions of the previous statement, we can find a single triangulated arrangement H of $O(r \log r)$ lines, such that H is good for at least a constant fraction of triples in T. This is obviously enough to prove the statement (we apply this $O(\log m)$ times).

Let G be the multiset of lines arising as the (multiset) union of the triples of T. Similarly as in the proof of the Cutting lemma 3.1, let $N \subseteq G$ be a $(1/4r)$-net for the range space with point set G and ranges defined by triangles, $|N| = O(r \log r)$. We let H be a triangulation of the arrangement of N. We know that any triangle in H is intersected by no more than $|G|/4r$ lines of G.

Let us say that a point $x \in X$ and a line $g \in G$ form a *visible pair*, if x lies in the zone of g in H. Any point x may contribute to at most $|G|/4r$ visible pairs, so the total number of visible pairs is at most $n|G|/4r$ and hence at least $3/4$ lines of G contribute to at most n/r visible pairs each. This implies that at least $1/12$ of the triples of T are good.   □

4. Removing Logarithms

The logarithmic factor in the bound of Theorem 1.3 is annoying. One cannot remove it in general, but what about the geometric applications? The most straightforward (and least successful) approach is to try to prove or disprove that for every fixed d, k, every finite $X \subset E^d$ and every $\varepsilon \in (0, 1)$, the range space $H_d^k(X)$ has an ε-net of size $O(1/\varepsilon)$. Pach and Wöginger [PW90] proved it for H_2, this is not too difficult. Seidel et al. [SWM90] proved it for H_3, but their proof does not generalize to higher dimensions at all, so this is an outstanding open problem. An alternative proof for H_3 was given by Clarkson [Cla89] in an unpublished note, and a similar proof is given in [Mat91d], where it easily follows from other results.

Some results were obtained for so-called *weak ε-nets*. If $(X, \mathcal{R})$ is a range space and $A \subseteq X$, a set $S \subseteq X$ is called a weak ε-net for A if it satisfies the ε-net condition for the range space induced by A, i.e. $S \cap R \neq \emptyset$ for every $R \in \mathcal{R}$ with $|A \cap R|/|A| > \varepsilon$. As an example in the plane, let us mention that a one-point weak $(2/3)$-net for the range space $H_2(X)$ always exists and it is usually called the *centerpoint* of X. In geometric range spaces, construction of weak ε-nets sometimes is a cheap trick only. Any finite set X in E^d admits a $d+1$ point weak 0-net in the range space H_d – it suffices to take the vertices of any simplex enclosing all points of X.

Seidel et al. [SWM90] established the existence of a weak ε-net of size $O(1/\varepsilon)$ for range spaces with ranges defined by balls in E^d. The proof is by a simple reduction to a known result about convex sets. Alon et al. [ABFK91] recently proved the existence of weak ε-nets of size $O(1/\varepsilon^{d+1-\alpha(d)})$ for convex sets in E^d, with certain small constant $\alpha(d) > 0$. This contrasts with the fact that ε-nets of size depending on ε only do not exist in these range spaces. This result has been applied by Alon and Kleitman [AK91] for a solution of a long-standing open problem in combinatorial geometry.

The effort to get rid of the logarithmic factor was successful in most of computational geometry applications of ε-nets; the use of ε-nets could have been refined or replaced by other methods.

A general method applicable to many situations was developed by Clarkson [Cla88]. The idea is to avoid worst-case estimates, but to use global expectations. The reader may get the best feeling of the difference by comparing this Clarkson's work with the older [Cla87]. The technique is enriched and used several times in [CEG+90]; see also [EGS90].

Another type of improvement was achieved for the Cutting lemma 3.1:

Theorem 4.1   [CF90]. (Improved Cutting Lemma.)   Let H be a collection of n hyperplanes in E^d and $r \leq n$ a parameter. One can find (in polynomial time) a $(1/r)$-cutting for H, consisting of $O(r^d)$ simplices.

The above bound on the number of simplices of the cutting is asymptotically optimal (since one simplex may contain at most $(n/r)^d$ vertices of the arrangement of H). We will sketch the proof of the Improved cutting lemma (but we have to omit a proof of a key lemma). First, let us pick a sample of r hyperplanes and triangulate its arrangement; call the resulting simplices the *primary simplices*. For every primary simplex s, consider the collection $H_s \subseteq H$ of hyperplanes intersecting it. We cannot exclude that some $|H_s|$ will be significantly bigger than n/r. Let us define the quantity $t_s = |H_s|r/n$, this is the "excess" of the simplex s and let n_t denote the number of simplices with $t_s \geq t$. A key result Chazelle and Friedman [CF90] prove is that the expected value of n_t is of order $O(r^d 2^{-t})$, i.e. the number of simplices with excess at least t decreases exponentially with t (more exactly, they prove that this is true if we use a specific way of triangulation for the arrangement of the sample).

To get a sufficiently fine subdivision, we will further subdivide each of the primary simplices; the simplices with a bigger excess will be subdivided more finely. The subdivision inside a simplex s is done using the original Cutting lemma 3.1. We apply that lemma on the collection H_s, with the value of parameter r equal to t_s. Then the whole space is subdivided into simplices intersected by at most $|H_s|/t_s = n/r$ hyperplanes of H_s. The simplices of this $(1/t_s)$-cutting for H_s are then intersected with the simplex s. The resulting cells are not simplices in general, but each of them can be triangulated using a constant number of simplices per cell, and each of the resulting simplices (*secondary simplices*) is obviously intersected by no more than n/r hyperplanes of H. This is done for every primary simplex s.

It remains to bound the number of secondary simplices used for all primary simplices s together. By Cutting lemma, we needed at most $O(t_s^d(\log t_s)^d)$ secondary simplices for a primary simplex s, hence the total number of secondary simplices is bounded by

$$\sum_s O(t_s^d(\log t_s)^d) \leq \sum_{t=1}^{\infty} n_t . O(t^d(\log t)^d),$$

and the expected value of this sum is bounded by

$$O(r^d) \sum_{t=1}^{\infty} t^d(\log t)^d 2^{-t} = O(r^d),$$

this finishes the proof. Note that we did not need the almost optimal bound on the number of simplices in Cutting lemma 3.1; any bound for the number of simplices of a $(1/r)$-cutting polynomially depending on r would do as well.

A different, more constructive proof of the Improved cutting lemma exists for the planar case ([Mat90]). Still another proof (for an arbitrary dimension) follows from a recent work of Chazelle [Cha91].

In geometric setting, the Short edge lemma 3.3 can also be improved by a logarithmic factor:

Lemma 4.2 (Geometric short edge lemma). *Given a set X of n points in E^d and a (multi)set Q of m hyperplanes in E^d, there exist two points $x, y \in X$ which are separated by at most $O(m/n^{1/d})$ hyperplanes of Q.*

(The reader may verify that a version of this lemma with an additional logarithmic factor indeed follows from Lemma 3.3.) Lemma 4.2 immediately follows from the Improved cutting lemma 4.1 applied on the collection Q with $r = \Omega(n^{1/d})$. We outline an alternative proof of Chazelle and Welzl [CW89] which does not use such a heavy machinery.

Define the *distance* of two points x, y (relative to Q) as the number of hyperplanes of Q separating x from y (where we wrap around infinity, i.e. of the two (projective) segments determined by x, y, we consider the one intersected by a smaller number of hyperplanes). This defines a metric on the set of full-dimensional cells of the arrangement of Q. The following lemma is not too hard to prove:

Lemma 4.3   [CW89].   *Any ball of radius $r \leq m/2$ in the above defined metric contains $\Omega(r^d)$ regions.*

Using this, one proves the geometric Short edge lemma by a volume argument. Consider a set R of regions of the arrangement of Q with the following properties:

(i) Any two regions of R have distance at least r.
(ii) No region can be added to R without destroying (i).

(Such a set R is usually called an r-net in the theory of metric spaces, but this notion has a little to do with our notion of ε-net.) The property (i) guarantees that if we construct a $(r/2)$-ball around every region of R, we get a collection of disjoint balls. Since the total number of regions in an arrangement of m hyperplanes is $O(m^d)$, we get from Lemma 4.3 the inequality $|R|\Omega(r^d) \leq m^d$ or $|R| = O((m/r)^d)$. Now choose r in such a way that $|R| < n$ (by the previous bound, one can take $r = O(m/n^{1/d})$). Assign every point of X to its nearest region in R; because of the property (ii), each point has distance less than r to its region. Since we have more points than regions, there must be two points x, y assigned to the same region, and such points are at distance at most $2r$ from each other, thus they form the desired short edge.   □

5. Removing the Randomization

We stated Theorem 1.3 as an *existence result* about ε-nets, and we did not say how one can find a small ε-net. From one point of view, this is very easy: the probabilistic proof shows that if we add some number c to the best existence bound for an ε-net and pick a random subset of this size, what we get will be an ε-net with probability about $1 - 2^{-c/\varepsilon}$. However, this is not quite satisfactory; what if we want to *verify* that our set is really an ε-net, let alone construct it deterministically?

Some results along these lines are known. There is a general method for de-randomizing probabilistic constructions, called the method of conditional probabilities (Raghavan [Rag86], Spencer [Spe87]). This method allows to convert probabilistic proofs of existence of some objects into polynomial time algorithms constructing such objects, provided that certain probabilities can be evaluated efficiently. The algorithms produced from "raw" probabilistic proofs have usually quite a big exponent in the complexity bound. Sometimes also the greedy algorithm for the set covering problem (usually attributed to Lovász) can be used to replace a simple probabilistic proof by a deterministic construction. The paper [CF90] contains, in addition to the proof of the Improved cutting lemma, several applications of the method of conditional probabilities and of an ingenious modification of the greedy algorithm, and as a result it is shown that most of randomized algorithms in computational geometry can be replaced by deterministic polynomial time algorithms. However, the complexity of a deterministic algorithm e.g. for the Improved cutting lemma in E^3 comes out as large as $O(n^{10}r)$.

A relatively efficient deterministic algorithm for the Improved cutting lemma in the plane (of complexity $O(nr^2 \log r)$) was given in [Mat90]. The running time was improved by Agarwal [Aga90a], where the complexity $O(nr \log n (\log r)^\omega)$ with $\omega < 3.3$ was attained. In a general dimension, an efficient deterministic algorithm was found in [Mat91b] and then improved and simplified in [Mat91a]. This algorithm has a good performance for all values of r, except for those very close to n ($r = n/\log n$, say). Recently Chazelle [Cha91] was able to remove this restriction and proved the following deterministic counterpart of the Improved cutting lemma:

Theorem 5.1 *Given a collection of n hyperplanes in E^d and a parameter r, $1 < r \leq n$, one can compute a $(1/r)$-cutting for H, consisting of $O(r^d)$ simplices, by a deterministic algorithm with running time $O(nr^{d-1})$.*

One can argue that this running time is optimal in the following sense: If one wants to compute not only the cutting, but also the collection of hyperplanes intersecting every simplex of the cutting, then already the output size is in general of order $\Omega(nr^{d-1})$. Thus this theorem is satisfactory in divide-and-conquer applications. If we only want to obtain the cutting, then this optimality argument is no longer valid, and indeed it is possible to find a $(1/r)$-cutting faster: for instance, for $r \leq n^{1/(2d-1)}$, a $(1/r)$-cutting with $O(r^d)$ simplices can be computed in $O(n \log r)$ time [Mat91c].

Berger et al. [BRS89] proved that the algorithm of [CF90] can be efficiently implemented in parallel (in NC). Most likely the methods could be applied also to show that also the algorithm of Theorem 5.1 admits an efficient parallel implementation.

The proof of a version of Theorem 5.1 given in [Mat91b] (with a worse running time) is given in geometric language. The method was then partially transferred to an abstract range space setting (and somewhat simplified) in [Mat91a], where a general result on the deterministic computation of ε-approximations is proved. Before we state the result, we give a remark concerning the form of input for such an algorithm.

If we assumed that the range space was given to us by a complete list of ranges, then already the input size for an algorithm computing an ε-approximation would be impractically large. However, the finite VC-dimension allows to compute a $(1/r)$-approximation without looking at all the ranges. We will assume that a range space $\Sigma = (X, \mathcal{R})$ is given by a *subspace oracle*, i.e. an oracle which given a set $A \subseteq X$, returns a list of all distinct sets of the form $R \cap A$ for some $R \in \mathcal{R}$ (in time $O(|A|k)$, where k is the number of sets returned). The point is that while there are many distinct ranges in $\mathcal{R}$, they can have only $\pi_{\mathcal{R}}(|A|) = |A|^{O(1)}$ distinct intersections with A. In geometric application such an oracle is usually easy to construct.

Theorem 5.2   [Mat91a]. *Let $\Sigma = (X, \mathcal{R})$ be a n-point range space with the shatter function satisfying $\pi_{\mathcal{R}}(m) = O(m^d)$ ($d \geq 1$ a constant). Having a subspace oracle for Σ, and given a parameter r, we can deterministically compute a $(1/r)$-approximation of size $O(r^2 \log r)$ for Σ, in time $O(n(r^2 \log r)^d)$.*   □

In particular, if r is a constant, we are able to compute a $(1/r)$-approximation in linear time. An algorithmic analog of Proposition 1.1 then allows us to compute a $(1/r)$-net of size $O(r \log r)$ for Σ within the same time bound; the method was sketched in the remark following Theorem 1.4. Note that this method becomes very inefficient for larger r; it would be very interesting to find a substantially better algorithm.

We will indicate how this theorem can be applied on the computation of ε-cuttings. Let H be a collection of n hyperplanes in $\mathbf{E}^d$ and let us consider the range space Σ with point set H and ranges of the form $\{h \in H;\ h \cap e \neq \emptyset\}$, where e is an arbitrary segment (let us remark that this range space is isomorphic to the range space $H_2^d(D(H))$, where $D(H)$ is the point set dual to H). It is not difficult to see that the shatter function of this range space is bounded by $O(n^{2d})$. The subspace oracle required by Theorem 5.2 is constructed as follows: Given a subset $A \subseteq H$, we construct the arrangement of A and for every pair of cells of this arrangement, we output the set of hyperplanes of A separating the cells of the pair. It is straightforward to achieve the running time $O(|A|^{2d+1})$. Hence by Theorem 5.2, we can compute a $(1/r)$-approximation of size $O(r^2 \log r)$ for Σ in time which depends linearly on n (and polynomially on r, but we will use this for a constant value of r anyhow).

The use of the $(1/r)$-approximation for the computation of an ε-cutting stems from the following simple lemma:

Lemma 5.3   *Let A be an ε-approximation for the above defined range space Σ and let Ξ be a δ-cutting for A. Then Ξ is a $d(\varepsilon + \delta)$-cutting for H.*

Corollary 5.4   *There exists an algorithm which for a collection of n hyperplanes and a parameter r bounded by a constant computes a $(1/r)$-cutting of size $O(r^d) = O(1)$ for H in time $O(n)$.*

Proof. We may compute a $(1/2dr)$-approximation A of constant size for the above defined range space Σ in time $O(n)$ and then compute a $(1/2dr)$-cutting of size $O(r^d)$ for A in constant time (e.g. by the polynomial-time algorithm due to [CF90]).   □

This basic result for a values of r bounded by a constant can be applied to compute $(1/r)$-cuttings for larger values of r by various methods. The most sophisticated one was discovered by Chazelle [Cha91] and it leads to Theorem 5.1.

Acknowledgement. Work on this chapter has been supported by the Hungarian Academy of Sciences during the computational geometry workshop in Visegrad, June 1989. I would also like to thank János Pach for overall support, and Emo Welzl for his lecture notes which helped me in writing this survey.

References

[Aga90a]   P. K. Agarwal: Partitioning arrangements of lines I: An efficient deterministic algorithm. Discrete & Computational Geometry 5 (1990) 449–483

[Aga90b]   P. K. Agarwal: Partitioning arrangements of lines II: Applications. Discrete & Computational Geometry, 5 (1990) 533–573

[ABFK91]   N. Alon, I. Bárány, Z. Füredi, and D. Kleitman: Point selections and weak ε-nets for convex hulls. Manuscript, 1991

[AHWW87] N. Alon, D. Haussler, E. Welzl, and G. Wöginger: Partitioning and geometric embedding of range spaces of finite Vapnik-Chervonenkis dimension. In: Proc. 3. ACM Symposium on Computational Geometry 1987, pp. 331–340

[AK91]   N. Alon and D. Kleitman: Piercing convex sets and the Hadwiger Debrunner (p, q)-problem. Manuscript, 1991

[Ass83]   P. Assouad: Densite et dimension. Ann. Inst. Fourier (Grenoble) 33 (1983) 233–282

[BEHW89] A. Blumer, A. Ehrenfeucht, D. Haussler, and M. Warmuth: Classifying learnable geometric concepts with the Vapnik-Chervonenkis dimension. Journal of the ACM 36(4) (1989) 929–965

[BRS89]   B. Berger, J. Rompel, and P. W. Shor: Efficient NC algorithms for set cover with applications to learning and geometry. In: Proc. 30. IEEE Symposium on Foundations of Computer Science 1989, pp. 54–59

[CEGS89]   B. Chazelle, H. Edelsbrunner, L. Guibas, and M. Sharir: Point-location in real-algebraic varieties and its applications. In: Proc. 16th International Colloquium on Automata, Languages and Programming 1989, pp. 179–192

[CEG$^+$90] K. L. Clarkson, H. Edelsbrunner, L. Guibas, M. Sharir, and E. Welzl: Combinatorial complexity bounds for arrangements of curves and spheres. Discrete & Computational Geometry 5 (1990) 99–160

[CF90]   B. Chazelle and J. Friedman: A deterministic view of random sampling and its use in geometry. Combinatorica 10(3) (1990) 229–249

[Cha89]   B. Chazelle: Lower bounds on the complexity of polytope range searching. J. Amer. Math. Soc 2(4) (1989) 637–666

[Cha91]   B. Chazelle: Cutting hyperplanes for divide-and-conquer. Tech. report
            CS-TR-335-91, Princeton University, 1991. Preliminary version: Proc.
            32. IEEE Symposium on Foundations of Computer Science, October
            1991, 29–38

[Cla87]   K. L. Clarkson: New applications of random sampling in computa-
            tional geometry. Discrete & Computational Geometry **2** (1987) 195–
            222

[Cla88]   K. L. Clarkson: Applications of random sampling in computational ge-
            ometry II. In: Proc. 4. ACM Symposium on Computational Geometry
            1988, pp. 1–11

[Cla89]   K. Clarkson, unpublished proof outline, December 1989.

[CS89]   K. L. Clarkson and P. Shor: New applications of random sampling in
            computational geometry II. Discrete & Computational Geometry **4**
            (1989) 387–421

[CSW90]   B. Chazelle, M. Sharir, and E. Welzl: Quasi-optimal upper bounds for
            simplex range searching and new zone theorems. In: Proc. 6. ACM
            Symposium on Computational Geometry, 1990, pp. 23–33

[CW89]   B. Chazelle and E. Welzl: Quasi-optimal range searching in spaces of
            finite VC-dimension. Discrete & Computational Geometry **4** (1989)
            467–490

[Dud78]   R. M. Dudley: Central limit theorems for empirical measures. Ann.
            Probab. **6** (1978) 899–929

[EGS90]   H. Edelsbrunner, L. Guibas, and M. Sharir: The complexity and con-
            struction of many faces in arrangements of lines and segments. Discrete
            & Computational Geometry **5** (1990) 161–196

[Fre81]   M. L. Fredman: A lower bound on the complexity of orthogonal range
            queries. Journal of the ACM **28** (1981) 696–705

[GP86]   J. E. Goodman and R. Pollack: Upper bounds for configurations and
            polytopes in $\mathbf{R}^d$. Discrete & Computational Geometry **1** (1986) 219–
            227

[HW87]   D. Haussler and E. Welzl: ε-nets and simplex range queries. Discrete
            & Computational Geometry **2** (1987) 127–151

[KPW92]   J. Komlós, J. Pach, and G. Wöginger: Almost tight bounds for epsilon-
            nets. Discrete & Computational Geometry 1992. To appear.

[Mat90]   J. Matoušek: Construction of ε-nets. Discrete & Computational Ge-
            ometry **5** (1990) 427–448

[Mat91a]   J. Matoušek: Approximations and optimal geometric divide-and-con-
            quer. In: Proc. 23. ACM Symposium on Theory of Computing 1991,
            pp. 506–511

[Mat91b]   J. Matoušek: Cutting hyperplane arrangements. Discrete & Compu-
            tational Geometry **6(5)** (1991) 385–406

[Mat91c]   J. Matoušek: Efficient partition trees. In: Proc. 7. ACM Symposium
            on Computational Geometry 1991, pp. 1–9. Also to appear in Discrete
            & Computational Geometry

[Mat91d]   J. Matoušek: Reporting points in halfspaces. In: Proc. 32. IEEE Sym-
            posium on Foundations of Computer Science, 1991, pp. 207–215

[MWW91]   J. Matoušek, E. Welzl, and L. Wernisch: Discrepancy and ε-approxi-
            mations for bounded VC-dimension. Combinatorica 1992. To appear;

also in Proc. 32. IEEE Symposium on Foundations of Computer Science (1991), pp. 424–430

[PW90]   J. Pach and G. Wöginger: Some new bounds for epsilon-nets. In: Proc. 6. ACM Symposium on Computational Geometry 1990, pp. 10–15

[Rag86]   P. Raghavan: Probabilistic construction of deterministic algorithms: approximating packing integer programs. In: Proc. 27. IEEE Symposium on Foundations of Computer Science 1986, pp. 10–18

[Sau72]   N. Sauer: On the density of families of sets. Journal of Combinatorial Theory Ser. A **13** (1972) 145–147

[Spe87]   R. Spencer: Ten lectures on the probabilistic method. CBMS-NSF, SIAM, 1987

[SWM90]   R. Seidel, E. Welzl, and J. Matoušek: Netting a lot with a little: Small ε-nets for disks and halfspaces. In: Proc. 6. ACM Symposium on Computational Geometry 1990, pp. 16–22

[VC71]   V. N. Vapnik and A. Ya. Chervonenkis: On the uniform convergence of relative frequencies of events to their probabilities. Theory Probab. Appl. **16** (1971) 264–280

[Wel88]   E. Welzl: Partition trees for triangle counting and other range searching problems. In: Proc. 4. ACM Symposium on Computational Geometry 1988, pp. 23–33

[YY85]   F. F. Yao and A. C. Yao: A general approach to geometric queries. In: Proc. 17. ACM Symposium on Theory of Computing 1985, pp. 163–168

Chapter IV. Complexity of Polytope Volume Computation

Leonid Khachiyan*

Abstract. We survey some recent results on the complexity of computing the volume of convex n-dimensional polytopes.

1. Jumps of the Derivatives

Let P be a bounded full-dimensional polyhedron in $\mathbb{R}^n$, and let $a \in \mathbb{R}^n$ be a fixed nonzero vector. Consider the "moving halfspace" $H(t) = \{x \in \mathbb{R}^n \mid < a, x > \le t\}$, sweeping P over the time interval $t \in (-\infty, +\infty)$. Our goal is to describe the behavior of the function $V(t) = \text{vol}_n[P \cap H(t)]$.

Let u be a vertex of P. We say that at the moment $\tau = < a, u >$ the halfspace $H(t)$ crosses u and call τ a *critical moment* for $V(t)$. Denote by $\tau_0 < \tau_1 < \cdots < \tau_N$ the critical set of $V(t)$, i.e., the set of all the instances at which $H(t)$ crosses at least one of the vertices of P. From the theory of mixed volumes it is well known (see, for example [16], Theorem 15.4) that in any time interval $t \in [\tau_k, \tau_{k+1}]$ containing no critical moments as interior points, $V(t)$ is a polynomial function of time $V(t) = p_{kn}t^n + \cdots + p_{k0}, \quad t \in [\tau_k, \tau_{k+1}]$, whose degree does not exceed n. Intuitively, we expect that in general position $V(t)$ is not analytic at any critical moment τ_k. To begin with consider the most simple case where P is an n-dimensional simplex $S_n = \text{conv.hull}\{u_0, u_1, \cdots, u_n\}$ defined by $n+1$ affine independent vertices. Suppose that $H(t)$ and $P = S_n$ are in general position i.e.,

$$P \text{ has no edges parallel to the boundary of } H(t). \tag{1.1}$$

In this case $V(t) = v_n(t) = \text{vol}_n[H(t) \cap S_n]$ has exactly $n+1$ critical moments $\tau_0 < \tau_1 < \cdots < \tau_n$. Let us show that *under the assumption* (1.1) *the first* $n-1$ *derivatives of* $v_n(t)$ *are continuous, and the jumps*

$$J_n(\tau_k) = \frac{d^n v_n(\tau_k + 0)}{dt^n} - \frac{d^n v_n(\tau_k - 0)}{dt^n}$$

---

* On leave from Computing Center of the USSR Academy of Sciences

of the n-th derivative have alternating signs:

$$(-1)^k J_n(\tau_k) > 0 , \quad k = 0, 1, \cdots, n. \tag{1.2}$$

Indeed, assume without loss of generality that $H(t)$ crosses the k-th vertex of S_n at the k-th critical moment, and write for $t > \tau_0$ the n-volume $v_n(t)$ of $H(t) \cap S_n$ as

$$v_n(t) = \frac{1}{n} \sum h(u_0, F_t) \, \mathrm{vol}_{n-1}[F_t]. \tag{1.3}$$

Here the sum is taken over the facets F_t of $H(t) \cap S_n$ and $h(u_0, F_t)$ is the height of the vertex u_0 with respect to aff.hullF_t. Clearly, only two heights in this sum are nonzero. The first one is a positive constant $h_n = n \, \mathrm{vol}_n[S_n] \, / \, \mathrm{vol}_{n-1}[S_{n-1}]$ and corresponds to the facet $H(t) \cap S_{n-1}$, where $S_{n-1} = \mathrm{conv.hull}\{u_1, \cdots, u_n\}$. By induction we may assume that the $(n-1)$-volume of this facet $v_{n-1}(t) = \mathrm{vol}_{n-1}[H(t) \cap S_{n-1}]$ has $n-2$ continuous derivatives, and the jump

$$J_{n-1}(\tau_k) = \frac{d^{n-1} v_{n-1}(\tau_k + 0)}{dt^{n-1}} - \frac{d^{n-1} v_{n-1}(\tau_k - 0)}{dt^{n-1}}$$

of the $(n-1)$-th derivative of $v_{n-1}(t)$ is positive at the first critical moment τ_1, negative at the second τ_2, and so on:

$$(-1)^k J_{n-1}(\tau_k) < 0, \quad k = 1, 2, \cdots, n. \tag{1.4}$$

Next, the second nonzero height in (1.3) is equal to $(t - \tau_0)/\|a\|$ and corresponds to the facet $S_n \cap \{x \in \mathbb{R}^n \mid \ < a, x >= t \ \}$, the section of S_n by the boundary of the moving halfspace $H(t)$. Clearly, the $(n-1)$-volume of that facet is equal to $\|a\| dv_n(t)/dt$. Therefore (1.3) can be written as

$$n v_n(t) = h_n v_{n-1}(t) + (t - \tau_0) \frac{dv_n(t)}{dt}.$$

The above recurrence implies that the first $n-1$ derivatives of $v_n(t)$ are continuous, and

$$J_n(\tau_k) = -\frac{h_n}{\tau_k - \tau_0} J_{n-1}(\tau_k), \quad k = 1, 2, \cdots, n.$$

Obviously $J_n(\tau_0) > 0$ and from (1.4) we obtain (1.2).

Thus, *if P is an n-dimensional simplex in general position with $H(t)$, and u is a vertex of P, then the jump $J(u)$ of the n-th derivative of $V(t)$ at the critical moment $\tau =< a, u >$ has the sign $(-1)^{e(u)}$, where $e(u)$ is the number of edges uv of P such that $< a, v > \ < \ < a, u >$.*

In fact, *one can also obtain the following explicit expression* [15] *for the jump of the n-th derivative of $V(t)$ at a vertex u :*

$$J(u) = \frac{(-1)^{e(u)}}{|\gamma_{i_1} \cdots \gamma_{i_n} \, \det[a_{i_1}, \cdots, a_{i_n}]|}. \tag{1.5}$$

Here $a_{i_1}, \cdots, a_{i_n} \in \mathbb{R}^n$ are the normal vectors of the active facets at u and $\gamma_{i_1}, \cdots, \gamma_{i_n} \in \mathbb{R}$ are the (unique) coefficients in the representation $a = \gamma_{i_1} a_{i_1} + \cdots + \gamma_{i_n} a_{i_n}$.

Observe that (1.5) is obvious for the critical moment τ_0. For $\tau_1, \cdots, \tau_n$ (1.5) follows from the fact that $|J(u)| = \text{const}$ for any of the 2^n "corners" obtained by cutting $\mathbb{R}^n$ by n hyperplanes through u.

Now let P be an arbitrary simple polyhedron, and let $H(t)$ be in general position with respect to P, see (1.1). Suppose that at a given instant τ the halfspace $H(t)$ crosses r vertices $u^1, \cdots, u^r$ of P. Since exactly n facets intersect at each vertex of P, we can cut out from the polyhedron r small simplices $S^1, \cdots, S^r$ with vertices at the points $u^1, \cdots, u^r$ and represent P as the disjoint union of the above r simplices and some polyhedron $P' = P \setminus (S^1 \cup \cdots \cup S^r)$ containing no vertices in the boundary of $H(\tau)$. This implies that *the jump of the n-th derivative of the function $V(t) = \text{vol}_n[H(t) \cap P]$ is equal to the sum of the corresponding jumps over the critical vertices:*

$$\frac{d^n V(\tau + 0)}{dt^n} - \frac{d^n V(\tau - 0)}{dt^n} = \sum_{<a,u>=\tau} \frac{(-1)^{e(u)}}{|\gamma_{i_1} \cdots \gamma_{i_n} \det[a_{i_1}, \cdots, a_{i_n}]|}. \quad (1.6)$$

Example 1. Let $P = C_3$ be the unit 3-dimensional cube and $a = (1, 1, -2)$. At the moment $\tau = 0$ the halfspace $H(t) = \{x \in \mathbb{R}^3 \mid x_1 + x_2 - 2x_3 \le t\}$ crosses two vertices $u = (0, 0, 0)$ and $v = (1, 1, 1)$ of the cube. However, $J(u) = -J(v) = -1/2$ and $V(t)$ is analytic at the critical moment $\tau = 0$.

Example 2. Let C_n be the unit n-cube and suppose that all the coordinates of the vector $a \in \mathbb{R}^n$ are positive. Then for any vertex $u \in \{0, 1\}^n$ of C_n one has $e(u) = |u| = $ the number of 1's in u.

Since in general position the first $n - 1$ derivatives of $V(t)$ are continuous, integrating (1.6) *we get the following formula [15] for the volume of the intersection of a simple polyhedron P with a halfspace $H(t)$*

$$V(t) = \frac{1}{n!} \sum (-1)^{e(u)} \frac{(\max(0, t - <a, u>))^n}{|\gamma_{i_1} \cdots \gamma_{i_n} \det[a_{i_1}, \cdots, a_{i_n}]|}, \quad (1.7)$$

where the sum is taken over all the vertices u of P.

In particular, *the polynomial*

$$\frac{1}{n!} \sum (-1)^{e(u)} \frac{(t - <a, u>)^n}{|\gamma_{i_1} \cdots \gamma_{i_n} \det[a_{i_1}, \cdots, a_{i_n}]|}$$

does not depend on t and is equal to the volume of P.

2. Exact Volume Computation is Hard

Consider the following well-known knapsack problem: given $(a, \tau) \in Z_+^{n+1}$ determine the solvability of the equation $< a, u > = \tau$ in Boolean variables $u \in \{0, 1\}^n$. This problem can be reformulated as follows: does the moving halfspace $H(t) = \{x \in \mathbb{R}^n \mid < a, x > \leq t\}$ cross a vertex of the unit n-cube C_n at the moment $t = \tau$?

Since the knapsack problem remains NP-complete under the additional assumption

$$u \in \{0, 1\}^n \ \& \ < a, u > = \tau \ \Rightarrow \ |u| = \text{const}, \qquad (2.1)$$

from (1.6) and *Example 2* it is easy to see that *computing the volume of the intersection of the unit n-dimensional cube with a rational halfspace is NP-hard* [12]. Indeed, if it were possible to compute the function $V(t) = \text{vol}_n[C_n \cap \{x \mid < a, x > \leq t\}]$ for rational t in polynomial time, then by means of interpolation one could verify the condition $d^n V(\tau + 0)/dt^n - d^n V(\tau - 0)/dt^n \neq 0$ in polynomial time as well. The latter condition, however, is equivalent to the solvability of the knapsack problem with the property (2.1).

In fact, it can be shown [3] that *the problem of computing the volume of the intersection of the unit n-cube with a rational halfspace is #P-hard*, also see [13].

It is essential for the validity of the last statement that the coefficients $(a, \tau) \in Z_+^{n+1}$ of the halfspace $H(\tau)$ be "large", since it is known [14] that the volume of the intersection of the unit cube with an arbitrary fixed number of rational halfspaces can be computed in pseudopolynomial time. However, it was observed in [12] that if $Q = \{1, \cdots, n; \prec\}$ is a partially ordered set and

$$P(Q) = \{x \in \mathbb{R}^n \mid 0 \leq x_i \leq 1, \ i = 1, \cdots, n; \ x_i \leq x_j \ \text{if} \ i \prec j \ \text{in} \ Q\}$$

is the order polyhedron of Q, then it is NP-hard to determine the volume of the intersection of $P(Q)$ with a rational halfspace defined by "small" (polynomial in n) coefficients $(a, \tau) \in Z_+^{n+1}$. Recently a much stronger result has been obtained in [2] as a direct corollary of the following important theorem : *the problem of computing the number of linear extensions of a given poset is #P -complete* (this theorem was conjectured in [17]). Since it is well-known, see, for example [22], that the number of linear extensions of a poset Q is equal to $n! \ \text{vol}_n[P(Q)]$, the latter result implies that determining the volume of order polyhedra $P(Q)$ is #P-hard. Thus,

computing the volume of rational polyhedra is strongly $\#P - hard$.

The following question was posed in [3]: can the volume of a rational polyhedron $P = \{x \in \mathbb{R}^n \mid Ax \leq b\}$ be always written as a reduced fraction whose denominator (hence, numerator) has the binary length bounded by a polynomial in the binary length of A and b? In other words, can the volume of a rational polyhedron be written in polynomial space? The answer to this question is negative [15]. This is shown by the example $P = T_a(C_n)$, where C_n

is the unit n-cube, $a = (2^{-1}, 2^{-2}, \cdots, 2^{-n})$, and T_a is the projective mapping $x \to x/(1+ < a, x >)$. Clearly, for a positive a the image $T_a(C_n)$ of the unit cube is defined by the $2n$ inequalities

$$x_i \geq 0, \quad x_i + < a, x > \leq 1, \quad i = 1, \cdots, n,$$

and it is easy to see that the jump (1.5) of the n-th derivative of the function

$$V(t) = \text{vol}_n \, [T_a(C_n) \cap \{x \mid < a, x > \leq t\}]$$

at a vertex $T_a(u)$, $u \in \{0, 1\}^n$, is given by

$$J(T_a(u)) = (-1)^{|u|} \frac{(1+ < a, u >)^{n-1}}{a_1 \cdots a_n}.$$

Now it follows from (1.7) that

$$V(t) = \frac{1}{n! \, a_1 \cdots a_n} \sum (-1)^{|u|} (1+ < a, u >)^{n-1} (\max(0, t - \frac{< a, u >}{1+ < a, u >}))^n,$$

where the sum is taken over the vertices $u \in \{0, 1\}^n$ of C_n. In particular, the volume of $T_a(C_n)$ can be written as

$$V(1) = \frac{1}{n! \, a_1 \cdots a_n} \sum \frac{(-1)^{|u|}}{1+ < a, u >}.$$

Substituting $a = (2^{-1}, 2^{-2}, \cdots, 2^{-n})$ we get

$$V(1) = \frac{2^{(n^2+3)/2}}{n!} \sum_{N=2^n}^{2^{n+1}-1} \frac{(-1)^{e(N)-1}}{N},$$

where $e(N)$ is the number of 1's in the binary expansion of N. Obviously the binary length of the denominator of the latter expression is not polynomial in n.

We now turn to the complexity of computing the volume of a polytope $P \subset \mathbb{R}^n$, given as the convex hull of a set of integer points $P = \text{conv.hull} \{u_1, \cdots, u_m\}$, $u_1, \cdots, u_m \in Z^n$. Can the volume of P be computed in polynomial time? [18, 8]. The following negative result [3], [13] considers the problem which is "polar" to determining the volume of the intersection of a cube and a halfspace:

Let $e_1, \cdots, e_n$ be the standard basis vectors in $\mathbb{R}^n$, and let $a \in Z^n$ be a given integer vector. Computing the volume of $O(a) = \text{conv.hull} \{+e_1, -e_1, \cdots, +e_n, -e_n, a\}$ is #P-hard.

Indeed, $O(a) = \text{conv.hull} \{O, a\}$, where $O = \{x \in \mathbb{R}^n \mid |x_1| + \cdots + |x_n| \leq 1\}$ is the unit n-octahedron. Hence the n-volume of $O(a)$ can be written as

$$\text{vol} \, O(a) = \text{vol} \, O + \sum_S \text{vol conv.hull} \{S, a\} = \frac{2^n}{n!} + \frac{1}{n!} \sum_\delta \max(0, < a, \delta > -1),$$

where the first sum is taken over the facets S of O that are visible from a, and the second sum is taken over the vectors $\delta \in \{-1,1\}^n$. Therefore

$$n!\{\text{vol}O(a+e_1)+\text{vol } O(a-e_1)-2 \text{ vol } O(a)\} = \#\{\delta \in \{-1,1\}^n | < a, \delta >= 1\}.$$

But the problem of determining the number of solutions $\delta \in \{-1,1\}^n$ to the equation $< a, \delta >= 1$ is well-known to be $\#P$-complete.

It would be interesting to show that the problem of computing $\text{vol}_n \text{conv.hull}\{u_1, \cdots, u_m\}$ is strongly $\#P$-hard. Another interesting problem is to prove the $\#P$-completeness for linear extension count for posets of height 2 [2]. One more question is: is it hard to compute the volume of a zonotope? [7]

Clearly, the problem of determining the volume of a rational polytope in facial or vertex descriptions remains $\#P$-hard if it is required to approximate the volume to a given absolute accuracy , say $|u - \text{vol}_n P| \leq 1$. In the next section we address the problem of approximating the volume to a given relative accuracy:

$$\left| \frac{u}{\text{vol}_n P} - 1 \right| \leq \varepsilon.$$

3. Volume Approximation

A convex body $P \subset \mathbb{R}^n$ is said to be ρ-*rounded* if $B \subseteq P \subseteq \rho B$, where $B = \{x \in \mathbb{R}^n \mid \|x\| \leq 1\}$ is the unit Euclidean ball. If $T : x \rightarrow Ax + b$ is an affine transformation such that the above inclusions hold for the image $T(P)$, we say that the transformation ρ-rounds P.

It is well-known [9] that for an arbitrary convex body $P \subset \mathbb{R}^n$ there exists an n-rounding affine transformation T and, in general, $\rho = n$ is the best possible value. Moreover, if $P \subset \mathbb{R}^n$ is a rational polytope in facial or vertex description, then for an arbitrary $\rho > n$ one can find a rational ρ-rounding affine transformation T in time polynomial in $\log(1/(\rho - n))$ and in the binary length of input data. In fact, one can "polynomially round" P in a more general case where P is given by a separation oracle (such an oracle can answer the question "$x \in P$?" for an arbitrary rational vector x and, if $x \notin P$ it gives a hyperplane that separates x from P). In this case it is still possible to find an $n^{3/2}$-rounding transformation $T = Ax + b$ in $O(n^4 \log \rho)$ arithmetic operations and in $O(n^2 \log \rho)$ calls to the oracle, provided that originally P is ρ-rounded, see [8]. Since vol $P = $ vol $T(P)/|\det A|$, we assume henceforth that P itself is $n^{3/2}$-rounded:

$$B \subseteq P \subseteq n^{3/2}B.$$

In particular, one can easily approximate the n-volume of convex bodies to relative accuracy $\varepsilon = n^{3n/4} - 1$. If P is given by a separation oracle, this exponential bound is close to the best possible in the class of deterministic methods with polynomial informational complexity [5, 1]. Specifically, *for any deterministic method which estimates the volume of convex bodies $P \subset \mathbb{R}^n$ in polynomial in n number of calls to a separation oracle,*

$$\varepsilon \geq \left(\frac{n}{\log n}\right)^{n/2} - 1$$

for some P, see [1].

However, Dyer, Frieze and Kannan [4] have recently shown that in the class of *randomized* algorithms the volume of convex bodies can be approximated in polynomial time to an *arbitrary* fixed relative accuracy, say $\varepsilon = 0.01$. More precisely, their remarkable theorem states that *there exists a randomized algorithm which for any given $\varepsilon > 0$ and $\beta > 0$ finds an approximation u to the volume of a convex body $P \subset \mathbb{R}^n$ to relative accuracy ε with probability of error less than β*

$$\text{Prob}\left\{ \left| \frac{u}{\text{vol}_n P} - 1 \right| \leq \varepsilon \right\} \geq 1 - \beta$$

in polynomial in n, $1/\varepsilon$, and $\log(1/\beta)$ number of arithmetic operations and membership tests "$x \in P$?"

We informally outline the main ideas of the algorithm. First the problem of approximating $\text{vol}_n P$ to the relative accuracy ε can be reduced to $k = O(n \log n)$ subproblems of estimating the ratios $\xi_i = \text{vol}_n P_i / \text{vol}_n P_{i-1}$, $i = 1, \cdots, k$, where

$$B = P_0 \subseteq \cdots \subseteq P_{i-1} \subseteq P_i \subseteq \cdots \subseteq P_k = P \subseteq n^{3/2} B$$

is the "tower of convex bodies" $P_i = (1 + 1/n)^i B \cap P$. Clearly, $\text{vol}_n P = \xi_1 \cdots \xi_k \text{vol}_n B$, where B is the unit Euclidean ball, and therefore it suffices to estimate each of the ξ_i's to the relative accuracy $\varepsilon_1 = O(\varepsilon/n \log n)$ with the probability of error $\beta_1 = O(\beta/n \log n)$. Since each ratio ξ_i in the tower is bounded from above

$$1 \leq \xi_i = \text{vol}_n P_i / \text{vol}_n P_{i-1} \leq (1 + 1/n)^n < e,$$

if it were possible to sample x uniformly from within P_i, then in $O(\varepsilon_1^{-2} \log(1/\beta_1))$ independent trials "$x \in P_{i-1}$?", we could estimate ξ_i as required. To generate $x \in P_i$ nearly uniformly, Dyer, Frieze and Kannan consider a discrete approximation CP_i to P_i made up of cubic "pixels"

$$C(m) = \{x \in \mathbb{R}^n \mid m_i \delta \leq x_i \leq (m_i + 1)\delta, \ i = 1, \cdots, n\}, \quad m \in Z^n$$

of a sufficiently small size δ. For the time being the reader may assume that CP_i consists of the cubes $C(m)$ that intersect *int* P_i. Since diam $C(m) = \delta n^{1/2}$ and P_i contains the unit ball, we have $CP_i \setminus P_i \subseteq (1 + \delta n^{1/2})P_i \setminus (1 - \delta n^{1/2})P_i$. This implies

$$| \xi_i - M_i/M_{i-1} | = O((1 + \delta n^{1/2})^n - (1 - \delta n^{1/2})^n) = O(\delta n^{3/2}),$$

where M_i is the number of cubes $C(m)$ in CP_i. Hence, with $\delta = O(\varepsilon_1 n^{-3/2})$ the problem of estimating ξ_i can be replaced by the problem of approximating the ratio M_i/M_{i-1} to the relative accuracy $O(\varepsilon_1)$ with the probability of error β_1. To pick randomly a cube $C(m)$ in CP_i, Dyer, Frieze and Kannan consider a simple random walk over the cubes $C(m)$ in CP_i, which converges to the uniform distribution. At the t-th step of the random walk, $t = 0, 1, ...$, we chose with probability $1/(2n)$ a facet of the present cube $C(m_t)$ and move to the cube $C(m')$ across the chosen facet $m_{t+1} = m'$ if we do not leave CP_i; otherwise we stay in the present cube: $m_{t+1} = m_t$. The convexity of P_i implies that the Markov chain of this random walk is connected i.e., for any two cubes $C(m)$ and $C(m')$ in CP_i the t step transition probability $p_{mm'}^{(t)}$ is positive for some t. In fact, $p_{mm'}^{(t)} > 0$ for all sufficiently large t, because some of the cubes $C(m)$ have self-loops with positive 1-step transition probabilities p_{mm}. In other words, the chain is ergodic and consequently, it has a unique stationary final distribution. Since the chain is symmetric $p_{mm'} = p_{m'm}$, the latter is easily seen to be uniform: $p_{mm'}^{(t)} \to p(\infty) = 1/M_i$, $t \to \infty$. Thus, for a sufficiently large t we can use the t-th cube in the random walk to sample nearly uniformly from CP_i. To bound t from above, Dyer, Frieze and Kannan use the following consequence of a recent result of Sinclair and Jerrum [21] on rapidly mixing Markov chain:

$$| p_{mm'}^{(t)} - p(\infty) | \le (1 - 0.5\,\Phi^2)^t, \qquad (3.1)$$

where Φ is the *conductance* of the chain. The latter quantity is defined as $\Phi = \min \Phi(A, B)$, where the minimum is taken over all the partitions (A,B) of the states of the chain, and (for symmetric chains)

$$\Phi(A, B) = \frac{\sum p_{mm'} : m \in A, m' \in B}{\min\{|A|, |B|\}}.$$

Intuitively, Φ measures the minimum relative connection strength between subsets of the states. In our case $\Phi(A, B)$ admits a simple geometric interpretation:

$$\Phi(A, B) = \frac{\delta}{2n} \frac{\mathrm{vol}_{n-1}[\partial C(A) \cap \partial C(B)]}{\min\{\mathrm{vol}_n C(A), \mathrm{vol}_n C(B)\}}, \qquad (3.2)$$

where $C(A) = \cup_{m \in A} C(m)$ and $C(B) = \cup_{m' \in B} C(m')$.

We now need the following isoperimetric inequality [19], [11]:

$$s(u, v) \ \mathrm{diam}\ Q > \min\{u, v\}, \qquad (3.3)$$

where $s(u, v)$ is the $(n-1)$-volume of the minimal surface partitioning a convex body $Q \subset \mathbb{R}^n$ into two n-volumes u and v.

Consider the definition (3.2). Since $C(A) \cup C(B) = CP_i$ is "close" to the convex body P_i, and the latter is contained in the ball of diameter $D = 2n^{3/2}$, it would be fine if we could apply the isoperimetric inequality (3.3) to bound the conductance $\Phi(A, B)$ from below

$$\Phi(A, B) > \frac{\delta}{2n} \frac{1}{D} \geq \text{const } \delta n^{-5/2} \tag{3.4}$$

for *any* partition (A, B). In view of the inequality of Sinclair and Jerrum this would immediately prove the polynomiality of the algorithm. Unfortunately, we can easily obtain (3.4) from (3.3) only for the case where $\min\{\text{vol}_n C(A), \text{vol}_n C(B)\}$ sufficiently exceeds $\text{vol}_n[CP_i \setminus P_i]$, or

$$\frac{\min\{|A|, |B|\}}{|A| + |B|} = \frac{\min\{|A|, |B|\}}{M_i} \geq \text{const } \delta n^{3/2}. \tag{3.5}$$

A possible way to bypass this difficulty is to replace P_i by its Euclidean α-neighborhood $P_i(\alpha)$ with $\alpha = O(\delta n^{3/2})$. Such a replacement is within the tolerance of estimating ξ_i and smoothes the boundary of P_i. At the same time it allows one to show that the conductance of the random walk on $CP_i(\alpha)$ is polynomially bounded away from zero, see [4]. Another approach was suggested by Lovasz and Simonovits [19]. They proved a generalization of the inequality of Sinclair and Jerrum, which allows to ignore very asymmetric partitions (A,B) in the definition of conductance and instead of pointwise convergence guarantees convergence for "big" subsets of states, provided that the initial distribution is sufficiently spread out. In particular, as a corollary for Markov chains with the uniform final distribution, Lovasz and Simonovits obtained the following result: *suppose that for every set of states C with at most M elements we have $|\text{Prob}\{m_t \in C\} - p(\infty)|C| \,| \leq \nu$ for $t = 0$. Then for any set of states D*

$$|\text{Prob}\{m_t \in D\} - p(\infty)|D| \,| \leq \nu + (1 - 0.5\Phi_M^2)^t [p(\infty)]^{-1/2} \quad \text{for all } t. \tag{3.6}$$

Here Φ_M is the M-conductance of the chain, defined as the largest number such that

$$\left\{ \sum \frac{p_{mm'} + p_{m'm}}{2} : m \in A, m' \in B \right\} \geq \Phi_M \left(\min\{|A|, |B|\} - M \right)$$

for any partition (A, B).

Since the latter definition actually ignores the partitions with $\min\{|A|, |B|\} \leq M$, we know that $\Phi_M \geq \text{const } \delta n^{-5/2}$ for $M = O(\delta n^{3/2} M_i)$, see (3.4) and (3.5). Suppose that we use as a starting distribution for the random walk over CP_i, a distribution over CP_{i-1} such that

$$|\,\text{Prob}\{m_0 \in D\} - |D|/M_{i-1}\,| \leq \nu_{i-1} \quad \text{for any set } D \text{ in } CP_{i-1}. \tag{3.7}$$

Then for every set C in CP_i with at most $M = O(\delta n^{3/2} M_i)$ elements we have

$$|\,\text{Prob}\{m_0 \in C\} - |C|/M_i\,| \leq \nu_{i-1} + M\left(\frac{1}{M_{i-1}} - \frac{1}{M_i}\right) = \nu_{i-1} + O(\delta n^{3/2}).$$

We know that $\Phi_M \geq \text{const } \delta n^{-5/2}$ and therefore the second term in (3.6) is exponentially small for t, polynomial in n and $1/\delta$. Hence we conclude that after a polynomial number of random moves on CP_i we can start the random walk on CP_{i+1} with $\nu_i = \nu_{i-1} + O(\delta n^{3/2})$, see (3.7). Since we can easily obtain a good starting distribution on the discrete approximation CP_0 to the unit ball $P_0 = B$, and $i \le k = O(n \log n)$, the latter recurrence implies $\nu_i = O(\delta n^{5/2} \log n)$ for all i. Recalling that the ratios M_i/M_{i-1} must be estimated to the relative accuracy ε_1 and therefore it suffices to have $\nu_i = O(\varepsilon_1) = O(\varepsilon/n \log n)$, we finally find $\delta \simeq \varepsilon/n^{7/2} \log^2 n$, and obtain the following crude lower bound on the conductance: $\Phi_M \ge \text{const } \varepsilon/n^6 \log^2 n$. We do not go into further analysis, since the polynomiality of the problem is already clear. We also skip another detail due to the fact that the question "$C(m) \cap \text{int } P_i \ne \emptyset$?" (or equivalently "is a given cube $C(m)$ in CP_i?") can be answered in polynomial time for explicitly given rational polytopes P, but not for convex bodies given by a membership oracle. In general, CP_i consists of the cubes $C(m)$ that *weakly* intersect P_i, see [4],[19].

Unfortunately, the complexity bound of the algorithm turns out to be very high: the algorithm requires $O(\varepsilon^{-4} n^{16} \log^6 n \log(n/\varepsilon) log(n/\beta))$ calls to a membership oracle for P, see [19]. Furthermore, it is clear that since the algorithm generates a random point in P by means of a diffusion process whose step distribution is centrally symmetric and *local*, it takes at least $(\text{diam } P/\delta)^2$ steps to achieve reasonable mixing on P. Thus, for radical improvements in the complexity we need more rapidly mixing "long-range" random walks on convex bodies. Sometimes we can also use a "method of finite elements" to simplify the problem of uniform generation for P. For example, if P is the order polyhedron of a poset $Q = \{1, \cdots, n; \preceq\}$, and Q has M linear extensions $m : m(1) \preceq m(2) \preceq \cdots \preceq m(n)$, we can decompose P into M simplices $S(m) = \{x \in \mathbb{R}^n | 0 \le x_{m(1)} \le x_{m(2)} \le \cdots \le x_{m(n)} \le 1\}$ and use the triangulation $P = \cup S(m)$ instead of cubic approximations to P. This gives us a simple random walk over the set of linear extensions of Q, which converges to the uniform distribution:$| \text{Prob}\{m_t = m\} - 1/M | \le (1 - 0.5\Phi^2)^t$. Since the triangulation $P = \cup S(m)$ is exact, $\text{vol}_{n-1}[\partial S(m)]/\text{vol}_n[S(m)] \simeq n^2$ and diam $P = n^{1/2}$, the isoperimetric inequality immediately gives us the bound $\Phi \ge \text{const } n^{-2.5}$ on the conductance of this random walk. In fact, we believe that the latter bound can be improved, and $\Phi \ge \text{const } n^{-2}$. The above almost uniform generator of linear extensions can be used for determining well-balanced comparisons [6, 10] in posets with encouraging computational results [11].

Note Added in Proof. Since the time of writing of this survey paper, there has been a substantial progress in reducing complexity bounds for polytope volume computation, obtained by Applegate and Kannan (1990), Dyer and Frieze (1991), and Lovasz and Simonovits (1991). The best currently known bound $O(\varepsilon^{-2} n^7 \log^2 n \log^3(\varepsilon^{-1}) \log(\beta^{-1}))$ is due to Lovasz and Simonovits [20].

References

[1] J. Barany and Z. Furedi: Computing the volume is difficult. Proc. 18th Symposium on Theory of Computing, 1986, pp. 442-447

[2] G. Brightwell and P. Winkler: Counting linear extensions is $\#P$-complete. Preprint, Bellcore, July 1990

[3] M. Dyer and A. Frieze: On the complexity of computing the volume of polyhedron. SIAM J. Computing **17** (1988) 967-974

[4] M. Dyer, A. Frieze and R. Kannan: A random polynomial time algorithm for estimating volumes of convex bodies. Proc. 21st Symposium on Theory of Computing, 1989, pp. 375-381

[5] G. Elekes: A geometric inequality and the complexity of computing volume. Discrete and Computational Geometry, **1** (1986) 289-292

[6] M. Fredman: How good is the information theory bound in sorting. Theoretical Computer Science, **1** (1976) 355-361

[7] P. Gritzmann: Private communication

[8] M. Grötschel, L. Lovasz, and A. Schrijver: Geometric Algorithms and Combinatorial Optimization. Springer, 1988

[9] F. John: Extremum problems with inequalities as subsidiary conditions. Studies and essays presented to R. Courant on his 60th birthday. Interscience Publ., New York 1948, pp. 187-204

[10] J. Kahn and M. Saks: Every poset has a good comparison. Proc. 16th Symposium on Theory of Computing, 1984, pp. 299-301

[11] A.Karzanov and L.Khachiyan: On the conductance of order Markov chains. Technical Report DCS TR 268, Rutgers University, June 1990

[12] L. Khachiyan: On the complexity of computing the volume of a polytope. Izvestia Akad. Nauk SSSR, Engineering Cybernetics **3** (1988) 216-217 (in Russian)

[13] L. Khachiyan: The problem of computing the volume of polytopes is $\#P$-hard. Uspekhi Mat. Nauk **44**, no. 3 (1989) 199-200

[14] M. Kozlov: Algorithms for volume computation, based on Laplace Transform. Preprint, Computing Center of the USSR Academy of Sciences, 1986

[15] J. Lawrence: Polytope volume computation. Preprint NISTIR 89-4123, U.S. Department of Commerce, National Institute of Standards and Technology, Center for Computing and Applied Mathematics, Galthersburg, October 1989

[16] K. Leichtweis: Convex sets. Springer, 1980

[17] N. Linial: Hard enumeration problems in geometry and combinatorics. SIAM J. on Algebraic and Discrete Methods **7**, no.2 (1986) 331-335

[18] L. Lovasz: Algorithmic aspects of combinatorics, geometry and number theory. Proc. Internat. Congress Math. Warsaw 1983, Vol. 1, Warsaw 1984, pp. 1591-1599

[19] L. Lovasz and M. Simonovits: The mixing rate of Markov chains, an isoperimetric inequality, and computing the volume. Preprint 27/1990, Mathematical Institute of the Hungarian Academy of Sciences, May 1990

[20] L. Lovasz and M. Simonovits: Random walks in a convex body and an improved volume algorithm. Preprint, Preliminary version, December 1991

[21] A. Sinclair and M. Jerrum: Approximate counting, generation and rapidly mixing Markov chains. Information and Computation **82** (1989) 93-133

[22] R. Stanley: Two order polytopes. Discrete and Computational Geometry **1** (1986) 9-23

Chapter V. Allowable Sequences and Order Types in Discrete and Computational Geometry

Jacob E. Goodman and Richard Pollack

1. Introduction

The allowable sequence associated to a configuration of points was first developed by the authors in order to investigate what combinatorial structure lay behind the Erdős-Szekeres conjecture (that any $2^{n-2} + 1$ points in general position in the plane contain among them n points which are in convex position). Though allowable sequences did not lead to any progress on this ancient problem, there did emerge an object that had considerable intrinsic interest, that turned out to be related to some other well-studied structures such as pseudoline arrangements and oriented matroids, and that had as well a combinatorial simplicity and suggestiveness which turned out to be effective in the solution of several other classical problems. These connections and applications are discussed in Sections 2, 3, and 4 of this paper.

In order to better understand the connection between allowable sequences and pseudoline arrangements, the authors proved a theorem which, when properly interpreted, suggested a natural generalization to higher dimension of the notion of "order" along a line, the concept of the "order type" of a set of points in $\mathbf{R}^d$. This idea plays a fundamental role in many geometric algorithms (in the plane it involves the decision about whether an ordered triple of points makes a left turn or a right turn, for example). The order type of a configuration, and many of its fundamental properties which have been explored in the past eight years, are discussed in the remaining sections.

Among the applications of the concept of order type is one which made possible the extension of Hadwiger's theorem on line transversals in the plane to hyperplane transversals in $\mathbf{R}^d$. We refrain from discussing this application in the present paper, since there is a complete discussion of it in Section 2 of the article "Geometric Transversal Theory" [51] which also appears in this volume, p. 163.

Acknowledgements. The work of the first author has been supported in part by NSA grant MDA904-89-H-2038, PSC-CUNY grant 666426, and the Center for Discrete Mathematics and Theoretical Computer Science (DIMACS). The work of the second author has been supported by NSF grant CCR-8901484, NSA grant MDA904-89-H-2030, and DIMACS.

2. Combinatorial Types of Configurations in the Plane and Allowable Sequences

The "combinatorial type" of a configuration of points in the plane was introduced in [36] for configurations in general position, and extended to arbitrary planar configurations in [40]. This classification is finer than the one given by order type but is very natural for describing certain problems in the plane; moreover, the concept of order type emerges from it quite naturally.

This classification is defined by associating to any configuration $C = \{p_1, \ldots, p_n\}$ of n labeled points in $\mathbf{R}^2$, with a fixed coordinate system, a certain sequence of permutations of $[1, n]$ and then identifying configurations which are asociated with the same sequence of permutations. Given a configuration $C = \{p_1, \ldots, p_n\}$ of n labeled points in $\mathbf{R}^2$ (we shall usually identify the point p_i with its index i), we project the points orthogonally onto a directed line L, thus obtaining a permutation of $[1, n]$ determined by the order in which the points fall on L. Let L rotate counterclockwise. A new permutation arises whenever L passes through a direction orthogonal to one determined by a *connecting line* of C, i.e. a line determined by two or more points of C. If the points $i_1, \ldots, i_k$ on such a connecting line appeared in the induced permutation in that order just before L passed orthogonally to the connecting line, they would appear in the reverse order $i_k, \ldots, i_1$ in the permutation induced on L just after. If $i_1, \ldots, i_k$ are all the points of C on the given line, the reversal of these indices in the *move* from one permutation to the next is called a *switch*. A move may consist of a single switch, or more than one (if there is more than one connecting line in the direction corresponding to that switch). Now allowing L to continue rotating, the doubly infinite sequence we obtain is called the *circular sequence of permutations* associated with C. This sequence is clearly periodic (the period corresponding to a full rotation of L), and is in fact determined by a half period (corresponding to L rotating though a half turn). For example, with the configuration in Fig. 2.1 we associate the following sequence.

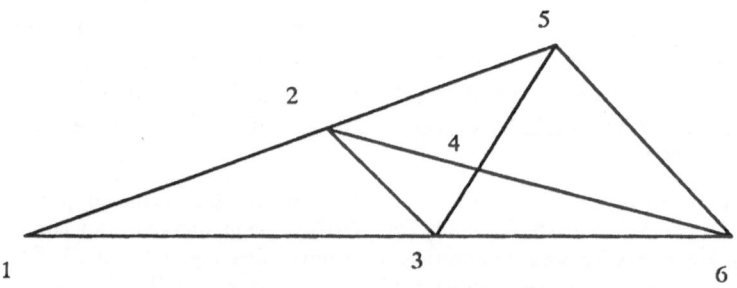

Fig. 2.1. A configuration of 6 points

```
. . . . . .
. . . . . .
. . . . . .
1 2 3 4 5 6
1 3 2 4 6 5
1 3 6 4 2 5
6 3 1 4′2 5
6 3 4 1 2 5
6 3 4 5 2 1
6 5 4 3 2 1
5 6 4 2 3 1
. . . . . .
. . . . . .
. . . . . .
```

Notice that many geometric features of a configuration C are encoded in the circular sequence associated with C. Among these are the following:

1. The extreme points of C are those which appear as initial (or terminal) terms of permutations in the circular sequence;
2. the number of directions determined by the connecting lines of C is half the period of the circular sequence;
3. a *semispace* (the intersection of a half-plane with C) is an initial or terminal segment of some permutation in the associated circular sequence;
4. a k-set (a semispace of cardinality k) is an initial or terminal segment of length k in some permutation;
5. points are collinear if they occur in the same switch;
6. the number of k-sets is the number of switches across the kth spot in a single period of the associated circular sequence.

Notice that for the dual arrangement A of lines (under the correspondence in which point (a, b) corresponds to line $y = -ax + b$ — see Fig. 2.2, which is the arrangement of lines dual to the configuration of Fig. 2.1), if we record the order of the lines from bottom to top as we scan the arrangement from left to right we obtain a half period (the one that starts with our directed line horizontal and pointing to the right) of the circular sequence associated with the original configuration C.

The features that we have just listed, that are encoded by the circular sequence of C, correspond to the following features of the dual arrangement A:

1. Indices which appear as initial (or terminal) terms of a permutation correspond to lines on the upper or lower envelope of A;
2. half the period of the circular sequence is the number of distinct abscissae of vertices of A;

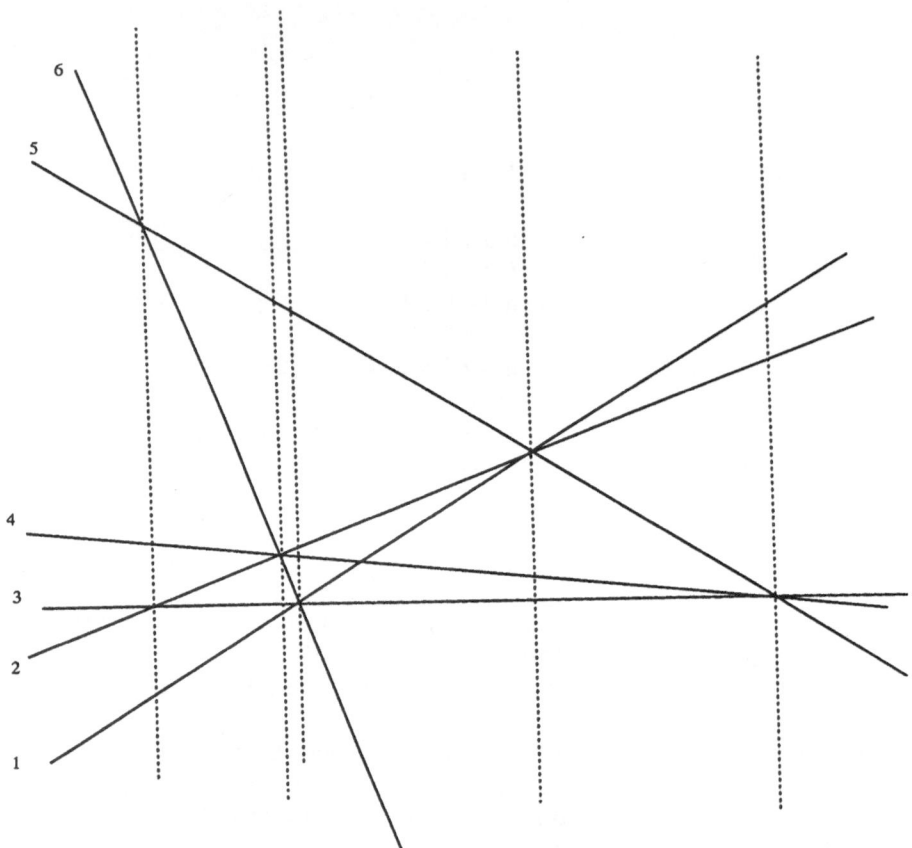

Fig. 2.2. The dual arrangement of 6 lines

3. an initial (or terminal) segment of a permutation corresponds to the set of lines above (or below) a cell of $\mathcal{A}$ (a connected component of the complement of the arrangement);
4. an initial or terminal segment of length k corresponds to a cell of $\mathcal{A}$ with k lines above (or below) it;
5. lines are concurrent if they occur in the same switch;
6. the number of vertices on the "k-hull" is the number of switches across the k-th spot.

Labeled configurations or arrangements that have the same associated circular sequence are called *combinatorially equivalent*. Unlabeled configurations which admit a labeling under which they or their mirror images are combinatorially equivalent are also called *combinatorially equivalent*.

Notice that a circular sequence

$$\Sigma = (\Pi_i)_{i \in Z}$$

has the following properties.

1. The move from Π_i to Π_{i+1} consists of reversing one or more nonoverlapping substrings of Π_i;

2. if a move reverses the pair i, j then every other pair of indices not reversing in that move is reversed in some move before the next move reversing the pair i, j.

It is natural to study sequences of permutations satisfying properties (1) and (2). Thus we call a doubly infinite sequence of permutations

$$\Sigma = (\Pi_i)_{i \in Z}$$

satisfying (1) and (2) an *allowable sequence* of permutations. Such an allowable sequence is necessarily periodic, with even period N, and permutations a half period apart are the reverse of each other:

$$\Pi_i = \tilde{\Pi}_{i+N/2}.$$

If each move in Σ corresponds to just the switch of a single pair of indices, Σ is called a *simple* allowable sequence; this would be the case if Σ arose from a configuration of points in general position (no three points collinear, no two connecting lines parallel).

It is natural to wonder, at this point, if every allowable sequence is *realizable* in the sense that there is a configuration of points whose allowable sequence is the associated circular sequence.

Circular sequences (at least in the *simple* case) had made their appearance in a slightly different guise more than a century ago. In 1881 R. Perrin [85] discussed what he called the *aspects* of a configuration. The aspect of a configuration from a point was the circular order in which the points appeared from that point, i.e., the sorted order of their polar angles from that point. He then considered the sequence of aspects as the point moved along a line which had all the intersection points of all the connecting lines of the configuration on one side. It is easy to see that this sequence is simply a half period of the associated circular sequence. Perrin went on to assert that (in our language) every simple allowable sequence is realizable, i.e., that properties (1) and (2) are necessary and sufficient for a sequence of permutations, each move being just a simple switch, to be the circular sequence of a configuration in general position. In 1980, V. Chvátal and G. Klincsek [16] also observed that properties (1) and (2) were necessary for realizability and asked whether these conditions were sufficient as well.

As it turns out, these properties are not sufficient for realizability. There is, in fact, up to a permutation of the indices, exactly one non-realizable simple allowable sequence of order 5. It is,

$$
\begin{array}{ccccc}
\cdot & \cdot & \cdot & \cdot & \cdot \\
\cdot & \cdot & \cdot & \cdot & \cdot \\
1 & 2 & 3 & 4 & 5 \\
2 & 1 & 3 & 4 & 5 \\
2 & 1 & 4 & 3 & 5 \\
2 & 1 & 4 & 5 & 3 \\
2 & 4 & 1 & 5 & 3 \\
4 & 2 & 1 & 5 & 3 \\
4 & 2 & 5 & 1 & 3 \\
4 & 2 & 5 & 3 & 1 \\
4 & 5 & 2 & 3 & 1 \\
5 & 4 & 2 & 3 & 1 \\
5 & 4 & 3 & 2 & 1 \\
\cdot & \cdot & \cdot & \cdot & \cdot \\
\cdot & \cdot & \cdot & \cdot & \cdot
\end{array}
$$

It is easy to see that a configuration with this allowable sequence as its circular sequence must be a pentagon whose sides and "parallel" diagonals meet as in Fig. 2.3; elementary considerations show that such a figure is impossible.

This example generalizes as follows:

Theorem 2.1 ([36]) *If $1, 2, \ldots, n$ are the vertices of a convex n-gon (numbered modulo n) listed in counterclockwise order, and if $a, k > 0$ and $a + 2k < n$, then it is impossible that the diagonals (or edges) i, $i + a$ and the "parallel" diagonals (or edges) $i - k$, $i + a + k$ intersect on the side of points $i + a$ and $i + a + k$ for all i.*

This provides many examples of non-realizable allowable n-sequences for $n \geq 5$.

For the case of 5 points all combinatorial classes are known, and they are summed up in the following theorem.

Theorem 2.2 ([36]) *There are precisely 20 combinatorial equivalence classes of simple allowable 5-sequences, all of which are geometrically realizable except for the class of the unrealizable pentagon.*

Let us note that there is a natural way to generalize the concept of the circular sequence of a planar configuration $\mathcal{C}$ to higher dimensions, discussed in [47]. It is easy to see that the combinatorial type of a configuration is determined—up to a reversal of its allowable sequence—by its associated *set* of permutations. This suggests the following definition of combinatorial equivalence in higher dimensions.

Definition 2.3 *Let $\mathcal{C} = \{p_1, \ldots, p_n\}$ be a labeled configuration of n points in $\mathbf{R}^d$. Let L be a directed line passing through O such that the points of $\mathcal{C}$ have distinct images under orthogonal projection to L, and let π_L be the associated permutation of the indices $1, \ldots, n$ induced by the direction on L. The set*

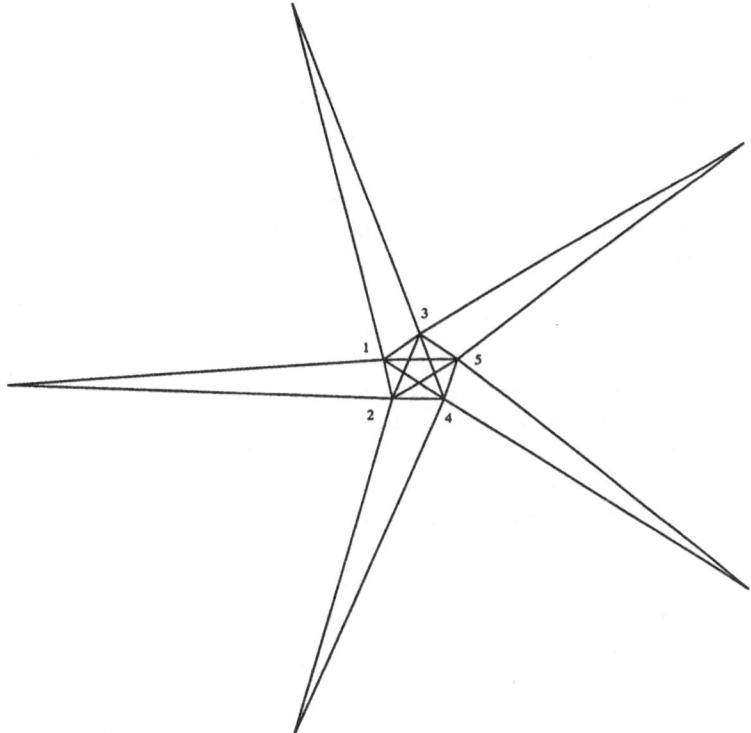

Fig. 2.3. The unrealizable pentagon

$\Pi(\mathcal{C})$ *of all permutations obtained in this way is called the* permutation set *of*
$\mathcal{C}$. *We say that two configurations,* $\mathcal{C}$ *and* $\mathcal{C}'$, *are* combinatorially equivalent
if $\Pi(\mathcal{C}) = \Pi(\mathcal{C}')$.

Just as in the plane the members of the permutation set of a configuration
$\mathcal{C}$ fall in a natural way into a circular sequence, so too, in $\mathbf{R}^d$, they form a
complex on the unit sphere $\mathbf{S}^{d-1}$, whose points correspond to the directed lines
L in the preceding definition. One gets, in this way, a much richer structure
than in the case of $d = 2$, one which has not yet been fully investigated.

Similar ideas have recently been explored by N. E. Mnëv [personal com-
munication].

3. Arrangements of Lines and Pseudolines

The study of arrangements of lines was initiated in 1826 by J. Steiner [94],
and that of pseudolines in 1926 by F. Levi [73]. An excellent survey of the
subject is due to B. Grünbaum [54]. The natural setting for arrangements
of lines is the real projective plane $\mathbf{RP}^2$ (although on occasion, we may be
sloppy about that detail and think of them as lying in the euclidean plane
$\mathbf{R}^2$.) An *arrangement* of lines is simply a finite collection of labeled lines in

$\mathbf{RP}^2$ which, to guarantee nondegeneracy, we require to have empty intersection. An arrangement has *vertices* which are the intersections of pairs of lines of the arrangement, *edges* which are maximal connected components of the complement of the vertices in the lines of the arrangement, and *faces* which are the maximal connected components of the complement of the lines of the arrangement in $\mathbf{RP}^2$. The vertices, edges and faces of the arrangement are the 0-, 1-, and 2-dimensonal cells (or faces) of the arrangement. The set of these cells (the *cell decomposition* induced by the arrangement) has a natural lattice structure (called the *face lattice* of the arrangement), and we consider two arrangements to be *isomorphic* if their face lattices are isomorphic under the correspondence induced by their labelings.

We saw in the last section that it is natural to associate a circular sequence with an arrangement of lines as well as with a configuration of points. It follows from the duality described in the last section that an allowable sequence can be realized as the circular sequence of a configuration of points if and only if it can be realized as the circular sequence of an arrangement of lines [40].

It has been observed by Grünbaum that in most arguments about the combinatorial properties of arrangements of lines it seems that the fact that the lines are straight is of no importance, i.e., all we need to know is that

1. each "line" of the arrangement is a continuous image of $\mathbf{RP}^1$ (the circle) which does not separate $\mathbf{RP}^2$, (i.e., whose complement in $\mathbf{RP}^2$ has only one connected component);
2. each pair of "lines" of the arrangement intersect exactly once (and therefore cross at their single point of intersection);
3. the intersection of the "lines" of the arrangement is empty.

A curve satisfying (1) is called a *pseudoline*, and a finite collection of curves satisfying (1), (2), and (3) is called an *arrangement of pseudolines*. Clearly, isomorphism of arrangements of pseudolines can be defined in exactly the same way as was done for arrangements of lines. Arrangements of pseudolines are particularly interesting for us because even though not every allowable sequence is *geometrically realizable* as the circular sequence of a configuration of points, or—equivalently—as the circular sequence of an arrangement of lines, it is nevertheless true that every allowable sequence is realizable as the circular sequence of an arrangement of pseudolines.

It is easy to see why this is so: it involves nothing more than coloring by numbers. For example, take the allowable sequence which corresponds to the bad pentagon of Fig. 2.3, and turn it on its side. Now connect all occurrences of the number i by a piecewise linear curve with a minimum number of bends. Finally, extend the ends of these paths by rays so that at the left end the slopes are decreasing as the index of the curve increases, and extend each curve to the right similarly by a ray whose slope is the same as the slope of the ray at the left end. It is easy to see that this family of curves is a pseudoline arrangement whose circular sequence is the given allowable sequence. (Notice that since these curves are all x-monotone, we can unambiguously record their order from bottom to top as we scan the arrangement from left to right.) This arrangement is called the *wiring diagram* of the allowable sequence [35] (see the figure below for the wiring diagram which corresponds to the unrealizable pentagon).

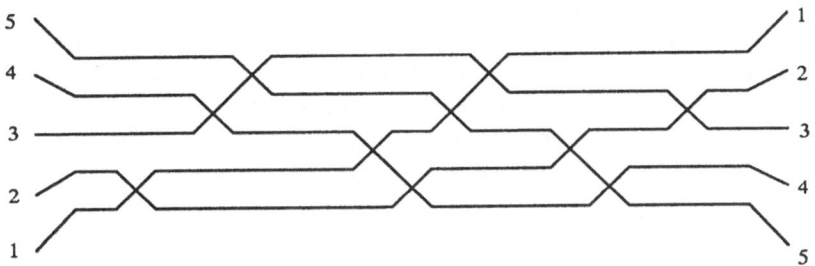

Fig. 3.1. Wiring diagram for the unrealizable pentagon

In a similar way, we can associate an allowable sequence with any arrangement of pseudolines in $\mathbf{RP}^2$. In fact there are many ways to do this. We first choose a point p in a 2-cell of the arrangement, which we will think of as the vertical point at infinity (when we think of the arrangement in $\mathbf{R}^2$). Next we provide an ordering of the vertices of the arrangement (which will correspond to the left-to-right ordering of the vertices of a line arrangement in $\mathbf{R}^2$ or to the order of the vertices by their polar angle about the pole p) by adjoining to the arrangement an auxiliary pseudoline joining each vertex of the arrangement to p. This can be accomplished by repeated use of the following important tool:

Lemma 3.1 (Levi Enlargement Lemma [73]) *Given an arrangement $\mathcal{A}$ of pseudolines in $\mathbf{RP}^2$ and two points p, q which do not both lie on the same pseudoline of $\mathcal{A}$, there is a pseudoline l which contains both p and q, such that $\mathcal{A} \cup \{l\}$ is still an arrangement of pseudolines.*

Now we construct an additional pair of pseudolines for each of the auxiliary lines. The order in which these pseudolines meet the pseudolines of the arrangement gives the permutations of the circular sequence which occur just before and just after the given vertex; see [45] for details.

The fact that so much of the geometry of configurations is captured by arbitrary allowable sequences or, dually, by arrangements of pseudolines is demonstrated by the fact that many of the fundamental theorems of plane combinatorial geometry (Radon's, Helly's, Kirchberger's, and Caratheodory's theorems, for example) also hold in this more general setting [38,39].

An arrangement of pseudolines is called *stretchable* if it is isomorphic to an arrangement of straight lines. An isomorphism of arangements can be extended to a homeomorphism of the entire plane by repeated application of Schoenflies' theorem; hence we see that isomorphism of pseudoline arrangements is precisely what corresponds, in the topological category, to equivalence of line arrangements.

Finally, we mention that pseudoline arrangements are identical with rank 3 acyclic oriented matroids, by the Folkman-Lawrence representability theorem [31], and that the duality between line arrangements and point configurations discussed earlier extends to a duality between pseudoline arrangements and "generalized configurations", consisting of points joined by pseudolines [35]. All one needs, therefore, is a suitable dictionary, to interpret any theorem or conjecture about oriented matroids as an equivalent theorem or conjecture about arrangements of pseudolines or generalized configurations, or vice-versa. See [6] for a very nice discussion of this connection.

4. Applications of Allowable Sequences

In the first subsection of this section we will apply what we know about realizability of allowable sequences of order ≤ 5 to show that all arrangements of at most 8 pseudolines are stretchable. The problems we will then consider in the remainder of this section are typical of many geometric problems dealing with point configurations or line arrangements in the plane, in that they have natural analogues as problems about allowable sequences. A solution to the problem for allowable sequences will guarantee that the corresponding geometric theorem is true. Since there are some allowable sequences that are not realizable (we shall see in Section 6 that in fact *most* circular sequences are not realizable), it is possible that the theorem may be false as a theorem about allowable sequences and yet true as a geometric theorem. Nevertheless, the combinatorial suggestiveness of allowable sequences has been effective in simplifying the effort to find solutions to several outstanding problems. Among these are:

1. the number of directions determined by $2n$ points not all on a line is at least $2n$ [98];
2. the number of k-sets among a set of n points is $O(nk^{1/2})$ [29,30,74];
3. the maximum number of "at-most-k-sets" is nk [2].

The success of this method illustrates the principle that when one finds the combinatorial essence of a problem in combinatorial geometry (i.e., makes the geometric problem a consequence of the "right" purely combinatorial problem), then one is well on the road to solving the problem.

In the second and third subsections we will discuss these problems.

4.1 Stretchability of Pseudoline Arrangements

Levi [73] observed that not all pseudoline arrangements are stretchable. One simply considers the Pappus configuration of a hexagon (6 lines) inscribed so that alternate vertices lie on each of two lines. Pappus' theorem asserts that the points of intersection of opposite sides of this hexagon are collinear, i.e., that they lie on a ninth line. If we now take this ninth line and bend it so that it fails to pass through one of these points of intersection it is clear that the resulting arrangement of 9 pseudolines (8 straight and 1 pseudo) is not stretchable. This arrangement, however, is not simple. Thirty years later G. Ringel [88] constructed a simple arrangement of nine pseudolines which is not stretchable, by modifying Levi's construction appropriately.

In 1971 R. J. Canham [14] and E. Halsey [57] independently gave a computer enumeration of all simple arrangements of 8 pseudolines, and showed that all simple arrangements of 7 or fewer pseudolines were stretchable. On the basis of the known evidence, Grünbaum conjectured [54] that every arrangement of 8 pseudolines, simple or not, is stretchable.

The fact that there is essentially only one non-realizable circular sequence of order 5 makes it possible to prove that all arrangements of 8 pseudolines are stretchable [37], settling Grünbaum's conjecture affirmatively. (Equivalently, every rank 3 oriented matroid on 8 points is realizable).

We outline the proof of this: Assume that the arrangement is simple (it will be evident from the proof of this case that otherwise the proof is even simpler). An easy inductive argument shows that any arrangement of pseudolines contains a triangle. Deform the arrangement so that this triangle becomes a point p. Using the Levi enlargement lemma we add pseusolines passing through p and through each of the other intersection points of the original arrangement. By Schoenflies' theorem there is a homeomorphism of the projective plane to itself in which all the pseudolines of this extended arrangement passing through p become straight lines. Now consider a model of the projective plane in which p is the vertical point at infinity so that these lines become a family of vertical lines in the euclidean plane and the remaining 5 pseudolines are equivalant to a wiring diagram for an allowable 5-sequence. If this 5-sequence doesn't come from the bad pentagon it is realizable and we are done. If it is an allowable sequence associated with the bad pentagon, say the sequence given in Section 2, then we can interchange the order of a certain pair of switches (say 3,5 and 1,4 in that example); this does not change the isomorphism class of the arrangement, but does make the allowable sequence realizable.

4.2 The Directions Problem

The directions problem, first posed by P. R. Scott in 1970 [90], asks for the determination of $f(n)$, the minimum number of directions (i.e., slopes) determined by n points in the plane which do not all lie on a line. He was able to prove:

Theorem 4.1 ([90])

$$(1 + \sqrt{8n - 7})/2 \leq f(n) \leq 2[n/2].$$

The upper bound comes from the vertices of the regular n-gon if n is even, and the vertices of the regular $(n-1)$-gon together with its center if n is odd.

In 1978, G. R. Burton and G. B. Purdy improved the lower bound, obtaining:

Theorem 4.2 ([13])

$$[n/2] \le f(n) \le 2[n/2].$$

Finally, by proving a more general theorem about allowable sequences, P. Ungar showed that the upper bound was indeed the correct answer. He proved:

Theorem 4.3 ([98]) *Let n be even. If an allowable sequence contains $12\ldots n$ and we do not get from it to $n(n-1)\ldots 1$ in a single move, then the number of moves required to reach $n(n-1)\ldots 1$ is at least n.*

His elegant proof depends upon counting how many moves are required to carry each index across the midline (the space between $n/2$ and $n/2+1$ in the initial permutation). A move in which some indices cross the midline is called a *crossing move*. At the ith crossing move an increasing string which straddles the midline is reversed. Let d_i denote the distance from the midline to the nearest end of this string. Hence at the ith crossing move exactly $2d_i$ indices cross the midline. Since every index must cross the midline, if there are t crossing moves all together, we have

$$2d_1 + 2d_2 + \cdots + 2d_t \ge n.$$

The proof is essentially completed by observing that between the ith crossing move and the one following, at least $d_i + d_{i+1} - 1$ non-crossing moves must occur, which follows from the following three facts.

1. A decreasing string can get shortened by no more than 1 at each end in one move;
2. an increasing string can get longer by at most 1 at each end in one move;
3. between the ith crossing move and the one following we must tear down a decreasing string of length d_i and build up an increasing string of length at least d_{i+1}.

R. E. Jamison has explored the configurations that are extremal from the point of view of the directions problem [64,66], and has also written an extensive survey of this problem [65].

4.3 k-Sets

Given a finite set of points $\mathcal{C} \subset \mathbf{R}^d$, a *k-set* of $\mathcal{C}$ is the intersection of $\mathcal{C}$ and an open half space H, with $\mathcal{C} \cap H$ of cardinality k. Let $f_d(k, \mathcal{C})$ denote the number of k-sets of $\mathcal{C}$ and $f_d(k, n) = \max_{|\mathcal{C}|=n} f_d(k, \mathcal{C})$. A related quantity,

$g_d(k,n)$, is defined as $\max_{|C|=n} g_d(k,C)$, where $g_d(k,C) = \sum_{0 \leq i \leq k} f_d(i,C)$ is the number of *at-most-k-sets* of C.

In what follows, when $d=2$ we will suppress the subscript d, i.e., we will write $f_2(k,n) = f(k,n)$ and $g_2(k,n) = g(k,n)$.

According to L. Lovász [74], G. Simmons was the first to raise the question of finding an upper bound for $f(n,2n)$, and gave a lower bound of $\Omega(n \log n)$. In 1971 Lovász found an upper bound for $f(n,2n)$:

Theorem 4.4 ([74])
$$f(n,2n) = O(n\sqrt{n}).$$

Lovász bounds the number of connecting lines of the set (called *halving lines*) which separate the remaining $2n-2$ points into two equal parts. (The segment joining the two points of the set on such a line is called a *halving segment*.)

The main idea of the proof is contained in a lemma which states that a line which separates the points so that there are k on one side and $n-k$ on the other side meets exactly $\min(k, n-k)$ halving segments.

All that is really needed of this lemma is the following, which is now known as Lovász' lemma.

Lemma 4.5 *If C is a set of n points in general position in the plane, then any line meets $O(n)$ halving segments.*

The ideas in this paper were elaborated and extended by Erdős, Lovász, Simmons, and Strauss, who were able to prove

Theorem 4.6 ([30]) *There are positive constants c_1, c_2 such that*
$$c_1 n \log k \leq f(k,n) \leq c_2 n \sqrt{k}.$$

The k-set problem (still only the planar version) was rediscovered in 1982 by H. Edelsbrunner and E. Welzl, because the function $f(k,n)$, and in particular $f([n/2],n)$, bounded the complexity of many natural problems in computational geometry.

It was easy to describe the (planar) problem as a problem about allowable sequences. Let $F(k,n)$ and $G(k,n)$ be the corresponding functions maximized over all allowable sequences. Thus, for an allowable sequence Σ, we define $F(k,\Sigma)$ to be the number of k-sets of Σ and $F(k,n) = \max_{|\Sigma|=n} F(k,\Sigma)$. Also, $G(k,n) = \max_{|\Sigma|=n} G(k,\Sigma)$, where $G(k,\Sigma) = \sum_{0 \leq i \leq k} F(i,\Sigma)$ is the number of *at-most-k-sets* of Σ.

From the observation that if n is an extreme point of the allowable sequence Σ then in a half period of Σ, n will switch at each of the first k spots exactly once (recall that each switch at one of these spots contributes 1 to $f(k,\Sigma)$), the authors proved

Theorem 4.7 ([43]) *For $2k < n$, $G(k,n) \leq 2nk - 2k^2 - k$.*

Independently, and by a purely geometric argument, J. Pach proved [personal communication] that $g(k,n) \leq 2nk$.

The exact values of $G(k,n)$ and $g(k,n)$ were finally found by N. Alon and E. Györi when they proved

Theorem 4.8 ([2]) *For* $2k < n$, $g(k,n) = G(k,n) = nk$.

That $nk \leq g(k,n) \leq G(k,n)$ is obvious. Thus it suffices to prove that $G(k,n) \leq nk$; this follows by observing that in any allowable sequence any index makes at most $4k$ switches among the first and last k spots, and if we count this for each index we have double-counted the switches at each spot. Hence there are at most $2nk$ switches at these spots, and this is twice the number of at-most-k-sets in the allowable sequence. (Alon and Györi also give an independent proof of the fact that $g(k,n) = nk$ along the lines followed by Erdős, Lovász, Simmons, and Strauss in their paper.) That $g(k,n) = nk$ was also proved independently by G.W. Peck [84].

What is surprising is that once again, as in the case of the directions problem, the truth turns out to be the same for both geometric configurations and the far more general allowable sequences.

Using entirely different methods (probabilistic ones), K. Clarkson and P. Shor have found asymptotically tight bounds for at-most-k-sets in all dimensions.

Theorem 4.9 ([17])

$$g_d(k,n) = \Theta(n^{\lfloor d/2 \rfloor} k^{\lceil k/2 \rceil})$$

as $n/k \to \infty$.

Shortly after Edelsbrunner and Welzl rediscovered the k-set problem, they found the same upper bound that had been found by Erdős, Lovász, Simmons, and Strauss. However, as in the proof of Alon and Györi, they bounded the number of switches at the kth spot and showed

Theorem 4.10 ([29])

$$F(k,n) = O(n\sqrt{k}).$$

Again, using allowable sequences, Welzl proved a curious generalization of the Alon–Györi theorem which also has as a special case the theorems of Edelsbrunner–Welzl and of Erdős, Lovász, Simmons, and Strauss.

For $K \subset \{1,2,\ldots,[n/2]\}$ and $\mathcal{C}$ (respectively Σ) a configuration (allowable sequence) of n points, define $g(K,\mathcal{C}) = \sum_{k \in K} f(k,\mathcal{C})$, $G(K,\Sigma) = \sum_{k \in K} F(k,\Sigma)$, and let $G(K,n), g(K,n)$ be the respective maxima over objects of size n. Welzl then proves

Theorem 4.11 ([101])

$$g(K,n) \leq G(K,n) \leq 2^{3/2} n \Big(\sum_{k \in K} k\Big)^{1/2}.$$

At the present time there is a slightly better lower bound on the number of $[n/2]$-sets in the case of allowable sequences than the $\Omega(n \log n)$ lower bound for configurations. M. Klawe, M. Paterson, and N. Pippenger construct an arrangement of pseudolines which is not known to be stretchable, to obtain

Theorem 4.12 ([70])

$$F(n, 2n) \geq n^{1 + \Omega(1 + \sqrt{\log n})}.$$

Following the methods developed by P. Edelman and C. Greene [26] to give an alternate proof of R. Stanley's theorem [93] on the number of allowable sequences, Edelman recently found the average number of k-sets exactly (averaged over allowable n-sequences).

Theorem 4.13 ([25]) *The average number of k-sets of an allowable n-sequence is*

$$\frac{(2k-1)(2(n-k)-1)}{2^{2n-5}} \binom{2k-2}{k-1} \binom{2(n-k-1)}{n-k-1}.$$

Apart from this last theorem of Edelman, allowable sequences have not played any role in the dramatic new results about this problem in the past two years. Rather, the higher dimensional results have come from following the initial idea in Lovasz' lemma, merged with extensions of a theorem of H. Tverberg. The first subcubic bound on $f_3(n, 2n)$ was found by I. Bárány, Z. Füredi, and L. Lovász [5], and they observed that their method would give a sub-n^d bound in dimension d if it were possible to prove a "colored" version of Tverberg's theorem (which they were able to do in dimension 2). They remark that in the notation of the following theorem, we obtain $f_d(n, 2n) = O(n^{d-\epsilon_d})$, where $\epsilon_d = N(d+1, d)^{-(d+1)}$. A proof of this theorem has just been claimed [99] by S. T. Vrećica and R. T. Živaljevic. Their result is:

Theorem 4.14 ([99]) *There exists a number $N = N(r, d)$ such that given $N(d+1)$ distinct points in $\mathbf{R}^d$ partitioned into $d+1$ classes (called "colors"), each of size N, we can choose $r(d+1)$ of these points to be the vertices of r multicolored simplices (each simplex having one vertex of each color) from among these points so that there is a point common to their convex hulls.*

In fact $N(r, d)$ can be chosen to be $2p(r) - 1$ where $p(r)$ is the smallest prime bigger than r.

The planar lower bound of Simmons extends easily to a lower bound of $n^{d-1} \log n$ in dimension d. In fact, a lower bound of $nf(n)$ in the planar case converts in general to a lower bound of $n^{d-1} f(n)$ in dimension d. The subcubic upper bound in dimension 3 of Bárány *et al* has been improved by Aronov, Chazelle, Guibas, Sharir, and Wenger [3] to show that $f_3(n, 2n) = O(n^{8/3} \log^{5/3} n)$. In the plane, the $O(n\sqrt{k})$ upper bound has been cracked by Pach, Steiger, and Szemerédi [83] when they showed that $f(k, n) = O(n\sqrt{k}/\log^* k)$.

Finally, we remark that in their paper Erdős *et al* ask whether their results can be extended to higher dimensions (we have seen that they can be, though only weakly), and whether they can be extended to other kinds of dividing curves and surfaces. This last has been accomplished by Sharir in [91], where he has extended many of the at-most-k-set results in this direction.

5. Order Types of Points in $\mathbf{R}^d$ and "Geometric Sorting"

As we have seen, isomorphic pseudoline arrangements may have different associated allowable sequences. As part of a paper elucidating the relationship between combinatorial equivalence (combinatorially equivalent pseudoline arrangements have the same associated allowable sequence) and isomorphism (isomorphic pseudoline arrangements decompose the plane into isomorphic lattices), the authors proved the following theorem.

Theorem 5.1 ([45]) *For $\Sigma = (\Pi_i)_{i \in \mathbf{Z}}$ and $\Sigma' = (\Pi_i')_{i \in \mathbf{Z}}$ nontrivial allowable n-sequences, the following are equivalent.*

1. *Σ and Σ' have the same sets of semispaces;*
2. *line $\overline{jk}$ separates i from m in Σ $\Longleftrightarrow$ the same holds in Σ';*
3. *each noncollinear triple ijk has the same orientation in Σ as in Σ' (or else this holds for all triples of Σ and the reverse $\overline{\Sigma'}$ of Σ');*
4. *the set of switch symbols in Σ is the same as that in Σ' (or else in $\widetilde{\Sigma'}$). (For each move, the switch symbol identifies which strings reverse in that move as well as their positions in the permutation just before the move takes place.)*

Two sequences which satisfied any of these equivalent conditions were called "semispace-equivalent". Identifying configurations whose allowable sequences are semispace-equivalent turns out to be the same as saying that they determine the same *oriented matroid* [6], and the different formulations of this equivalence correspond to some of the different ways in which oriented matroids (at least those of rank 3) have been defined. Two arrangements sitting in the euclidean plane are isomorphic if and only if any allowable sequences of each are semispace-equivalent. (If we look at arrangements in the projective plane, a slightly coarser equivalence relation, *local equivalence*, plays the same role.)

Buried in these equivalent properties was a rather surprising theorem about configurations of points: the implication (4) $\Rightarrow$ (3) amounts to saying that for a configuration $\mathcal{C}$ of points, knowing for every directed connecting line l of $\mathcal{C}$ *how many* points are to the left of l determines for every such l *which* points are to its left. In other (though less precise) words, these sets are determined by their cardinalities.

It is helpful to think of the analagous (trivial) theorem on the line. Given a set $\mathcal{C} = \{p_1, p_2, \ldots, p_n\} \subset \mathbf{R}$, knowing—for each i—the number of elements of

C less than (to the left of) p_i determines which elements of C are less than p_i. This is obvious, since knowing the number of elements of C less than p_i means knowing the position of p_i once C has been sorted. For this reason, one might think of the process of finding the switch symbols of a planar configuration as a definition of "sorting" in the plane. In fact, it turns out that the implication (4) $\Rightarrow$ (3) extends to any dimension d, and to point sets of nearly arbitrary degeneracy in $\mathbf{R}^d$. But first a few definitions.

As we have seen, the linear order of the set C on the line is determined by knowing for each i, j the sign of $p_i - p_j$, i.e., knowing the vector

$$\left(\text{sgn det} \begin{pmatrix} p_i & 1 \\ p_j & 1 \end{pmatrix} \right)_{1 \le i < j \le n}$$

This is a collection of $O(n^2)$ pieces of information, i.e., $O(n^2)$ numbers, which reduce to $O(n)$ numbers once we sort C.

So too, in the plane, a configuration $C = \{p_1, \ldots, p_n\}$ (with $p_i = (x_i, y_i)$), determines a *planar ordering* given by the orientation (clockwise or counterclockwise) of each triple of points in C:

$$\left(\chi_C(i,j,k) = \text{sgn det} \begin{pmatrix} x_i & y_i & 1 \\ x_j & y_j & 1 \\ x_k & y_k & 1 \end{pmatrix} \right)_{1 \le i < j < k \le n}$$

This is a collection of $O(n^3)$ numbers. The planar ordering is clearly equivalent to knowing

$$\Lambda_C(i,j) = \{k \mid \chi(i,j,k) > 0\}.$$

The implication (4) $\Rightarrow$ (3) says that Λ_C is determined by;

$$\lambda_C(i,j) = \text{card } \Lambda_C(i,j),$$

which encodes the $O(n^3)$ numbers stored in Λ_C into space $O(n^2)$. Thus, computing λ_C amounts to sorting the configuration C, just as it did on the line.

The preceding discussion generalizes in a natural way to a d-*ordering* of a configuration $C = \{p_1, \ldots, p_n\} \subset \mathbf{R}^d$ as the map χ_C from ordered $(d+1)$-tuples of $[1, n]$ to $\{-1, 0, +1\}$ given by;

$$\chi_C(i_o, \ldots, i_d) = \left(\text{sgn det} \begin{pmatrix} p_{i_o}^1 & \cdots & p_{i_o}^d & 1 \\ \vdots & \vdots & \ddots & \vdots \\ p_{i_d}^1 & \cdots & p_{i_d}^d & 1 \end{pmatrix} \right)_{1 \le i_o < \cdots < i_d \le n}$$

We define the *Lambda-matrix* of C by

$$\Lambda_C(i_1, \ldots, i_d) = \{k \mid \chi_C(i_1, \ldots, i_d, k)\}$$

(this is actually a d-dimensional array), and the *lambda-matrix* of C by

$$\lambda_C(i_1, \ldots, i_d) = \text{card} \left(\Lambda_C(i_1, \ldots, i_d) \right).$$

An inductive argument then yields the so-called "fundamental theorem of geometric sorting":

Theorem 5.2 ([41]) Λ_C *is determined by* λ_C.

Configurations C and C' are said to have the same *order type* if $\Lambda_C = \Lambda_{C'}$ or, equivalently, if $\lambda_C = \lambda_{C'}$.

A natural question to ask is: How fast can we compute the lambda-matrix of a configuration of n points in $\mathbf{R}^d$? Trivially we can do it in time $O(n^{d+1})$, and since the output size is $\Omega(n^d)$ it cannot be done faster than that.

Dual to a configuration of points in $\mathbf{R}^d$ is an arrangement of hyperplanes. Such arrangements have played a central role in computational geometry for the past decade. In a seminal paper [28], H. Edelsbrunner, J. O'Rourke, and R. Seidel found an $O(n^d)$ algorithm to "construct" an arrangement $\mathcal{A}$ of n hyperplanes in $\mathbf{R}^d$, in the sense that it constructs the full lattice of faces of $\mathcal{A}$ of all dimensions. It is easy to use this algorithm to calculate the lambda-matrix of the dual configuration in the same time. Hence the lambda-matrix can be calculated in time $O(n^d)$, which is optimal.

6. The Number of Order Types in $\mathbf{R}^d$

How many different order types are there on n points in $\mathbf{R}^d$? Certainly no more than $3^{\binom{n}{d+1}}$ (respectively $2^{\binom{n}{d+1}}$ for simple order types), since there are that many mappings of $(d+1)$-subsets of an n-set to $\{-1, 0, +1\}$ (respectively to $\{-1, +1\}$ for a simple order type). It is obvious, however, that not all such mappings correspond to an order type, i.e., that there are mappings which are not realized by the mapping χ_C for any configuration C of n points in $\mathbf{R}^d$. (Incidentally, one definition of an oriented matroid is in terms of precisely such mappings; i.e., an oriented matroid is simply a mapping $\chi : [1, n] \to \{-1, 0, +1\}$ which satisfies an obvious condition in common with a map χ_C arising from a configuration C of n points in $\mathbf{R}^d$, namely, that χ restricted to any $2d$-element subset of $[1, n]$ is realizable. Under some additional natural conditions (that $|\chi|$ is a matroid) it is possible to replace $2d$ by $d + 3$.)

It would appear that we could compute the number of distinct simple order types of n points as follows. Let $f(n)$ be this number. Now k points in general position in the plane determine $\binom{k}{2} \sim \frac{k^2}{2}$ lines, which in turn determine $\sim \frac{k^4}{8}$ cells [105], and the order type on these k points can be extended to a simple one on $k + 1$ points by placing an additional point in any one of these cells. Moreover, placing the point in different cells will give different order types because the point will be on different sides of any connecting line which separates the two cells. Thus $f(k+1) \sim \frac{k^4}{8} f(k)$, and it follows (using Stirling's formula) that there are essentially

$$\frac{2^4}{8} \frac{3^4}{8} \cdots \frac{n^4}{8} = n^{4n + O(\frac{n}{\log n})}$$

order types of n points in the plane; by the same argument in dimension d, it seems that there are essentially

$$\frac{d^{d^2}}{(d!)^{d+1}} \frac{(d+1)^{d^2}}{(d!)^{d+1}} \cdots \frac{n^{d^2}}{(d!)^{d+1}} = n^{d^2 n + O(\frac{n}{\log n})}$$

order types of n points in $\mathbf{R}^d$.

Unfortunately, this argument is fatally flawed, since different realizations of the same order type may not have corresponding cells in which to place an additional point (see Figure 6.1: we can move point 4 about so that line $\overline{34}$ passes above or below the intersection of lines $\overline{12}$ and $\overline{56}$ without changing the order type of these six points). For this reason it provides only a lower bound on the number of simple order types on n labeled points in $\mathbf{R}^d$. If $f_s(d, n)$ is this number, we have

Theorem 6.1 $f_s(d, n) \geq n^{d^2 n (1 + O(1/\log n))}$

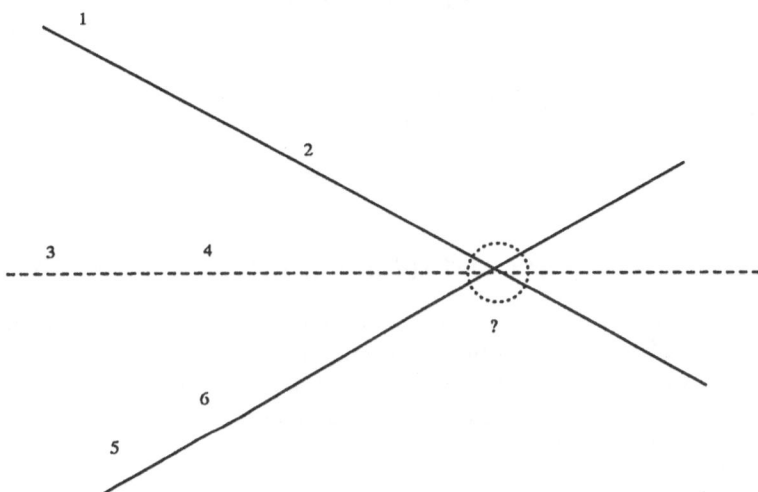

Fig. 6.1. Is the line $\overline{34}$ above or below the vertex?

Nevertheless, it is clear that the number of simple order types is no greater then the number of sign patterns of the sequence of polynomials

$$p_1(x), p_2(x), \ldots, p_{\binom{n}{d+1}}(x),$$

where

$$x = (x_1^1, x_1^2, \ldots, x_1^d, x_2^1, \ldots, x_n^d) \in \mathbf{R}^{dn}$$

and the polynomials p_i are all the $\binom{n}{d+1}$ determinants

$$\left(\det \begin{pmatrix} 1 & x_{i_0}^1 & \cdots & x_{i_0}^d \\ \vdots & \vdots & \ddots & \vdots \\ 1 & x_{i_d}^1 & \cdots & x_{i_d}^d \end{pmatrix} \right)_{1 \leq i_0 < \cdots < i_d \leq n}$$

One way to bound this number is to apply the following theorem proved independently by J. Milnor and R. Thom:

Theorem 6.2 ([79,97]) *If $X \subset \mathbf{R}^m$ is defined by polynomial inequalites of the form*

$$f_1 \geq 0, \ldots, f_p \geq 0$$

of total degree $d = \deg f_1 + \cdots + \deg f_p$, then the sum of the Betti numbers of X (and in particular the 0th Betti number, which is the number of connected components of X) is $\leq \frac{1}{2}(2+d)(1+d)^{m-1}$.

Using this method to bound the number of sign sequences of polynomials, the authors were able to prove

Theorem 6.3 ([47]) *Let $f_s(d,n)$ be the number of simple order types on n labeled points in $\mathbf{R}^d$. Then*

$$f_s(d,n) \leq n^{d(d+1)n}.$$

Shortly thereafter, Alon used the technique of introducing slack variables and modified this argument, using a related theorem of Milnor's from the same paper, to obtain an improved bound on all order types (simple or not):

Theorem 6.4 ([1]) *Let $f(d,n)$ be the number of order types on n labeled points in $\mathbf{R}^d$. Then*

$$(n/d)^{d^2 n (1+O(\frac{\log d}{\log n}))} \leq f_s(d,n) \leq f(d,n) \leq (n/d)^{d^2 n (1+O(\frac{1}{\log(n/d)} + \frac{\log\log(n/d)}{d\log(n/d)}))}.$$

The theorems of Thom and Milnor bound the sum of the Betti numbers and thus do not give a sharp bound on the number of connected components or on the number of sign sequences. In 1968, however, H. E. Warren counted the number of connected components in the complement of the zero set of a collection of polynomials to obtain:

Theorem 6.5 ([100]) *The number of sign patterns of m polynomials in n variables, each of degree at most d, is at most $\left(\frac{4edm}{n}\right)^n$.*

Applying Warren's theorem, we get a slightly stronger bound, with the error term in the lower bound matched in the upper bound:

Theorem 6.6

$$f(d,n) = \left(\frac{n}{d}\right)^{d^2 n (1+\Theta(\frac{1}{\log(n/d)}))}.$$

An immediate consequence of each of these theorems is a corresponding bound on the number of combinatorially distinct d-dimensional polytopes with n vertices, since the vertex configurations of combinatorially distinct polytopes must have different order types (though the converse is false). This give a bound far smaller than the best previously known bound of $n^{O(n^{\lfloor d/2 \rfloor})}$ coming from the upper bound theorem for polytopes [71,75].

Theorem 6.7 ([47]) *Let $g(d, n)$ be the number of combinatorially distinct convex polytopes with n vertices. Then*

$$g(d, n) \leq (\frac{n}{d})^{d^2 n (1 + O(\frac{1}{\log(n/d)}))}.$$

The techniques of this section can also be used to count the number of different combinatorial types of configurations in the plane (i.e., the number of distinct realizable allowable sequences).

Theorem 6.8 ([47]) *Let $h(n)$ be the number of distinct combinatorial equivalence classes of numbered configurations of n points in $\mathbf{R}^2$ or, equivalently, the number of realizable allowable sequences. Then*

$$h(n) \leq n^{8n}.$$

This bound is particularly surprising when we compare it with the total number of allowable sequences. P. Edelman and C. Greene [26] and R. Stanley [93] have shown that there are precisely

$$\frac{(n-2)!\binom{n}{2}!}{1^{n-1}3^{n-2}\cdots(2n-3)^1}$$

of these. A comparison of this constant with the bound in Theorem 6.8 shows that (in a very strong sense) most allowable sequences are not realizable.

In a similar spirit, G. Kalai [67] has shown that there are far more triangulated spheres than there are simplicial polytopes.

Theorem 6.9 ([67]) *Let $s(d, n)$ be the number of distinct triangulations of $\mathbf{S}^{d-1}$ with n labeled vertices. Then*

$$\log s(d, n) \geq c(d) n^{\lfloor (d-1)/2 \rfloor}.$$

When we compare this result with the upper bound on the number of simplicial polytopes given in Theorem 6.7, we see that almost all triangulated spheres are not polytopal.

7. Isotopy and Realizability Questions

The isotopy conjecture had been considered by many authors including Bokowski-Sturmfels [10], Dress, Lovász, and Goodman-Pollack [42,46], and was first formulated by Ringel [88] in the dual form for simple arrangements of lines. In this form it can be stated as follows: Given isomorphic simple arrangements $\mathcal{A}$ and $\mathcal{A}'$, it is possible to continuously move the lines of $\mathcal{A}$ into the corresponding lines of $\mathcal{A}'$ so that during the motion the arrangement remains simple (and hence the isomorphism class does not change). For point configurations in $\mathbf{R}^d$ this can be stated as: Given simple configurations $\mathcal{C}$ and $\mathcal{C}'$ which have the same order type, it is possible to continuously move the points of C into the corresponding points of C' so that during the motion the configuration remains simple, i.e., the order type of the configuration does not change. When first considered, this conjecture has usually seemed "obvious" to many researchers. Nevertheless, it resisted proof for many decades.

(One very naive attempt in the plane is the following. With $C = \{p_1, \ldots, p_n\}$ and $C' = \{p'_1, \ldots, p'_n\}$, let $p_i(t) = p_i + t(p'_i - p_i)$, i.e., each point in C moves at constant speed to arrive at the corresponding point in C' in unit time. Now we only have to check that this preserves the orientation of every triple, i.e., if $\chi_C(i, j, k) = \chi_{C'}(i, j, k) > 0$ then $\chi_C(t)(i, j, k) > 0$ for all $t \in [0, 1]$. This turns out to be false, however, and for an amusing reason, the same reason that explains the so-called "baseball paradox", which is that there can be two baseball players, Joe and Dom, so that in both the first and second halves of the baseball season Joe has a better batting average than Dom, but nevertheless Dom has a better batting average than Joe over the whole season).

The isotopy conjecture has a very simple statement if we consider the *realization space* of a configuration. Given a configuration C of n points in $\mathbf{R}^d$, which we identify (by concatenating coordinates) with a point of $\mathbf{R}^{dn}$, let $\mathcal{R}(C)$ (the realization space of C) be the set of all points in $\mathbf{R}^{dn}$ which (regarded as configurations in $\mathbf{R}^d$) have the same order type as C.

The isotopy conjecture then states that for every configuration of points $C \subset \mathbf{R}^d$, the realization space $\mathcal{R}(C)$ is connected.

There was considerable surprise in the spring of 1987 when N. White found a counterexample to the isotopy conjecture, at least in the non-simple case:

Theorem 7.1 ([104]) *There is a (non-simple) configuration of 42 points with disconnected realization space.*

The surprise was somewhat mitigated by the fact that his example was for a rather highly degenerate configuration, and there had been less confidence in the conjecture in this case. It seemed as though the story was completely resolved a few months later, however, when B. Jaggi and P. Mani-Levitska found an arrangement of 17 lines with disconnected realization space, which they were able to modify to obtain another arrangement, this time simple, consisting of 17 lines with disconnected realization space.

Though their arrangement was simple, their argument to show that its realization space was disconnected was not. While studying their paper, B. Sturmfels and N. White observed that any *constructible* configuration of n points with disconnected realization space could be modified to obtain a simple configuration of about $4n$ points with disconnected realization space. (A configuration $C = \{p_1, \ldots, p_n\} \subset \mathbf{RP}^2$ is constructible if the points can be labeled so that p_1, p_2, p_3, p_4 are in general position and p_t lies on at most two connecting lines determined by $p_1, \ldots, p_{t-1}$ for $t = 5, \ldots, n$.) An application of this method to the non-simple examples of White and of Jaggi and Mani-Levitska then yielded

Theorem 7.2 ([63]) *There is a simple configuration with disconnected realization space.* [1]

Unbeknownst to these authors, the isotopy problem was meanwhile being pursued by a group in Leningrad under the leadership of A. Vershik. In his Ph.D dissertation, N. E. Mnëv proved a startling theorem. It was announced in a Doklady article [80] (that, unfortunately, received little notice), and finally in 1988 his full paper [81] appeared in the West. As Vershik remarks in the introduction, Mnëv showed that isotopy for configurations in the plane (as well as for isomorphism classes of polytopes) is false in the strongest possible sense. He proved:

Theorem 7.3 ([80,81]) *Any semi-algebraic variety is homotopy equivalent to the realization space of a suitable simple planar order type.*

In particular, therefore, since there are disconnected semi-algebraic varieties, there are disconnected realization spaces of configurations.

The idea behind Mnëv's argument (later simplified by P. Shor [92]) is to replace any semi-algebraic variety (the solution set of a system of polynomial equalities and inequalities) with another one which is *stably equivalent* to it and in which the variables are totally ordered, i.e., $0 < x_1 < x_2 < \ldots < x_n$. (Spaces X and Y are stably equivalent if X is homeomorphic to $Y \times \mathbf{R}^d$ for some d, or vice-versa.) Mnëv and Shor accomplish this in different ways; essentially, Mnëv finds an equivalent variety in the neighborhood of some point $(x_1, x_2, \ldots, x_n)$, while Shor finds one in the neighborhood of infinity. Given this reduction, a configuration is constructed whose realization space is essentially the solution set of the given sytem. The construction is roughly as follows: We start with three lines in the projective plane which we interpret as the x-axis, the y-axis, and the line at infinity. We introduce points on the x-axis corresponding to the variables x_i. Then any polynomial expression in these variables can be constructed by a succession of the usual ruler-and-compass constructions we learned from Euclid to do geometric algebra. In this way we introduce many new points which are the intersection points of connecting lines joining previously determined points. The algebraic inequalities are simply the expression that certain points lie to the left or right of others. There is the additional observation that if we make each new point constructed on the y-axis much closer to the line at infinity than the one before, then the resulting order type (really an oriented matroid, because the extra inequalities might not yield a realization by points) is entirely determined.

The resulting configuration is not simple, however. But it is constructible, in the sense described above. Now, starting from the last point constructed, we replace each point lying on two connecting lines by four points which lie

---

[1] The record for the smallest counterexample to the isotopy conjecture is held by P. Suvorov [96], who gives a configuration of 13 points with disconnected realization space.

very close to the given point and surround it. Thus, "scattering" each point into about 4 points (an idea that seems to have been used first by M. Las Vergnas [72]), we obtain a simple configuration of about four times as many points and whose realization space is essentially the same.

The time for constructing this configuration and its size are linear in the complexity of the input semi-algebraic set. As a result it follows that whether or not a given system of polynomial equations and inequalities has a solution (this question is what is known as the decision problem for the Existential Theory of the Reals, usually referred to as ETR) is linearly reducible to deciding whether a given oriented matroid (or, equivalently. a given arrangement of pseudolines) is geometrically realizable (stretchable). Since ETR is known to be NP-hard (as a consequence of the NP-hardness of quadratic programming [89]), it follows that the problem of deciding whether an oriented matroid is realizable by a configuration of points or whether an arrangement of pseudolines is stretchable is also NP-hard.

Independently of the work of Mněv, Shor also proves in [92] that stretchability of pseudolines is NP-hard. He accomplishes this by reducing a variant of 3-SAT known as monotone 3-SAT [34] to the stretchability of certain pseudoline arrangements obtained by cleverly linking together distorted Pappus and Desargues configurations of pseudolines which turn out to be stretchable precisely when the given boolean formula they model is satisfiable. (The 3-SAT problem is to determine whether a boolean expression in conjunctive normal form which contains at most 3 variables per clause has a truth assignment of the variables which renders the boolean expression true, and for monotone 3-SAT the clauses contain either all non-negated variables or all negated variables.) Shor's paper also gives a clear exposition of the NP-hardness result using Mněv's methods.

The problem of the realizability, or "coordinatizability", of oriented matroids (or equivalently the stretchability of pseudoline arangements) was explored by J. Bokowski and his students [8,9,10,11,95]. They developed an algorithm which might generate a *solvability sequence* for an arrangement of pseudolines (for an oriented matroid), which if found would guarantee that the arrangement was stretchable (that the oriented matroid was coordinatizable). They also showed that a large class of stretchable pseudoline arrangements possessed a solvability sequence, and that any order type having a solvability sequence would necessarily satisfy the isotopy conjecture [10]. Bokowski and Sturmfels suspected that there existed order types which did not have solvability sequences. J. Richter [8] found such an example, with 9 points; interestingly, the corresponding order type still possessed the isotopy property.

There is a natural extension of the concept of stretchability for an arrangement of pseudolines to *d-stretchability*, which means that the arrangement is isomorphic to an arrangment of graphs of polynomials of degree at most d. Let $d(n)$ be the smallest d such that every simple arrangement of n pseudolines is d-stretchable. Thus $d(8) = 1$ and $d(9) > 1$. Unfortunately, there is a very large gap in what is known about $d(n)$.

Theorem 7.4 ([44]) *There exist constants c_1, c_2 such that,*

$$c_1 n^{1/2} \leq d(n) \leq c_2 n^2.$$

The lower bound follows from a counting argument on the number of d-stretchable arrangements using Milnor's theorem [79], while the upper bound follows from a construction which uses a theorem of D. J. Newman on comonotone polynomial approximation [82].

8. Lattice Realization of Order Types and the Problem of Robustness in Computational Geometry

A natural question to ask is: How compactly can we store an order type of n points in $\mathbf{R}^d$? A naive answer is that we need only nd pieces of information, where each "piece" is one of the coordinates of one of the points in a particular realization. This is naive because it doesn't take into account the size of the number describing this coordinate. What if the coordinate is very large? What if it is an irrational number, and has no compact representation? So it would seem that storing (representing) an order type by giving a coordinate representation might not be such a good idea.

An alternative is to use the λ-matrix, which takes space $O(n^d)$ (or $O(n^d \log n)$ if we take into account the size of the integer entries). This, at one time, appeared nearly optimal when we had only a bound of $2^{O(n^d)}$ on the number of such order types (since we clearly need at least $\log K$ bits even to name K different objects). When we found that there were many fewer such order types, only about $n^{d^2 n}$ (see Section 6), it became reasonable to ask if there was a much more compact way to encode them than by using the λ-matrix.

It makes sense to look again at this question in terms of a coordinate representation of an "optimal" point set having a given order type: We would like to know how small a number $b = b(d, n)$ will suffice so that every order type of n points in $\mathbf{R}^d$ has a representative so that every coordinate uses at most b bits. Suppose (looking ahead to a negative answer!) we relax our problem so that we consider only simple order types. The question we will now address was posed rather succinctly by B. Chazelle when he asked [personal communication, 1986], at the first ACM Symposium on Computational Geometry, "How many bits does it take to know an order type?".

The relevance of this question to computational geometry is twofold. First, the decisions made during the running of a typical geometric algorithm are largely, often exclusively, based on order type (or partial order type) information about a set of points given as part of the input data: whether a point is on the positive or negative side of a hyperplane spanning a subset of the input data or, dually, whether a vertex of an arrangement is on one side or another of an input hyperplane. This is in fact the basis of the combinatorial side of computational geometry [27], and our knowledge of this side seems deep and appears to be advancing quite rapidly.

The second connection is to the problem of *robustness* in geometric algorithms. Most algorithms are described and their complexity analyzed assuming that the computer is an "infinite precision" machine, i.e., that it calculates perfectly with real numbers and that a single real number occupies a single unit of space in the machine. But this is not the way real computers work at all. They are finite precision machines: arithmetic is perfectly accurate only up to a certain number of bits, and beyond that various kinds of roundings are done which produce a variety of errors. If these errors simply produced output that deviated from the truth to an extent based on the size of the errors themselves, the problem would be serious enough. However, the problem in geometric algorithms is much worse. During the running of an algorithm, many decisions are typically based on whether a determinant is positive or negative (an order type decision). If the value of this determinant is too small for the precision of the computer to distinguish from zero, what decision should be made? Should it simply be rounded to zero? Chosen to be positive or negative randomly, or on the basis of some heuristic information? All known choices are capable of leading to the same possible catastrophe: the calculation produces results that contradict the geometry, and the program crashes. There is no output (let alone incorrect output), or else the output may be pure nonsense. A *robust* algorithm is one that manages to avoid these pitfalls. (We refrain from being more precise here because part of the problem is to find a suitable definition of robustness.) These issues have been pursued quite actively (but with limited success so far) during the past few years [21,32,33,52,55,59,60,61,62,76,77].

From the point of view of providing a possible solution to the robustness problem, the answer to Chazelle's question is particularly disappointing. Goodman, Pollack, and Sturmfels showed in 1989 that an exponential number of bits is both necesary and sufficient to give a coordinate representation of any planar order type on n points. More precisely, they proved;

Theorem 8.1 ([49]) *Let $f(d, n)$ be the smallest integer N such that every configuration C of n points in general position in $\mathbf{R}^d$ can be realized, up to order type, on the grid $G(d, N) = \{(i_1, \ldots, i_d) \mid -N \leq i_j \leq N\}$. Then there exist constants $c_1 = c_1(d), c_2 = c_2(d)$ such that*

$$2^{2^{c_1 n}} \leq f(d, n) \leq 2^{2^{c_2 n}}.$$

The construction of a configuration giving the lower bound proceeds using some of the same machinery used by Mnëv (see Section 7). One first finds a highly degenerate configuration by successive squaring so that the largest x_1-coordinate is doubly exponential in the number points constructed. The same "scattering" technique mentioned above is then used to convert this configuration into one with about four times as many points that is in general position. Finally, the projective invariance and other properties of the cross-ratio lead eventually to a proof of the lower bound.

The proof of the upper bound uses the following lemma of D. Yu. Grigor'ev and N. N. Vorobjov:

Lemma 8.2 ([53]) *Suppose that the polynomials* $h_1, \ldots, h_k \in Z[X_1, \ldots, X_m]$ *satisfy the bounds* $\deg h_i < d$ *and* $l(h_i) \leq M$, *where* $l(h)$ *is the maximum bit length of the coefficients of* h. *If* W *is any connected component of the semi-algebraic set defined by the system* $h_1 \geq 0, \ldots, h_k \geq 0$, *then* W *meets the ball of radius* $R = \exp((M + \log k)(d^{qm}))$ *centered at the origin for some natural number* q.

The $k = \binom{n}{d}$ inequalities are taken to be the (after scaling) assertions that each of the determinants in the $m = nd$ variables whose positivity determine the order type is at least 1. The lemma then guarantees a realization of this order type in a disk of doubly exponential radius, with the property that the area of any triangle is at least 1. Another scaling then permits each point to "snap" to a grid point without changing the order type, and the upper bound follows.

It turns out that this result can be expressed in another form, which involves a seemingly less "computational" concept than bit size.

Let us define the spread $\sigma(\mathcal{C})$ of a configuration $\mathcal{C} = \{p_1, \ldots, p_n\}$ in general position in $\mathbf{R}^d$ to be

$$\sigma(\mathcal{C}) = \frac{\max_{1 \leq i_0 < \cdots < i_d \leq n} \text{vol} \langle p_{i_0}, \ldots, p_{i_d} \rangle}{\min_{1 \leq i_0 < \cdots < i_d \leq n} \text{vol} \langle p_{i_0}, \ldots, p_{i_d} \rangle},$$

where $\langle p_{i_0}, \ldots, p_{i_d} \rangle$ denotes the simplex spanned by the points $p_{i_0}, \ldots, p_{i_d}$.

If $\mathcal{C} = \{p_1, \ldots, p_n\}$ is a configuration of points on a line, and we want to find a configuration $\mathcal{C}'$ having the same order with the smallest spread, it is clear that we should spread the points so they are equally spaced. Thus the *intrinsic spread* $\tilde{\sigma}$ of a configuration $\mathcal{C}$ of n points on a line, in this sense, is n.

In the same way, the intrinsic spread of a labeled configuration $\mathcal{C}$ in $\mathbf{R}^d$ is defined to be the minimum, over all configurations $\mathcal{C}'$ having the same order type as $\mathcal{C}$, of the spread $\sigma(\mathcal{C}')$. (It is not difficult to see that this minimum is actually attained by some configuration.) Thus the intrinsic spread of $\mathcal{C}$ is simply a measure of how "evenly" it is possible to realize the order type of $\mathcal{C}$, and it is natural to ask just how large this instrinsic spread may be for configurations of n points, i.e., how "badly spread out" some order types may be in this measure.

It is not hard to see that Theorem 8.1 can then be rephrased as:

Theorem 8.3 ([49]) *Let* $g(d, n)$ *be the maximum of the intrinsic spread* $\tilde{\sigma}(\mathcal{C})$ *over all simple configurations* $\mathcal{C}$ *of* n *points in* $\mathbf{R}^d$. *Then there exist constants* $c_1 = c_1(d), c_2 = c_2(d)$ *such that*

$$2^{2^{c_1 n}} \leq g(d, n) \leq 2^{2^{c_2 n}}.$$

References

[1] N. Alon: The number of polytopes, configurations, and real matroids. Mathematika **33** (1986) 62–71

[2] N. Alon and E. Győri: The number of small semispaces of a finite set of points in the plane. J. Comb. Theory, Ser. A **41** (1986) 154–157

[3] B. Aronov, B. Chazelle, H. Edelsbrunner, L. Guibas, M. Sharir and R. Wenger: Points and triangles in the plane and halving planes in space. Discrete Comput. Geom. **6** (1991) 435–442

[4] B. Aronov, J. E. Goodman, R. Pollack and R. Wenger: Hadwiger's theorem for line transversals does not generalize to higher dimensions. Unpublished manuscript, 1989

[5] I. Bárány, Z. Füredi and L. Lovász: On the number of halving planes. In: Proc. Fifth Annual ACM Sympos. on Computational Geometry. 1989, pp. 140–144

[6] A. Björner, M. Las Vergnas, B. Sturmfels, N. White and G. Ziegler: Oriented Matroids. Cambridge Univ. Press, Cambridge 1992

[7] R. G. Bland and M. Las Vergnas: Orientability of Matroids. J. Comb. Theory, Ser. B **23** (1978) 94–123

[8] J. Bokowski, J. Richter and B. Sturmfels: Nonrealizability proofs in computational geometry. Discrete Comput. Geom. **5** (1990) 333–350

[9] J. Bokowski and B. Sturmfels: Oriented matroids and chirotopes–problems of geometric realizability. TH Darmstadt, 1985

[10] J. Bokowski and B. Sturmfels: On the coordinatization of oriented matroids. Discrete Comput. Geom. **1** (1986) 293–306

[11] J. Bokowski and B. Sturmfels: Computational Synthetic Geometry. Lecture Notes in Mathematics, vol. 1355. Springer, Berlin Heidelberg 1989

[12] E. Boros and Z. Füredi: The number of triangles covering the center of an n-set. Geometriae Dedicata **17** (1984) 69–77

[13] G. R. Burton and G. B. Purdy: The directions determined by n points in the plane. J. London Math. Soc. (2) **20** (1979) 109–114

[14] R. J. Canham: Arrangements of hyperplanes in projective and Euclidean spaces. University of East Anglia, Norwich 1971

[15] S. E. Cappell, J. E. Goodman, J. Pach, R. Pollack, M. Sharir and R. Wenger: The Combinatorial Complexity of Hyperplane Transversals. Adv. Math., to appear

[16] V. Chvátal and G. Klincsek: Finding largest convex subsets. Cong. Num. **29** (1980) 453–460

[17] K. L. Clarkson and P. W. Shor: Applications of random sampling, II. Discrete Comput. Geom. **4** (1989) 387–421

[18] R. Cordovil: Sur les matroïdes orientés de rang trois et les arrangements de pseudodroites dans le plan projectif réel. European J. Comb. **3** (1982) 307–318

[19] R. Cordovil: Oriented matroids and geometric sorting. Canad. Math. Bull. **26** (1983) 351–354

[20] L. Danzer, B. Grünbaum and V. Klee: Helly's Theorem and its Relatives. In: Convexity. Proc. Sympos. Pure Math., Vol. 7, (1963) pp. 100–181. Amer. Math. Soc., Providence

[21] D. Dobkin and D. Silver: Recipes for geometry and numerical analysis, Part 1: an empirical study. In: Proc. Fourth Annual ACM Sympos. on Computational Geometry, 1988, pp. 93–105

[22] A. Dreiding, A. Dress and A. Haegi: Classification of Mobile Molecules by Category Theory. Studies in Physical and Theoretical Chemistry **23** (1982) 39–58

[23] A. Dreiding and K. Wirth: Classification of Finite Ordered Point Sets in Oriented d-Dimensional Space. Math. Chemistry **8** (1980) 341–352

[24] A. Dress: Chirotopes and Oriented Matroids. Bayr. Math. Schriften **21** (1986) 14–68

[25] P. Edelman: On the average number of k-sets. Discrete Comput. Geom. **8** (1992) 209–213

[26] P. Edelman and C. Greene: Combinatorial correspondences for Young tableaux, balanced tableaux and maximal chains in the Bruhat order of S_n. In: Combinatorics and Contemporary Math., vol. 34. Amer. Math. Soc., Providence 1984, pp. 42–99

[27] H. Edelsbrunner: Algorithms in Combinatorial Geometry. Springer, Berlin 1987

[28] H. Edelsbrunner, J. O'Rourke and R. Seidel: Constructing arrangements of lines and hyperplanes with applications. SIAM J. Comput **15** (1986) 341–363

[29] H. Edelsbrunner and E. Welzl: On the number of line-separations of a finite set in the plane. J. Comb. Theory, Ser. A **38** (1985) 15–29

[30] P. Erdős, L. Lovász, A. Simmons and E. G. Strauss: Dissection graphs of planar point sets. In: A Survey of Combinatorial Theory. North-Holland, Amsterdam 1973, pp. 139–149

[31] J. Folkman and J. Lawrence: Oriented Matroids. J. Combin. Theory, Ser. B **25** (1978) 199–236

[32] S. Fortune: Stable maintenance of point-set triangulation in two dimensions. In: Proc. 30th Annual IEEE Sympos. on the Foundations of Computer Science, 1989, pp. 494–499

[33] S. Fortune and V. Milenkovic: Numerical stability of algorithms for line arrangements. In: Proc. Seventh Annual ACM Sympos. on Computational Geometry, 1991, pp. 334–341

[34] M. R. Garey and D. S. Johnson: Computers and Intractability: A Guide to the Theory of NP-Completeness. W. H. Freeman and Co., New York 1979

[35] J. E. Goodman: Proof of a conjecture of Burr, Grünbaum, and Sloane. Discrete Math. **32** (1980) 27–35

[36] J. E. Goodman and R. Pollack: On the combinatorial classification of non-degenerate configurations in the plane. J. Comb. Theory, Ser. A **29** (1980) 220–235

[37] J. E. Goodman and R. Pollack: Proof of Grünbaum's conjecture on the stretchability of certain arrangements of pseudolines. J. Comb. Theory, Ser. A **29** (1980) 385–390

[38] J. E. Goodman and R. Pollack: Convexity theorems for generalized planar configurations. In: Proc. of the Conf. on Convexity and Related Combinatorial Geometry, Univ. of Okla., 1980. Marcel Dekker, New York 1982, pp. 73–80

[39] J. E. Goodman and R. Pollack: Helly-type theorems for pseudoline arrangements in $\mathbf{P}^2$. J. Comb. Theory, Ser. A **32** (1982) 1–19

[40] J. E. Goodman and R. Pollack: A theorem of ordered duality. Geometriae Dedicata **12** (1982) 63–72

[41] J. E. Goodman and R. Pollack: Multidimensional Sorting. SIAM J. Comput. **12** (1983) 484–507

[42] J. E. Goodman and R. Pollack: Isotopy problem for configurations. In: List of Problems, Tagungsbericht Oberwolfach. Tagung über kombinatorische Geometrie. September 1984

[43] J. E. Goodman and R. Pollack: On the number of k-subsets of a set of n points in the plane. J. Comb. Theory, Ser. A **36** (1984) 101–104

[44] J. E. Goodman and R. Pollack: Polynomial realization of pseudoline arrangements. Comm. Pure Appl. Math. **38** (1984) 725–732

[45] J. E. Goodman and R. Pollack: Semispaces of configurations, cell complexes of arrangements. J. Comb. Theory, Ser. A **37** (1984) 257–293

[46] J. E. Goodman and R. Pollack: A combinatorial version of the isotopy conjecture. In: Discrete Geometry and Convexity. Ann. New York Acad. Sci., 1985, pp. 12–19

[47] J. E. Goodman and R. Pollack: Upper bounds for configurations and polytopes in $\mathbf{R}^d$. Discrete Comput. Geom. **1** (1986) 219–227

[48] J. E. Goodman and R. Pollack: Hadwiger's Transversal Theorem in Higher Dimensions. J. Amer. Math. Soc. **1** (1988) 301–309

[49] J. E. Goodman, R. Pollack and B. Sturmfels: Coordinate representation of order types requires exponential storage. In: Proc. 21st Annual ACM Sympos. on the Theory of Computing 1989, pp. 405–410

[50] J. E. Goodman, R. Pollack and B. Sturmfels: The Intrinsic Spread of a Configuration in $\mathbf{R}^d$. J. Amer. Math. Soc. **3** (1990) 639–651

[51] J. E. Goodman, R. Pollack and R. Wenger: Geometric Transversal Theory. In: Recent Advances in Discrete and Computational Geometry. Springer, Berlin Heidelberg 1991. This volume, pp. 163–198

[52] D. H. Greene and F. F. Yao: Finite-resolution computational geometry. In: Proc. 27th Annual IEEE Sympos. on the Foundations of Computer Science 1986, pp. 143–152

[53] D. Yu. Grigor'ev and N. N. Vorobjov: Solving systems of polynomial inequalities in subexponential time. J. Symbolic Com. **5** (1988) 37–64

[54] B. Grünbaum: Arrangements and Spreads. Amer. Math. Soc., Providence 1972

[55] L. Guibas, D. Salesin and J. Stolfi: Epsilon geometry: building robust algorithms from imprecise computations. In: Proc. Fifth Annual ACM Sympos. on Computational Geometry, 1989, pp. 208–217

[56] H. Hadwiger: Über Eibereiche mit gemeinsamer Treffgeraden. Portugal. Math. **16** (1957) 23–29

[57] E. Halsey: Zonotopal complexes on the d-cube. University of Washington, Seattle 1971

[58] E. Helly: Über Mengen Konvexer Körper mit Gemeinschaftlichen Punkten. Jber. Deutsch. Math. Verein **32** (1923) 175–176

[59] C. Hoffman: The problems of accuracy and robustness in geometric computation. Computer **22** (1989) 31–42

[60] J. Hopcroft and P. Kahn: A paradigm for robust geometric algorithms. Technical Report TR 89-1044, Cornell University 1989

[61] M. Iri and K. Sugihara: Geometric algorithms in finite precision arithmetic. Technical Report RMI 88-10, University of Tokyo 1988

[62] M. Iri and K. Sugihara: Construction of the Voronoi diagram for one million generators in single precision arithmetic. First Canadian Conference on Computational Geometry, 1989

[63] B. Jaggi, P. Mani-Levitska, N. White and B. Sturmfels: Uniform oriented matroids without the isotopy property. Discrete Comput. Geom. **4** (1989) 97–100

[64] R. E. Jamison: Structure of slope critical configurations. Geometriae Dedicata **16** (1984) 249–277

[65] R. E. Jamison: A survey of the slope problem. In: Discrete Geometry and Convexity. Ann. New York Acad. Sci., 1985, pp. 34–51

[66] R. E. Jamison and D. Hill: A catalogue of sporadic slope-critical configurations. Cong. Num. **40** (1983) 101–125

[67] G. Kalai: Many triangulated spheres. Discrete Comput. Geom. **3** (1988) 1–14

[68] M. Katchalski: Thin Sets and Common Transversals. J. Geom. **14** (1980) 103–107

[69] P. Kirchberger: Über Tschebyschefsche Annäherungsmethoden. Math. Ann. **57** (1903) 509–540

[70] M. Klawe, M. Paterson and N. Pippenger: Inversions with $n^{1+\Omega(1+\sqrt{\log n})}$ transpositions at the median. Unpublished manuscript, 1982

[71] V. Klee: The number of vertices of a convex polytope. Canad. J. Math. **16** (1964) 701–720

[72] M. Las Vergnas: Order properties of lines in the plane and a conjecture of G. Ringel. J. Comb. Theory, Ser. B **41** (1986) 246–249

[73] F. Levi: Die Teilung der projectiven Ebene durch Gerade oder Pseusogerade. Ber. Math.-Phys. Kl. sächs. Akad. Wiss. Leipzig **78** (1926) 256–267

[74] L. Lovász: On the number of halving lines. Ann. Univ. Sci. Budapest, Eötvös, Sect. Math. **14** (1971) 107–108

[75] P. McMullen: The maximum number of faces of a convex polytope. Mathematika **17** (1970) 179–184

[76] V. Milenkovic: Verifiable implementations of geometric algorithms using finite precision arithmetic. Artificial Intelligence **37** (1988) 377–401

[77] V. Milenkovic: Double precision geometry: a general technique for calculating line and segment intersections using rounded arithmetic. In: Proc. 30th Ann. IEEE Sympos. on the Foundations of Computer Science, 1989

[78] V. Milenkovic and L. R. Nackman: Finding compact coordinate representations for polygons and polyhedra. In: Proc. Sixth Annual ACM Sympos. on Computational Geometry 1990, pp. 244–259

[79] J. Milnor: On the Betti numbers of real varieties. Proc. Amer. Math. Soc. **15** (1964) 275–280

[80] N. E. Mněv: On manifolds of combinatorial types of projective configurations and convex polyhedra. Soviet Math. Dokl. **32** (1985) 335–337

[81] N. E. Mněv: The universality theorems on the classification problem of configuration varieties and convex polytope varieties. In: Topology and Geometry-Rokhlin Seminar, Lecture Notes in Mathematics, vol. 1346. Springer, Berlin Heidelberg, 1988, pp. 527–543

[82] D. J. Newman: Efficient co-monotone approximation. J. Approx. Theory **25** (1979) 189–192

[83] J. Pach, W. Steiger and E. Szemerédi: An upper bound on the number of planar k-sets. In: Proc. 30th Ann. IEEE Sympos. on the Foundations of Computer Science 1989, pp. 72–79

[84] G. W. Peck: On k-sets in the plane. Discrete Math. **56** (1985) 73–74

[85] R. Perrin: Sur le probléme des aspects. Bull. Soc. Math. France 10 (1881) 103–127

[86] R. Pollack and R. Wenger: A family of generalizations of Hadwiger's theorem to hyperplane transversals. Unpublished manuscript

[87] R. Pollack and R. Wenger: Necessary and sufficient conditions for hyperplane transversals. Combinatorica 10 (1990) 307–311

[88] G. Ringel: Teilungen der projectiven Ebene durch Geraden oder topologische Geraden. Math. Z. 64 (1956) 79–102

[89] S. Sahni: Quadratic Programming is NP-hard. SIAM J. Computing 3 (1974) 262–279

[90] P. R. Scott: On the sets of directions determined by n points. Amer. Math. Monthly 77 (1970) 502–505

[91] M. Sharir : On k-sets in arrangements of curves and surfaces. Discrete Comput. Geom. 6 (1991) 593–613

[92] P. W. Shor: Stretchability of pseudoline arrangements is NP-hard. In: Applied Geometry and Discrete Mathematics–The Victor Klee Festschrift, American Math. Soc., Providence 1991, pp. 531–554

[93] R. Stanley: On the number of reduced decompositions of elements of Coxeter groups. Europ. J. Combinatorics 5 (1984) 359–372

[94] J. Steiner: Einige Gesetze über die Teilung der Ebene und des Raumes. J. Reine Angew. Math. 1 (1826) 349–364

[95] B. Sturmfels: Computational Synthetic Geometry. Ph.D. dissertation, University of Washington, Seattle 1987

[96] P. Suvorov: Isotopic but not rigidly isotopic plane systems of straight lines. In: Topology and Geometry–Rokhlin Seminar, Lecture Notes in Mathematics, 1346. Springer, Berlin Heidelberg 1988, pp. 545–555

[97] R. Thom: Sur l'homologie des variétés algébriques réelles. In: Differential and Combinatorial Topology, Princeton University Press, Princeton 1965, pp. 255–265

[98] P. Ungar: $2N$ noncollinear points determine at least $2N$ directions. J. Comb. Theory, Ser. A 33 (1982) 343–347

[99] S. T. Vrećica and R. T. Živaljević: The colored Tverberg's problem and complexes of injective functions, to appear

[100] H. E. Warren: Lower bounds for approximation of nonlinear manifolds. Trans. Amer. Math. Soc. 133 (1968) 167–178

[101] E. Welzl: More on k-sets of finite sets in the plane. Discrete Comput. Geom. 1 (1986) 95–100

[102] R. Wenger: A Generalization of Hadwiger's Theorem to Intersecting Sets. Discrete Comput. Geom. 5 (1990) 383–388

[103] R. Wenger: Geometric Permutations and Connected Components. Technical report TR-90-50, DIMACS 1990

[104] N. White: A nonuniform matroid which violates the isotopy conjecture. Discrete Comput. Geom. 4 (1989) 1–2

[105] T. Zaslavsky: Facing Up to Arrangements: Face Count Formulas for Partitions of Space by Hyperplanes. Memoirs Amer. Math. Soc. (1), 154 Amer. Math. Soc., Providence 1975

[106] T. Zaslavsky: Bilateral Geometries. Unpublished manuscript, 1980

Chapter VI. Hyperplane Approximation and Related Topics

Nikolai M. Korneenko and Horst Martini

Abstract. This chapter is a survey on problems related to the approximation of finite weighted point sets in d-space by hyperplanes with respect to certain metrics. From the viewpoint of Computational Geometry, we shall discuss the orthogonal and vertical L_1-fit problem as well as the orthogonal and vertical L_∞-fit problem. Though we shall lay emphasis on the othogonal L_1-fit problem, suitable generalizations to the other problems are given as well. For example, the reader will find informations on k-flats, $k \in \{0, \ldots, d-2\}$, being extremal with respect to finite point sets and leading also to the well-known Fermat-Torricelli point and related subjects.

1. Introduction

The recent and rapid progress in computational geometry, high-lighted by the books [M84, PS85, E87, M88], has provided new insights into various research areas and enriched them with new methods and problems as well as with new points of view on old questions.

As an example, we shall consider the problem of approximating a set of n points in $\mathbb{R}^d$ by a linear function (the linear fit problem), which is of fundamental importance in statistics and numerical mathematics [R64]. Furthermore, the lower-dimensional cases of this problem also have gained attention in operations research as point/line facility location problems [LMW88]. Thus, there are several numerical methods and algorithms for this class of problems, of varying efficientcy and satisfying different goodness-of-fit criteria. However, it turned out that a fairly large subclass of this problem class is more amenable to combinatorial rather than numerical treatment.

Consider n arbitrary points $p_i = (x_{1i}, \ldots, x_{di})$, $i = 1, \ldots, n$, in Euclidean space $\mathbb{E}^d$ with corresponding positive weights w_i. In the context of the (facility) location problems they are called *demand points*. The object which is to be placed optimally with respect to this point set is referred to as *facility*, i.e. in this chapter we might speak of hyperplane facilities.

For a hyperplane $H : (b_0 + b_1 x_1 + \ldots + b_d x_d = 0)$ to be optimal, the following criteria are most often considered:

$$\text{MINMAX}: \qquad \underset{H}{\text{minimize}} \ \max_i \ w_i \text{Dist}(p_i, H), \qquad\qquad (1)$$

$$\text{MINSUM}: \qquad \underset{H}{\text{minimize}} \ \sum_i \ w_i Dist(p_i, H), \qquad\qquad (2)$$

$$\text{MINSQUARES}: \qquad \underset{H}{\text{minimize}} \ \sum_i \ w_i (\text{Dist}\,(p_i, H))^2, \qquad\qquad (3)$$

where Dist ($\cdot$, $\cdot$) is a distance or a deviation function. The common metrics are the Euclidean, the Manhattan, and the Chebyshov metrics. For the Euclidean distance we obtain

$$D_{\mathrm{E}}(p_i, H) = \left(\sum_{j=1}^{d} b_j^2 \right)^{-\frac{1}{2}} \left| b_0 + \sum_{j=1}^{d} b_j x_{ji} \right|. \qquad\qquad (4)$$

For the Manhattan interpoint distance

$$D_{\mathrm{M}}(p_i, p_j) = \sum_k |x_{ki} - x_{kj}|$$

one may easily verify that

$$D_{\mathrm{M}}(p_i, H) = \left(\max_{j=1,\dots,d} b_j \right)^{-1} \left| b_0 + \sum_{j=1}^{d} b_j x_{ji} \right|,$$

and the original least-distances linear fit problem is split into d instances of a problem with vertical absolute deviation (5) as Dist ($\cdot$, $\cdot$):

$$D_{\text{vert}}(p_i, H) = \left| x_{1i} - \left(b_0 + \sum_{j=2}^{d} b_j x_{ji} \right) \right|. \qquad\qquad (5)$$

The Chebyshov distance

$$D_{\mathrm{C}}(p_i, p_j) = \min_k \{ |x_{ki} - x_{kj}| \}$$

in $\mathbb{E}^2$ is known to be isometric to the Manhattan one. However, this isometry is unique among all L_p-norms [CLW79], and in higher dimensions the D_{C}-problem is reduced to 2^{d-1} instances of a problem with (5) as distance.

Sometimes (1), (2), (3) are called L_∞-, L_1- and L_2-fit, respectively, with (4) corresponding to the orthogonal L_p-fit and with (5) corresponding to the vertical one. E.g. ((2), (4)) is said to be the orthogonal L_1-fit problem.

Cost functions of the class (3) are commonly used in statistics and numerical mathematics, because (3) together with (4) or (5) is a smooth function having simple closed-form solutions obtainable by the famous least-squares method. We shall exclude them here, i.e. we restrict our attention to the four remaining criteria.

As is common in discussions on the computational complexity of problems from computational geometry (see [PS85, M84]), we shall use an infinite precision real random access machine [AHU78] with unit-cost operations $(+, -, *, /$ and square root) as a model of computation.

On algorithms appearing throughout the paper, the dimension d is assumed to be fixed. The set $P = \{p_i | i = 1, \ldots, n\}$ of demand points and the set $W = \{w_i\}$ of weights are also assumed to be fixed, and possible notational dependence on P will be suppressed.

For a hyperplane H let H^+, H^-, H^0 denote the sets of demand points lying on either side of H and in H, respectively. Similarly, W^+, W^-, W^0 will denote the corresponding sets of weights. $|X|$ is the cardinality of a set X.

The chapter is arranged as follows: A brief discussion of the orthogonal L_1-fit is given in Sect. 2, while the vertical L_1-fit is described in Sect. 3. The corresponding L_∞-fits follow in Sects. 4 and 5, respectively. Finally some related problems are discussed in Sect. 6. Once again we note that these linear fit problems have been widely discussed in numerical analysis and other areas; in this chapter, we are particularly interested in their combinatorial computational complexity in the spirit of the books [RND77, AHU78, M84, PS85, M88], and we shall try to demonstrate applications of new approaches to these classical problems which have emerged in computational geometry.

2. MINSUM Problem: Orthogonal L_1-Fit

2.1 Basic Facts

A combinatorial discussion of this problem was given in [NM80], and the computational complexity of its two-dimensional version was treated in [We75, MN80, MN83, MT83, D84, DKM84, DKM85, LC85, K88, K89, YKII88]. For the general case, we refer to [IKY88, HR88, KM89, KM90, HIIRY90]. For the sake of convenience, we state the problem once more:

$$\underset{b_0, \ldots, b_d}{\text{minimize}}\ S_E(H) = \left(\sum_{j=1}^d b_j^2\right)^{-\frac{1}{2}} \sum_{i=1}^n w_i \left| \left(b_0 + \sum_{j=1}^d b_j x_{ji}\right) \right|. \tag{6}$$

In view of robust statistics, a hyperplane satisfying (6) shall be called a *median hyperplane* with respect to the point set P.

The following auxiliary problem might be of interest in itself.

Anchored Median Hyperplane Problem. In $\mathbb{E}^d$, $d \geq 2$, find a hyperplane $H(b_0, \ldots, b_d)$ passing through a fixed point q (the so-called *anchor*) and satisfying (6) with respect to a set of demand points p_i with positive weights w_i, $i = 1, \ldots, n$. $\qquad\qquad\square$

The two-dimensional case of this problem was considered e.g. in [MN83]. For general d, its connection with zonotopes (i.e. vector sums of line segments) was first observed by Martini [M87]. Using support functions of zonotopes, it is easy to prove

Theorem 1 [KM89, KM90]. *In $\mathbb{E}^d$, $d \geq 2$, let there be given a set P of n arbitrary points with corresponding positive weights, and let $q \in \mathbb{E}^d$ be some anchor. If the set P and q together cannot be placed in one (median) hyperplane, then each q-anchored median hyperplane of P contains $d-1$ points which together with q are affinely independent.*

From Theorem 1 one may simply deduce the general statement on unanchored median hyperplanes, namely

Theorem 2 [KM89, KM90]. *Let P be a set of n arbitrary points with positive weights and $k = \dim \mathrm{aff}(P) \leqq d$ in Euclidean d-space, $d \geqq 2$. Then the following holds:*

For $k < d$ each median hyperplane with respect to P contains P.
For $k = d$

(a) each median hyperplane is the affine hull of d affinely independent points from P and

(b) for each median hyperplane $W^+ < \frac{W}{2}$ and $W^- < \frac{W}{2}$.

This Theorem is stronger than similar claims from [MN80, NM80, DKM84, MT83,YKII88], which only refer to the existence of a median line satisfying the conditions of Theorem 2, while Theorem 2 concerns each median hyperplane.

We shall say that a hyperplane through d affinely independent points of a weighted set P is a *blocked* hyperplane. A blocked hyperplane satisfying the inequalities of Theorem 2(b) is called *halving*. Thus, each median hyperplane for P is a halving one, but not vice versa.

2.2 General Case: $O(n^d \log n)$ Time and Linear Space Algorithm

By Theorem 2(a) one can find the median hyperplanes in $O(n^{d+1})$ time, namely by enumerating all $C_n^d = O(n^d)$ blocked candidate hyperplanes and computing the corresponding weighted distance sums. (Enumeration algorithms spending constant time per candidate-k-subset can be taken from [RND77], Sect. 5.2.2.)

For improving this time complexity, it is better to use a kind of "coherence" between "neighbouring" hyperplanes based on the piecewise linearity of (6) (noted in [MT83, DKM84, DKM85]) rather than calculating the cost function for every candidate hyperplane from the scratch:

Lemma 1. *Let $H1$, $H2$ be two crossing hyperplanes such that one of the open double wedges determined by them contains no demand points. Then, given $S_{\mathrm{E}}(H1)$ from (6), we may compute $S_{\mathrm{E}}(H2)$ in $O(1)$ time.*

The proof is evident if we rewrite $S_E(H)$ in (6) in the form

$$\left(\sum b_i^2\right)^{-\frac{1}{2}} \left[\sum_{i \in H^+} w_i\left(b_0 + \sum b_j x_{ji}\right) - \sum_{i \in H^-} w_i(b_0 + \ldots)\right]. \qquad (7)$$

Therefore, one can write a subroutine which is of interest in itself: the *Two-Dimensional Anchored Median Line Algorithm.*

Algorithm 1. (Input: point count n, point set P, weight set W, anchor q; Output: anchored median line)

Step 1: Sort all the blocked candidate lines $H_i = \mathrm{aff}(q, p_i)$ with $p_i \neq q$ and $i = 1, \ldots, n$ by their slopes. Without loss of generality, assume that the indices give the order.

Step 2: Starting from the horizontal position, rotate the anchored line all the way around while stopping it where it hits the demand points to update $S_E(H)$ and its current minimum. Since the demand points are sorted by slope, then after computing the initial $S_E(H_{\text{horiz}})$ by Lemma 1 the i-th update of (7) can be accomplished in time $O(|H_i^0| + |H_{i+1}^0|)$, by keeping track of the sets H^+, H^-, H^0 and sigmas in (7) during the rotation.

Step 1 takes $O(n \log n)$ time, Step 2 needs $O(\sum H_i^0) = O(n)$ time. As for the original (unanchored) problem in $\mathbb{E}^2$, take all the demand points as anchors, spending in total $O(n^2 \log n)$ time and linear space.   □

Now it is easy to handle the d-dimensional case: One can anchor the candidate hyperplanes by all $(d-1)$-point sets in turn leaving them with the single rotational degree of freedom to proceed as above, arriving at the claimed $O(n^d \log n)$ time algorithm. A short description is the following.

Algorithm 2. (Input: d, n, P, W;   Output: median hyperplane)
    begin
        for   ind := 1   to $\binom{n}{d-1}$ do
          begin
            Generate the next $(d-1)$-point subset P_{ind} (cf. [RND77], Sect. 5.2.2).
            It determines a $(d-2)$-flat F_{ind}. Take its orthogonal complement
            C_{ind} (being a 2-flat) and project P orthogonally onto it, with P_{ind}
            mapping onto a point q_{ind}.
            Call Algorithm 1 to solve the instance of the anchored median line
            problem in the 2-plane C_{ind} with anchor q_{ind}.
            Update the current minimum of (6).
          end
    end.   □

2.3 General Case: $O(n^d)$ Time

Further improvements can be obtained by exploiting another kind of geometric coherence resting in geometric sorting. Lee and Ching [LC85] investigated a 2-dimensional problem of repeated use in various situations, namely the

Simultaneous Angular Sort.   Given a point set P in the plane. For each $p_i \in P$ find an ordering of $P \backslash \{p_i\}$ by the polar angle around p_i.   □

In [LC85] it is shown how to perform this in $O(n^2)$ time within the space $O(n^2)$. In particular, the concept of geometric duality between planar point configurations and line arrangements is used.

There are various kinds of duality transforms convenient in different situations [E87]. One possible kind of such a transform T is presented in the following (we prefer it because of its symmetry with its inverse):

$$
\begin{array}{c|ccc|c}
\text{object} & (x, y) - \text{plane} & \xrightarrow{\quad T \quad} & (k, b) - \text{plane} & \text{parameter} \\
\text{space} & L : (y + kx + b = 0) & \longleftrightarrow & L_T = (k, b) & \text{space} \\
& p = (x, y) & \longleftrightarrow & P_T : b + xk + y = 0) &
\end{array}
$$

The transform T maps points into non-vertical lines and vice versa. Assuming that no two demand points share their x-coordinates (this can be assured in $O(n \log n)$ time by an appropriate rotation), it is easy to see that the lines sharing a common point p and *sorted by slope* are mapped via T into points lying on and *sorted along the line* p_T (cf. [E87].

Thus, the configuration P of demand points shall be transformed by T into some line arrangement P_T. It turns out (see [CGL83, EOS86, E87] that the full description of (the planar subdivision generated by) P_T can be constructed in $O(n^2)$ time and space. Moreover, for each anchor p from P one may spend time $O(n)$ to extract from this description both the relevant line p_T and the set of the vertices of P_T *sorted along* p_T and corresponding to the p-anchored line set H_i, $i = 1, \ldots, n$ (see Algorithm 1), which is *sorted by slope* [LC 85]. Having obtained the set of sorted lines H_i, one proceeds as in Sect. 2.2. The whole procedure now uses only $O(n^2)$ time and space.

Immediately, one may generalize this to d dimensions. The extended form of the duality transform T is given by

$$
\begin{aligned}
H : (y + a_1 x_1 + \ldots + a_{d-1} x_{d-1} + a_d = 0) & \xrightarrow{T} H_T = (a_1, \ldots, a_d) \\
p = (x_1, \ldots, x_{d-1}, y) & \longmapsto p_T : (a_d + x_1 a_1 + \ldots \\
& \qquad\qquad + x_{d-1} a_{d-1} + y = 0)
\end{aligned}
$$

which also preserves several positional relations as incidence, sideness, order and others. One can extend it for all non-vertical flats in $\mathbb{E}^d$:

$$F_{d-k} = \begin{cases} H_1 : (y + \ldots = 0) \\ \quad\vdots \\ H_k : (y + \ldots = 0) \end{cases} \overset{T}{\underset{}{\rightleftarrows}} F_{k-1} = \mathrm{aff}((H_1)_T, \ldots, (H_k)_T)$$

(k simultaneous
linear equations)

Here again T casts the configuration P of demand points into the hyperplane arrangement P_T, and the whole description of the related cell complex may be constructed in $O(n^d)$ time and space (see again [EOS86, E87]). Moreover, the object space hpyerplanes, anchored by a $(d-1)$-point anchor Q and ordered by a turn angle, are mapped via T into the point set lying on and ordered along the 1-flat Q_T. On the lines of [LC85] one may extract the desired ordered point set, transform it back into the anchored hyperplane set sorted by a turn angle and perform the rotate-and-update method using (7) as in Sect. 2.2, in total time and space $O(n^d)$.

Now we shall present the reduction of the space requirements to $O(n)$ (cf. [KM89]). Let us start with the two-dimensional case and reconsider the simultaneous sort of Lee and Ching [LC85]. In [EG86] it was shown that linear space requirements are sufficient. For many problems one need not construct the whole description of the arrangement (in the dual plane) but only to traverse all its vertices in a partial order consistent with their orderings along the lines of the arrangement. We may perform this traverse using *topological sweep*. This is the generalization of the well-known line-sweep paradigm (see [PS85]) and may be considered as sweeping the plane from the left to the right hand side, not by a usual vertical line but by a vertically monotone curve (topological line or *pseudoline*). Here it is important that the vertices on a line of the arrangement should be traversed in the left-to-right sorted order [EG86].

Therefore, searching the median lines we need not presort all p_i-anchored line bundles with the application of Algorithm 1 to them in turn. Instead of this we simply "interleave" n "processes" of Algorithm 1 in accordance with the transitions of the pseudo-line sweeping.

Algorithm 3. (Median lines via dual plane topological sweep) For simplicity, we shall assume that no three demand points are collinear.

1. Initialize n candidate median lines L_i, $i = 1, \ldots, n$, as verticals through each demand point.
2. During the topological sweep, consider all pseudoline transition points in the dual plane (cf. [EG86]), corresponding to a line $\mathrm{aff}(p_i, p_j)$ in the object plane. For each such point, update candidate lines L_i and L_j as in Algorithm 1 and maintain a pointer to the current minimum.   □

Obviously, $O(n^2)$ time in linear space is needed.

In d-space we may group the candidate blocked hyperplanes into two-variate bundles to apply $\binom{n}{d-2}$ times the described planar pseudoline sweep (see [KM89, KM90]). More formally, a $(d-2)$-point set P_{d2}, or the $(d-3)$-flat $Q = \mathrm{aff}(P_{d2})$, shall be an anchor. Then the Q-anchored hyperplanes are mapped via T into points on the 2-flat Q_T. A point $p_i \in P \backslash Q$ is mapped via T into the line L_i in Q_T being the intersection of Q_T and the hyperplane $(p_i)_T$. If some additional $p_i \in P \backslash Q$ is fixed to obtain a $(d-1)$-point anchor $P_{d1} = P_{d2} \cup \{p_i\}$ (to be used in Algorithm 2), then the P_{d1}-anchored hyperplanes are mapped via T into the points along the line L_i in Q_T. Therefore, a planar situation is obtained which is amenable to a slightly extended Algorithm 3. We apply this $\binom{n}{d-2}$ times. Each such major step takes $O(n)$ time to map $P \backslash Q$ onto a line set $\{L_i\}$ in the plane Q_T plus $O(n^2)$ time and linear space for the extension of Algorithm 3. Hence, the claimed $O(n^d)$ time linear space complexity is verified.

2.4 Restrictions and Extensions

As seen, the algorithms in Sect. 2.3 do not exploit the statement (b) from Theorem 2. It was shown in [DKM84, YKII88] that in the planar weighted case the worst-case complexity cannot be improved by Theorem 2(b), since there may be as many as Cn^2 halving lines: Take a medium-sized odd number k, with $\frac{n}{3} < k < \frac{n}{2}$, say. Place k demand points evenly on a circle of a very large radius R and give them very large equal weights. Take now a concentric circle with very small radius r and place the remaining $n-k$ points on it with very small weights. It is easy to see that then all the lines through pairs of points lying on different circles are halving ones, and there may be as many as $\frac{n^2}{12}$ of them. In [YKII88] one may find a basically similar example in dual terms of weighted arrangements.

In view of this fact we turn to the unweighted case.

2.4.1 Unweighted Case, $d = 2$.
Throughout Sect. 2.4 multiple demand points are obviously excluded, and all points shall be given in general position, i.e., for any subset S one has $\dim \mathrm{aff}(S) = \min(|S| - 1, d)$. (If this is not the case, one might apply the method of "simulation of simplicity" from [E86, EM88, E87] incrementing conceptually the coordinates of the demand points:

$$x_{ij}^* = x_{ij} + eps^{2^{d*i+j-1}}; \quad j = 1, \ldots, d, \ i = 1, \ldots, n.$$

Here "eps" is thought to be sufficiently small to preserve all relevant relations and properties, e.g. to preserve the halving lines. In fact, it is not necessary to seek the adequate "eps". One only has to perform computations symbolically with polynomials in "eps".)

The following subquadratic bounds on the numer $h(n)$ of halving lines to a planar set of n points are known:

$h(n) < n^{\frac{3}{2}}$ ([L71, ELSS73], see also [EW83, E87]) and

$$h(n) < \frac{n^{\frac{3}{2}}}{\log^{\frac{1}{100}} n} \quad \text{(cf. [PSS89])}.$$

Hence there are possibilities for improvements. In [K88, K89] it was proposed to implement the halving line rotation procedure due to Lovász [L71, ELSS73] in a computational geometric manner: Starting with an arbitrary halving line $H_1 = \text{aff}(p_1, p_2)$, which has the initial orientation from p_1 to p_2, one rotates it clockwise around p_2 (while preserving the orientation as intrinsic) until it hits some demand point p_3 to obtain $H_2 = \text{aff}(p_2, p_3)$. Then H_2 is rotated clockwise around p_3 to obtain H_3, etc., until one returns ot the starting position.

It turns out that for odd n all lines H_i are halving ones, and for even n the line H_i is halving if and only if it is oriented from p_i to p_{i+1}. Otherwise, H_i is an $(\frac{n}{2}, \frac{n}{2} - 2)$-divider line. Now one may computationally reconsider this process, as noted in [K88, K89].

It is obvious that H_{i+1} is the tangent from p_i to one of the convex polygons

$$CH_i^+ = \text{conv}((H_i)^+) \quad \text{or} \quad CH^- = \text{conv}((H_i)^-).$$

Furthermore, in the course of this rotation, CH_i^- and CH_i^+ are being changed by inserting p_{i-1} into one of them and deleting p_{i+1} from one of them. Now one may recall the dynamic convex hull maintenance algorithm due to Overmars and van Leeuwen [OvL81], taking $O(\log^2 n)$ time per update (and $O(n \log n)$ time to initialize). The data structure from [OvL81] is based on balanced trees, and it allows the computation of point-polygon tangents by binary search in $O(\log n)$ time. Therefore, the line rotation procedure of Lovász may be implemented in time $O(h(n) \log^2 n)$, bearing in mind that the number of $(\frac{n}{2}, \frac{n}{2} - 2)$-dividers for even n has asymptotically the same upper bound as the number of halving lines. With the current estimates of $h(n)$ [PSS89] we obtain the overall time complexity $O(n^{\frac{3}{2}} \log^{2-a} n)$, where a is a small constant, which e.g. can be supposed as $\frac{1}{100}$.

For this case, a subquadratic algorithm was also proposed by Yamamoto et al. [YKII88]. It operates in the dual plane, where so-called $\lceil \frac{n}{2} \rceil$-belts correspond to the halving lines. To construct these belts, Yamamoto et al. refer to [EW86]. As seen, in the latter paper the belts are constructed with the aid of the dynamical maintenance procedure for the halfplane intersection from [OvL81], which is basically the dual way of the dynamical maintenance of the convex hull from the same paper. Therefore, it turns out that the algorithm from [YKII88] is basically the dual version of that in [K88, K89] described here. The question for the time optimal algorithm remains to be answered yet. The only lower bound for this is $\Omega(n \log n)$, proved in [YKII88] by reduction from the

Uniform Gap on a Circle Problem   [YKII88, LW86]: Consider n points on the unit halfcircle, with two of them being diametral. It is asked whether they are spaced evenly along the circle.   □

There are two ways to improve the time complexity of the line rotation procedure. First, the best lower bound for the worst case $h(n)$ is still $\Omega(n \log n)$ (see [ELSS73, EW83, E87]). Second, as for the dynamic convex hull maintenance problem, the "batched dynamic" version rather than the on-line setting is needed, because all the points to be inserted and deleted are known in advance. Here one may also carry out some kind of amortized complexity analysis to get rid of at least one logarithm.

Here the importance from the simplicity assumption for the point set and hence of the "simulation of simplicity" must be stressed: In the rotating line algorithm the transition from H_i to H_{i+1} takes at least $\theta(|H_i^0| + |H_{i+1}^0|)$ time to update the sets H^+, H^-, H^0 necessary to maintain sigmas in (7). For the non-simple configuration this would contribute time

$$T_{HHH}(n) = O(|H_1^0| + \ldots + |H_{h(n)}^0|) = O\left(\operatorname{Incid}(n, h(n))\right),$$

which is the number of incidences between demand points and halving lines. For a set of n points and t lines the upper bound on this quantity is given in [ST83]:

$$\operatorname{Incid}(n, t) = O(n^{\frac{2}{3}} t^{\frac{2}{3}}).$$

It might possibly be improved for the particular case of halving lines, but currently from [ST83, PSS89] one sees that the term

$$T_{HHH} = O(n^{\frac{5}{3}} \log^c n)$$

would clearly dominate the overall time complexity.

2.4.2 Unweighted Case, $d > 2$.
The question for the number $h(n)$ of halving hyperplanes in d-space, $d > 2$, was posed in [ELSS73]. Until now, very little is known about this problem. The first non-trivial upper bound is

$$h(n) = O(n^{3-eps}), \quad d = 3 \quad \text{(cf. [BFL89])},$$

with "eps" being a small constant, for instance $\frac{1}{343}$ or $\frac{1}{64}$. A still more recent bound is

$$h(n) = O(n^{\frac{8}{3}} \log^{\frac{5}{3}} n), \quad d = 3 \quad \text{(see [ACEGSW90])}.$$

It also remains to be answered whether Cn^d is the worst case number of halving hyperplanes for weighted point sets in $\mathbb{E}^d$, which would justify the separate treatment of the unweighted case.

2.4.3 Median k-Flats.
In an analogous manner one may consider median k-flats with $k \in \{0, \ldots, d-2\}$. A natural starting point would be the case

$d = 2$, $k = 0$. It is the so-called simple facility location problem, identical with the question of the well-known Fermat-Torricelli point of a given point set. The history of this problem (see e.g. [K74, K76, Sc86]) goes back to the 17th century, and a lot of numerical methods were proposed for it (cf. additionally [P61, K73, P76]). However, except for some special cases, any discussion of its computational complexity is quite non-trivial. First, it is not so obvious whether the infinite precision real RAM with unit-cost square roots is a fully adequate computational model for some kinds of geometric optimization problems, because sometimes it hides in itself too much computational complexity as shall be explained here.

When discussing the computational complexity of an optimization problem "Minimize F(X)" (e.g. for proving the NP-hardness of this problem) it is common to consider the related recognition problem: "Is $F(X) < L$?". Such a recognition problem is said to be in the class NP (cf. [GJ79]) if (informally) for a feasible input X one can check in polynomial time whether the problem has a positive answer. For the Fermat-Torricelli problem one would have to verify in polynomial time whether $\sum (m_i)^{\frac{1}{2}} < L$, in terms of the number of bits in all m_i, L. However, such an algorithm is not known [G84, AF86]. Thus the corresponding recognition version is not known to be in NP. (Clearly, this is not the case for the median hyperplane problem.)

This fact presents no obstacle to the existence of a polynomial time algorithm for finding the optimum. An example is the known Euclidean Minimum Spanning Tree problem with any greedy algorithm for it (see [GGJ76]). Here the crucial fact is that the problem at hand must allow the combinatorial characterization of optimum solution in terms of its input.

As it turns out, this may hardly be the case for the Fermat-Torricelli problem, despite its convexity. Namely, Bajaj showed that its solution (i.e. the coordinates of the point with the least distance sum to a given point set) in general cannot be expressed in radicals over the field of rationals in terms of the input points, even for simple 5-point configurations in the plane (cf. [B86, B88]). He applies Galois theory using the symbolic manipulation software tool MACSYMA. Hence the Fermat-Torricelli problem turns out to be highly intractable: It does not allow exact algorithms under models of computation with arithmetic operations and the extraction of k-th roots. This leaves us only numerical of symbolic approximation methods. Nevertheless, by the standard ellipsoid method an ε-approximative solution of the Fermat-Torricelli problem can be constructed in polynomial time (cf. [CT90]).

Finally, we shall refer to several generalizations of this problem, certainly yielding new starting points for research in the sense of Computational Geometry. The extension of the classical question to more general metric spaces was investigated e.g. in [F63, H66, ILM88, C67, CG85, C88]. In particular, a natural generalization of the Euclidean distance was considered by [P86, MPV87]. An interesting discussion of the original problem in $\mathbb{E}^d$ was recently given by Neumann [N87].

The question remains for median k-flats of n weighted points in $\mathbb{E}^d$ with $k \in \{1, \ldots, d-2\}$.

As the hyperplane problem is connected with zonotopes [M87, KM89, KM90], the case of median lines ($k = 1$) relates to vector sums of $(d-1)$-balls. Unfortunately, despite this geometric approach an analogy to the statement (a) in Theorem 2 does not hold. Moreover, for each $k \in \{1, \ldots, d-2\}$ the problem is at least as difficult as the Fermat-Torricelli problem in $\mathbb{E}^{d-k}$ (cf. [GM90]). The case $d = 1$, $k = 0$, solvable in linear time, is discussed in [BFPRT72] and [KM90].

3. MINSUM Problem: Vertical L_1-Fit

The problem is

$$\underset{b_1, \ldots, b_d}{\text{minimize}} \sum_{i=1}^{n} w_i \left| x_{1i} - \left(b_0 + \sum_{j=2}^{d} b_j x_{ji} \right) \right|. \tag{8}$$

It is easy to show that there is an optimal hyperplane being the affine hull of d affinely independent demand points (all other cases are trivial). Hence, the $O(n^d)$ algorithm from Sect. 2.3 remains valid. However, the case here is more favourable due to its piecewise linearity and convexity properties. An $O(n \log^2 n)$ algorithm for this problem in $\mathbb{E}^2$ was given by Megiddo and Tamir [MT83], but Zemel [Z84] proposed a linear time algorithm for any fixed dimension. He extended Megiddo's multidimensional prune-and-search method [Me84] and solved the following linear program, also called

Multiple-Choice Linear Programming Problem   (MCLPP):

$$\max \sum_{j \in n} c_j x_j \tag{9}$$

subject to

$$\sum_{j \in N} a_{ji} x_j = a_{01}, \quad i = 1, \ldots, d,$$

$$\sum_{j \in J_k} b_j x_j = b_{0k}, \quad k = 1, \ldots, r,$$

where the sets J_k are mutually disjoint, $x_j \geq 0$, $j \in N$.   □

If one takes the linear program which is dual to (9), then some manipulations yield the equivalent problem

$$\text{minimize} \sum_{i=1}^{d} a_{0i} z_i + \sum_{k=1}^{r} b_{0k} v_k \tag{10}$$

subject to

$$v_k \geq s_j - \sum_{i=1}^{d} t_{ji} z_i, \quad j \in J_k^+, \quad k = 1, \ldots, r,$$

$$v_k \leq -s_j + \sum_{i=1}^{d} t_{ji} z_i, \quad j \in J_k^-, \quad k = 1, \ldots, r,$$

$$J_k^+ \cup J_k^- = J_k,$$

see also [Z84].

Now one may apply Megiddo's recursion and prune-and-search in each separate block J_k, lowering concurrently down the dimension d and the sizes of all J_k. At the bottom level of the recursion, an algorithm of Zemel [Z80] is applicable. This algorithm has $O(n \log \max |J_k|)$ time, which is $O(n)$ if all J_k are pruned down to a constant size. Therefore, the total time is $O(n)$ [Z84]. Further, in [Z84] (8) is rewritten to the form

$$\text{minimize} \sum_{i=1}^{n} t_i$$

subject to

$$t_i \geq w_i \left(x_{1i} - \left(b_0 + \sum_{j=2}^{d} x_{ji} \right) \right),$$

$$t_i \leq -w_i \left(x_{1i} - \left(b_0 + \sum_{j=2}^{d} x_{ji} \right) \right),$$

which shows the format of (10) and thus can be solved in linear time for any fixed dimension. Moreover, the following extension of the fitting problem may be stated as (10), too.

Vertical L_1-Fit for Clusters. Let $K = \{P_1, \ldots, P_k\}$ be a partition of the set of demand points into k subsets. Find a hyperplane minimizing $\sum_i \text{dist}(P_i, H)$, where

$$\text{dist}(P_i, H) = \max_{p_j \in P_i} D_{\text{vert}}(p_j, H) \quad (\text{cf. (5)}). \qquad \square$$

Thus, this problem is also solvable in linear time for any fixed d. In [YKII88] one may find an independent application of prune-and-search with respect to (8) and $d = 2$ (see also [IKY89]).

Here we have to note that the hidden constants in the time complexity estimate for the prune-and-search method grow polyexponentially with d. Therefore one has to remember that (8) is a linear program, and hence it is solvable in time which is polynomial in terms of the input length (with variable dimension).

4. MINMAX Problem: Orthogonal L_∞-Fit

Here the problem is given by

$$\underset{H}{\text{minimize}} \ \max_i w_i D_E(p_i, H) \,. \tag{11}$$

The planar case was discussed e.g. in [MN83].

4.1 Unweighted Case

For the unweighted version the problem is equivalent to the

Point Set Width Problem. Find a minimum width layer between two parallel hyperplanes containing a given point set in $\mathbb{E}^d$. □

The planar case was solved repeatedly in [D84, E85, HT85, LW86] (see also [CO85]) with optimal $\theta(n \log n)$ time (cf. [LW86]) by basically the same procedure which was first used by Shamos [S78] for the problem (12) below and which was named "rotating calipers" by Toussaint [T83]: In the plane, an edge of the minwidth strip must span a side of the polygon conv (P), while the other edge in general passes through a vertex of conv (P). Such pairs (edge, vertex) lying on parallel supporting lines are called antipodal. After constructing conv (P) one may enumerate all antipodal pairs in $O(n)$ time, first by constructing any initial strip and then "rotating" it around conv (P) fitting it tight as with the calipers. (This method is useful for many geometric problems in the plane.) This $O(n \log n)$ time algorithm turns out to be worst-case optimal, as it was proved by Lee and Wu [LW86] with a reduction from the Maximum Gap on a Circle Problem. (Note that various Gaps-on-a-circle are useful for location problems, see also [L86].)

To solve the unweighted problem in higher dimensions, it would be sufficient to enumerate all the blocked antipodal pairs for conv (P) (e.g. in $\mathbb{E}^3$ there are in general the two types "facet-vertex" and "skew edge-edge"). An efficient way to do this for $d = 3$ was given by Houle and Toussaint [HT88]. Namely, consider conv (P) with no vertical faces (otherwise rotate it). Then one may divide it into upper and lower hulls UH and LH consisting of faces facing up and down, respectively. Then, for any pair of non-vertical parallel supporting planes, one plane touches UH and the other LH. All parallel pairs of supporting planes basically comprise a two-parametric set. Therefore, one may consider a geometric transform mapping a plane into its slope parameters [HT88]:

$$H : (z = ax + by + c = 0) \ \xrightarrow{T} \ H_T = (a, b) \in \mathbb{R}^2 \,.$$

When applied to UH, this yields a planar subdivision UPS where the facets of UH are mapped into vertices of UPS, the vertices of UH are mapped into faces of UPS and edges into edges. LPS is defined analogously. Overlaying

UPS onto LPS, the antipodal pairs will fit onto each other. For example, the blocked pair (facet, vertex) will correspond to a vertex of UPS (or LPS) falling into the face of LPS (resp. UPS). To accomplish this, one may use an algorithm of Guibas and Seidel [GS86] spending $O(n + I)$ time, where I is the number of edges in the resulting overlay. (Clearly, this is $O(n^2)$ in the worst case.) Together with computing conv(P), the total time complexity is $O(n \log n + I)$.

Recently, an algorithm for the weighted case (see the next section) was proposed in [HIIR89, HIIRY90] with optimal $\theta(n \log n)$ time complexity for $d = 2$.

Though leaving the concept of fixed dimension, we shall also mention recent results of Gritzmann and Klee [GK90], namely: For variable d, the computation of the width of conv(P) is NP-complete (even if conv(P) is a simplex).

4.2 Weighted Case

For the plane Morris and Norback [MN83] proved a blockedness property as in the unweighted case: Each optimal line with respect to (11) is at a maximum weighted distance from at least three demand points. It is easy to extend their proof to d dimensions. Thus, every optimal hyperplane is at the maximum weighted distance from at least $d + 1$ affinely independent demand points. This yields an $O(n^{d+1})$ time algorithm: Enumerate all $\binom{n}{d+1}$ corresponding sets of points. Each set gives $d + 1$ candidate hyperplanes, one for each choice of d points to determine slope parameters, with the remaining point to determine the intercept of the coordinate axis chosen to be vertical.

In the planar case, [DKM84, DKM85] and [LW86] resulted in different $O(n^2 \log n)$ time $O(n^2)$ space algorithms, while [K89] showed how the algorithm from [DKM84] may be improved to $O(n^2)$ time via the simultaneous sort from [LC85] discussed in Sect. 2.3.

By a geometric transformation to a three-dimensional convex hull problem, [E85] showed that $O(n \log n)$ time suffices to find a transversal of n circles in the plane if it exists. This method can be generalized to find the optimal line in the sense of (11) in the same amount of time.

An improvement for the general case was recently given in [HIIR89, HIIRY90]. It was shown that, unless all demand points lie in the common hyperplane, then the following problem turns out to be equivalent to (11):

$$\text{maximize}\left(\sum_{i=1}^{d} b_i^2\right)^{\frac{1}{2}} \tag{11a}$$

subject to

$$1 \geq w_i \left(b_0 + \sum_{j=1}^{d} b_j x_{ji} \right), \quad i = 1, \ldots, n, \tag{11b}$$

$$1 \geq -w_i \left(b_0 + \sum_{j=1}^{d} b_j x_{ji} \right), \quad i = 1, \ldots, n. \tag{11c}$$

Furthermore, any optimal solution of (11) must be a vertex of the feasibility region (11b, 11c). This region is the intersection of $2n$ semispaces in $\mathbb{E}^{d+1}$, and therefore its construction is known to be equivalent to the construction of a convex hull in the dual space of the same dimension (see e.g. [E87]. Hence, the computational complexity of the proposed approach is dominated by the complexity of the convex hull problem. It is known to be in $O(n \log n)$ time and linear space for $d = 2$ (which is optimal) and $O(n^{\lfloor \frac{d}{2} \rfloor + 1})$ time, $O(n^{\lfloor \frac{d+1}{2} \rfloor})$ space (cf. [E87]) for $d > 2$. As M. Houle pointed out, other techniques may be used to generate the vertices of the feasibility region, such as [S86] or the randomized convex hull construction algorithm (expected-time optimal for higher dimensions of R. Seidel [S90]).

4.3 MINMAX Problem: k-Flats

Considering the analogous problem for k-flats, the unweighted version for $k = 1$ and $d = 2$ is a well-known problem referring to minimal enclosing circles (see e.g. [S78]). It was finally solved in $O(n)$ time in [Me84] for any fixed dimension (minimal enclosing ball.) The weighted case ($k = 1$, $d = 2$) is solvable in $O(n \log n)$ time (cf. [C84]). For other dimensions k this seems to be tractable (at least in unweighted cases), unlike the minsum k-flats.

5. MINMAX Problem: Vertical L_∞-Fit

This is the well-known Chebyshov approximation problem

$$\operatorname*{minimize}_{b_0, \ldots, b_d} \max_i w_i \left| x_{1i} - \left(b_0 + \sum_{j=2}^{d} b_j x_{ji} \right) \right| \tag{12}$$

(cf. [R64]). For the planar unweighted case Shamos [S78] proposed the method of rotating calipers mentioned above. However, (12) is known to be reducible to a linear program in dimension $d + 1$. Therefore, one gets a linear time algorithm (see [Me84]) for any fixed dimension (and a polynomial time algorithm irrespective to d). In the plane, this problem was also discussed in the dynamic setting [RW87]:

Dynamic Vertical L_∞-Fit. Maintain the Chebyshov line dynamically while inserting and deleting the demand points of $P \in \mathbb{E}^2$.   □

Rey and Ward [RW87] presented a method based on the dynamic convex hull maintenance from [OvL81] and the monotonicity properties of the convex hull, leading to $O(n \log n)$ preprocessing time, $O(\log^2 n)$ update and search time. In [Bh88] the insertion time was improved to $O(\log n)$.

6. Related Issues

Obviously, the least-distances problems are closely connected with various combinatorial geometric notions: convex hulls, halving lines, etc.; therefore any progress in these areas might promote research on least-distances topics. We shall now discuss some problems related to least-distances questions in a more general sense and, in addition, refer to [K89] and Table 2 here.

6.1 Stabbing Lines/Hyperplanes

One might consider the recognition version of (1): Is there a hyperplane such that

$$\max_i w_i \, \text{Dist}\,(p_i, H) \leqq D \; ?$$

Introducing n spheres with corresponding radii $\frac{D}{w_i}$ and centers p_i, one can formulate this latter problem as a *transversal* or *stabbing* one: Is there a hyperplane intersecting all these spheres? Problems of this kind for various types of stabbed objects were briefly discussed in the survey [K89]. For more recent advances we refer to [AD88, AV88, R88, W88, ARW89, EW89, H89, HR88, HIIR89, CGPPSW90, ES90].

6.2 k-Hyperplane Problems

A natural way of generalizing the location problems discussed is the consideration of k facilities to be located in an optimal manner.

It turns out (cf. [MT82]) that, for the case discussed here, such k-line/hyperplane problems are NP-hard in a strong sense [GJ 79] for any isotone cost function based on any reasonable metric, if one allows k to be a part of the input. The reason is the strong NP-completeness of the following problem (see once more [MT83]):

Point Line Cover Problem. For any n and k, decide whether n given points can be covered by k straight lines.   □

From this one can also conclude the NP-completeness of the multi-stabbing line problem: For any n and k, decide whether n given objects may be stabbed by k lines.

However, a lot of open problems remain to be solved, e.g.: For any n and fixed k locate k lines optimally for n points (or find a k-stabber for n objects). In particular, what can be said for $k = 2$?

The following tables will give once more a survey of the main results regarding the problems discussed in this chapter.

Table 1. Chronology for the computational complexity of basic problems (W = Weighted, U = Unweighted, O = Orthogonal, V = Vertical)

Function	Dimension	Complexity: T = Time S = Space	Lower bound for time
minsum WO	1	$T = O(n)$, [KM89, KM90], implicitly in [BFPRT72]	$\Omega(n)$
	2	$T = O(n^3)$, [MN80] $T = O(n^2 \log n)$, $S = O(n^2)$, [MT83] $T = O(n^2 \log n)$, $S = O(n)$, [LC85, DKM85] $T = O(n^2)$, $S = O(n^2)$, [LC85]	
	d	$T = O(n^{d+1})$, [NM80] $T = O(n^d)$, $S = O(n)$, [IKY88, KM89, KM90]	
minsum UO	2	$T = O(h(n) \log^2 n)$, [K88, K89, YKII88], where $C_1 n \log n < h(n) < C_2 n^{3/2} \log^{.01} n$ [ELSS73, EW83, PSS89]	$\Omega(n \log n)$, [YKII88]
minsum WV	1	$T = O(n)$, [KM89, KM90], implicitly in [BFPRT72]	$\Omega(n)$
	2	$T = O(n)$, [Z84, YKII88, IKY89] $T = O(n \log^2 n)$, [MT83]	$\Omega(n)$
	d	$T = O(n)$, [Z84]	$\Omega(n)$
minmax UO	2	$T = O(n^2)$, [MN83] $T = O(n \log n)$, [KD82, D84, E85, HT85, LW86, HIIRY90, HIIR89]	$\Omega(n \log n)$, [LW86]
	3	$T = O(n \log n + I)$ with $I = O(n^2)$, [HT88]	
	3	$T = O(n \log n)$, [HR88, HIIR89, HIIRY90]	
minmax WO	1	$T = O(n)$, [HR88, HIIR89, HIIRY90]	
	2	$T = O(n^3)$, [MN83] $T = O(n^2 \log n)$, $S = O(n^2)$, [DKM84, DKM85, LW86] $T = O(n^2)$, $S = O(n^2)$, [K89] $T = O(n \log n)$, [HR88, HIIR89, HIIRY90, E85]	
	3	$T = O(n \log n)$, [HR88, HIIR89, HIIRY90]	
	≥ 3	$T = O(n^{\lceil \frac{d+1}{2} \rceil})$, $S = O(n^{\lceil \frac{d}{2} \rceil})$, [HR88, HIIR89, HIIRY90]	
minmax UV	2	$T = O(n \log n)$, [S78]	$\Omega(n)$
minmax WV	d	$T = O(n)$ via linear programming [Me84]	

Completion: For variable dimension in the cases minsum WV and minmax WV one obtains polynomial time via linear programming.

Table 2. Some extensions and related topics

---

1. *Anchored minsum hyperplane problem*
   Dim = 2 $O(n^2)$, [MN83]
                $O(n \log n)$, [KM89, KM90], see also [MT83, DKM84]
   Dim = d $O(n^{d-1})$, by incremental rotation [MT83, DKM84]

2. *Anchored minsum ray problem*
   Dim = 2 $O(n^2)$, [MN83]
                $O(n \log n)$, by incremental rotation [MT83, DKM84]

3. *Minsum linear facility with trunk*
   Dim = 2 $O(n^2 \log n)$, [DKM84, DKM85]
                $O(n^2)$ by combination of [DKM84, DKM85] and [LC85, EG86]
                (as for minsum WO)

4. *Minsum k-dimensional flats*
                $k = 0$: combinatorially unsolvable [B86, B88];
                        ε-approximation from it: polynomial time[CT90]
                $k \in \{1, \ldots, d-2\}$: at least as difficult as the Fermat-Torricelli
                problem in E^{d-k} [GM90]

5. *Minmax k-dimensional flats*
   Dim = 2 $k = 1$ (weighted): $O(n \log n)$, [C84]
   Dim = d $k = 1$ (unweighted): $O(n)$ [Me84] (Minimal enclosing ball)

6. *Minimum width strip*
                Equivalent to minmax UO

7. *Hyperplane stabbing hyperspheres*
                Reduced to minmax WO

8. *Dynamic* minmax *UV*
   Dim = 2   Preprocessing time: $O(n \log n)$
                Search and update: $O(\log^2 n)$, [RW87]
                Insertion update: $O(\log n)$, [Bh88]

9. *Optimal k-hyperplanes, various criteria*
                NP complete in a strong sense [MT82]

10. Maxmin *unweighted separating hyperplane*
    Dim = 2   (empty corridor): $O(n^2)$, [HM88]

11. *Maxmin k-flat*
    Dim = 2 $k = 1$ (unweighted): $O(n \log n)$, [T83] (maximum empty circle)

12. *Minmax circular fit* (narrowest ring)
                $O(n \log n + I)$, [W86], where I denotes the intersection count
                for the closest-point and farthest-point Voronoi diagrams.

13. *Exact fit or maximal stabber*
                One has to find a line passing through m of n given points or
                through the maximal number of points [GOR91]

14. *Shortest transversal*
                Find al shortest line which intersects all given objects [BT91].

Some more information on line/hyperplane placement and geometric location problems can be found in the surveys [L86, K89].

Acknowledgement. The authors wish to thank H.-D. Hecker for triggering their interests in higher dimensional location problems and M. Houle for critical discussion.

References

[AHU78]   A. Aho, J. Hopcroft and J. Ullmann: The Design and Analysis of Computer Algorithms. Addison-Wesley 1978

[AF86]   V. Akman and W.R. Franklin: On the question "is $\sum \sqrt{a_i} < L$?". Bull. EATCS **28** (1986) 16–20

[ACEGSW90] B. Aronov, B. Chazelle, H. Edelsbrunner, L.J. Guibas, M. Sharir and R. Wenger: Points and triangles in the plane and halving planes in space. Proc. 6th Annual Symp. Comput. Geom., Berkeley, CA, 1990 (see also: Tech. Rep. DIMACS Center, 90/9, 7 pages), pp. 112–115

[AD88]   D. Avis and M. Doskas: Algorithms for high dimensional stabbing problems. Theoret. Found. Computer Graph. CAD, Ed. R.A. Earnshaw. Springer, Berlin Heidelberg New York 1988, pp. 199–208

[ARW89]   D. Avis, J.M. Roberts and R. Wenger: Lower bounds for line stabbing. Inform. Process. Lett. **33** (1989) 2, 59–62

[AW88]   D. Avis and R. Wenger: Polyhedral line transversals in space. Discrete Comput. Geom. **3** (1988) 257–265

[B86]   C. Bajaj: Proving geometric algorithms nonsolvability: An application of factoring polynomials. J. Symbolic Comput. **2** (1986) 1, 99–102

[B88]   C. Bajaj: The algebraic degree of geometric optimization problems. Discrete Comput. Geom. **3** (1988) 177–191

[BFL89]   I. Bárány, Z. Füredi and L. Lovász: On the number of halving planes. Proc. 5th ACM Symp. Comput. Geom., Saarbrücken 1989, pp. 140–144

[Bh88]   B.K. Bhattacharya: An efficient on-line Chebyshov approximation algorithm for a finite planar point set. Proc. 26th Allerton Conf. Commun., Control. Comput. 1, 1988 107–108

[BFPRT72]   M. Blum, R. Floyd, V. Pratt, R. Rivest and R. Tarjan: Time bounds for selection. J. Computer Syst. Sci. **7** (1972) 448–461

[BT91]   B. Bhattacharya and G.T. Toussaint: Computing shortest transversals. Computing **46** (1991) 93–119

[CGPPSW90] S.E. Cappell, J.E. Goodman, J. Pach, R. Pollack, M. Sharir and R. Wenger: On the combinatorial complexity of hyperplane transversals. Proc. 6th Annual Sympos. Comput. Geom., Berkeley, CA, 1990

[CT90]   R. Chandrasekaran and A. Tamir: Algebraic optimization: The Fermat-Weber location problem. Math. Progr. **46** (1990) 219–224

[CGL83]   B. Chazelle, L. Guibas and D.T. Lee: The power of geometric duality. Proc. 24th Ann. IEEE Sympos. Found. Computer Sci., 1983, pp. 217–225 (see also BIT **25** (1985) 76–90)

[CG85]   G.D. Chakerian and M.A. Ghandehari: The Fermat problem in Min-
           kowski spaces. Geom. Dedicata **17** (1985) 227–238
[C88]   D. Cieslik: The Fermat-Steiner-Weber problem in Minkowski spaces.
           Optimization **19** (1988) 485–489
[C86]   K.L. Clarkson: Linear programming in $O(n3^{d^2})$ time. Inform. Pro-
           cess. Lett. **22** (1986) 21–24
[C67]   E.J. Cockayne: On the Steiner problem. Canad. Math. Bull. **10**
           (1967) 431–450
[C84]   R. Cole: Slowing down sorting networks to obtain faster sorting al-
           gorithms. 25th Ann. IEEE Sympos. Found. Computer Sci., 1984,
           pp. 255–260 (see also J. ACM **34** (1987) 1, 200–208)
[CO85]   D. Comer and M. O'Donnell: Geometric problems with application
           to hashing. SIAM J. Comput. **11** (1985) 217–226
[CLW79]   D. Coppersmith, D.T. Lee and C.W. Wong: An elementary proof of
           non-existence of isometries between $L(k,p)$ and $L(k,q)$. IBM J. Res.
           Devel. **26** (1979) 696–699
[D84]   N.N. Doroshko: The solution of two mainline problems in the plane.
           Manuscript deposited in VINITI USSR, 14.06.1984, Reg.no. 39614
           (1984), 15 pp.
[DKM84]   N.N. Doroshko, N.M. Korneenko and N.N. Metel'skij: Optimizing
           line location with respect to a planar point system. Insitute Math.
           Acad. Sci. BSSR, Minsk, USSR, Preprint No. 23 (1984), 19pp.
[DKM85]   N.N. Doroshko, N.M. Korneenko and N.N. Metel'skij: Mainline prob-
           lems in the plane. Doklady AN BSSR **29** (1985) 872–873
[D86]   M.E. Dyer: On a multidimensional search technique and its appli-
           cation to the Euclidean one-centre problem. SIAM J. Comput. **15**
           (1986) 725–738
[E85]   H. Edelsbrunner: Finding transversals for sets of simple geometric
           figures. Theoret. Computer Sci. **35** (1985) 55–69
[E86]   H. Edelsbrunner: Edge-skeletons in arrangements with applications.
           Algorithmica **1** (1986) 1, 93–109
[E87]   H. Edelsbrunner: Algorithms in Combinatorial Geometry. Springer,
           Berlin Heidelberg New York (1987)
[EG86]   H. Edelsbrunner and L. Guibas: Topologically sweeping an arrange-
           ment. Proc. 18th ACM Sympos. Theory of Computing 1986, pp. 389–
           403 (see also J. Computer Syst. Sci. **38** (1989) 165–194)
[EM88]   H. Edelsbrunner and E. Mücke: Simulation of simplicity: A technique
           to cope with degenerate cases in geometric algorithms. Proc. 4th
           ACM Sympos. Comput. Geom., Baltimore, MD, 1988, pp. 118–133
[EOS86]   H. Edelsbrunner, J. O'Rourke and R. Seidel: Constructing arrange-
           ments of lines and hyperplanes, with applications. Proc. 24th Ann.
           IEEE Sympos. Found. Computer Sci., 1983, pp. 83–91 (see also SIAM
           J. Comput. **15** (1986) 341–363)
[ES90]   H. Edelsbrunner and M. Sharir: A hyperplane incidence problem with
           applications to computing distances. Preprint, Dept. Computer Sci.,
           Univ. of Illinois at Urbana-Champaign, 1990, 9 pp.
[EW83]   H. Edelsbrunner and E. Welzl: On the number of equal-sized semi-
           spaces of sets of points in the plane. Lect. Notes Computer Sci. **154**
           (1983) 182–187

[EW86]   H. Edelsbrunner and E. Welzl: Constructing belts in two-dimensional arrangements, with applications. SIAM J. Comput. **15** (1986) 1, 271–284

[ELSS73]   P. Erdős, L. Lovász, A. Simmons and E. Strauss: Dissection graphs of planar point sets. In: A Survey of Combinatorial Theory, Eds. J. Srivastava et al., 1973, pp. 139–149

[EW89]   P. Egyed and R. Wenger: Stabbing disjoint translates in linear time. Proc. 5th Ann. Sympos. Comput. Geom., 1989, pp. 364–369

[F63]   R.L. Francis: A note on the optimum location of new machines in existing plant layouts. J. Indust. Engrg. **14** (1963) 57–59

[GGJ76]   M.G. Garey, R.L. Graham and D.S. Johnson: Some NP-complete geometric problems. Proc. 8th ACM Sympos. Theory of Comput., 1976, pp. 10–22 (see also Theoret. Computer Sci. **1** (1976) 237–267)

[GJ79]   M.G. Garey and D.S. Johnson: Computers and Intractability. Freeman 1979

[G84]   R.L. Graham: Unsolved problem P73. Bull. EATCS **26** (1984) 205–206

[GK90]   P. Gritzmann and V. Klee: Computational complexity of inner and outer j-radii of polytopes in finite-dimensional normed spaces. Anwendungsbezogene Optimierung und Steuerung, Schwerpunktprogramm DFG, Report No. 222, 1990, 44 pp., to appear in. Math. Progr.

[GM90]   P. Gritzmann and H. Martini: A note on the generalized Fermat-Torricelli problem. Manuscript, 1990

[GOR91]   L. Guibas, N. Overmars and J.-M. Robert: The exact fitting problem for points. Proc. 3rd Canad. Conf. Comput. Geom. 1991, pp. 171–174

[GS86]   L. Guibas and R. Seidel: Computing convolutions by reciprocal search. Proc. 2nd ACM Sympos. Comput. Geom., 1986, pp. 90–99 (see also Discrete Comput. Geom. **2** (1987) 175–193)

[H66]   M. Hanan: On the Steiner problem with rectilinear distance. SIAM J. Appl. Math. **14** (1966) 255–265

[H89]   J. Hershberger: Finding the upper envelope of n line segments in $O(n \log n)$ time. Inform. Process. Lett. **33** (1989) 169–174

[HIIR89]   M.E. Houle, H. Imai, K. Imai and J.-M. Robert: Weighted orthogonal linear L_∞-approximation and applications. Lect. Notes Computer Sci. **382** (1989) 183–191

[HIIRY90]   M.E. Houle, H. Imai, K. Imai, J.-M. Robert and P. Yamamoto: Orthogonal weighted linear L_1- and L_∞-approximation and applications. Submitted to Discrete Appl. Math., 1990

[HM88]   M.E. Houle, and A. Maciel: Finding the widest empty corridor through a set of points. In: Snapshots of Computational and Discrete Geometry, Ed. G. Toussaint, School Computer Sci., McGill Univ., Montreal, TR SOCS-88.11, 1988, pp. 201–214

[HR88]   M.E. Houle and J.-M. Robert: Orthogonal weighted linear approximation and applications. Tech. Report, School Computer Sci., McGill Univ., Montreal, No.SOCS-88.13, 1988, 24 pp.

[HT85]   M.E. Houle and G.T. Toussaint: Computing the width of a set. Proc. ACM Sympos. Comput. Geom. 1985, pp. 1–7

[HT88]   M.E. Houle and G.T. Toussaint: Computing the width of a set. IEEE
            Trans. Pattern Anal. Mach. Intell. **10** (1988) 761–765

[ILM88]   H. Idrissi, P. Loridan, C. Michelot: Approximation of solutions for
            location problems. J. Optim. Theory Appl. **56** (1988) 1, 127–143

[IKY88]   H. Imai, K. Kato and P. Yamamoto: Algorithms for orthogonal L_1
            linear approximation for points in two and higher dimensions. Proc.
            13th Int. Sympos. Math. Progr., Tokyo, 1988

[IKY89]   H. Imai, K. Kato and P. Yamamoto: A linear time algorithm for
            linear L_1 approximation of points. Algorithmica **4** (1989) 77–96

[K88]   N.M. Korneenko: Optimal straight lines for planar point sets. 33rd
            Int. Wiss. Koll. TH Ilmenau, Posterbeitr. Sekt. C1, 1988, pp. 15–18

[K89]   N.M. Korneenko: Optimal lines in the plane: a survey. Geobild
            '89, Mathematical Research, Vol. 51, Akademie-Verlag, Berlin 1989,
            pp. 43–52

[KM89]   N.M. Korneenko and H. Martini: The minsum hyperplane problem.
            Inst. Math. Acad. Sci. BSSR, Minsk, USSR, Preprint No. 35, (1989),
            26 pp.

[KM90]   N.M. Korneenko and H. Martini: Approximating finite weighted
            point sets by hyperplanes. Lect. Notes Computer Sci. **447** (1990)
            276–286

[K73]   H.W. Kuhn: A note on the Fermat problem. Math. Progr. **4** (1973)
            98–107

[K74]   H.W. Kuhn: Steiner's problem revisited. In: Studies in Optimization,
            eds. G.B. Dantzig and B.C. Eaves, MAA Studies in Mathematics,
            Vol. 10, Math. Assoc. of America, 1974, pp. 52–70

[K76]   H.W. Kuhn: Nonlinear programming: a historical view. In: Nonlinear
            Programming, eds. R.W. Cottle and C.E. Lemke, SIAM-AMS Proc.,
            Vol. 9, Amer. Math. Soc., 1976, pp. 1–26

[KD82]   Y. Kurozumi and W. Davis: Polygonal approximation by the mini-
            max method. Computer Graph. Image Process. **19** (1982) 248–264

[L86]   D.T. Lee: Geometric location problems and their complexity. Lect.
            Notes Computer Sci. **233** (1986) 154–167

[LC85]   D.T. Lee and Y.T. Ching: The power of geometric duality revisited.
            Inform. Process. Lett. **21** (1985) 117–122

[LW86]   D.T. Lee and Y.F. Wu: Geometric complexity of some location prob-
            lems. Algorithmica **1** (1986) 193–211

[L71]   L. Lovász: On the number of halving lines. Ann. Univ. Eötvös
            Loránd, Budapest, Sekt. Math. **14** (1971) 107–108

[LMW88]   R.F. Love, J.G. Morris and G.O. Wesolowsky: Facility Location:
            Models and Methods. North-Holland 1988

[M87]   H. Martini: Some results and problems around zonotopes. Coll. Math.
            Soc. J. Bolyai 48, Intuitive Geometry, eds. K. Böröczky and G. Fejes
            Toth, North-Holland 1987, pp. 383–418

[Me84]   N. Megiddo: Linear programming in linear time when the dimension
            is fixed. J. ACM **31** (1984) 114–127

[MT82]   N. Megiddo and A. Tamir: On the complexity of location linear fa-
            cilities in the plane. Oper. Res. Lett. **1** (1982) 194–197

[MT83]   N. Megiddo and A. Tamir: Finding least-distances lines. SIAM J.
            Algebr. Discrete Meth. **4** (1983) 207–211

[M84]   K. Mehlhorn: Data Structures and Algorithms, Vol. 3: Multi dimen-
            sional Search and Computational Geometry. Springer, Berlin Heidel-
            berg New York 1984
[M88]   K. Mehlhorn: Datenstrukturen und effiziente Algorithmen, Band 3;
            Mehrdimensionale Datenstrukturen und Algorithmische Geometrie,
            Teubner, Stuttgart 1988
[MPV87]   D.S. Mitrinović, J.E. Pečarić and V. Volenec: The generalized Fermat-
            Torricelli point and the generalized Lhuilier-Lemoine point. C. R.
            Math. Rep. Acad. Sci. Canada 6 (1987) 2, 95–100
[MN80]   J.G. Morris and J.P. Norback: A simple approach to linear facility
            location. Transp. Sci. 14 (1980) 1–8
[MN83]   J.G. Morris and J.P. Norback: Linear facility location – solving ex-
            tensions of basic problems. Eur. J. Oper. Res. 12 (1983) 90–94
[N87]   O. Neumann: On a problem raised by Fermat in Euclidean geometry
            (minimal sums of distances). Manuscript, Sekt. Math., FSU Jena,
            1987, 16 pp.
[NM80]   J.P. Norback and J.G. Morris: Fitting hyperplanes by minimizing
            orthogonal deviations. Math. Progr. 19 (1980) 103–105
[OvL81]   M.H. Overmars and J. van Leeuwen: Dynamically maintaining con-
            figurations in the plane. J. Comput. Syst. Sci. 23 (1981) 166-204
[PSS89]   J. Pach, W. Steiger and E. Szemeredi: An upper bound on the number
            of planar k-sets. Proc. 30th IEEE Sympos. Found. Computer Sci.,
            1989, pp. 72–79
[P61]   F.P. Palermo: A network minimization problem. IBM J. Res. Dev. 5
            (1961) 335–339
[P86]   P. Penning: Expoints. Nieuw Archief voor Wiskunde (3), 4 (1986)
            19–31
[P76]   B.I. Poljak: On the Bertsekas method for minimization of composite
            functions. Lect. Notes Control Inform. Sci. 14 (1976) 179–186
[PS85]   F.P. Preparata and M.I. Shamos: Computational Geometry: An In-
            troduction. Springer, Berlin Heidelberg New York 1985 (see also 3rd
            corrected printing, 1990)
[RND77]   E.M. Reingold, J. Nievergelt and N. Deo: Combinatorial Algorithms:
            Theory and Practice. Prentice Hall, Englewood Cliffs, NJ, 1977
[RW87]   C. Rey and P. Ward: On determining the on-line minmax fit to a
            discrete point set in the plane. Inform. Process. Lett. 24 (1987) 97–
            101
[R64]   J. Rice: The Approximation of Functions. Vol. 1: The Linear Theory.
            Addison-Wesley 1964
[R88]   J.-M. Robert: Stabbing hyperspheres by a hyperplane. In: Snap-
            shots of Computational and Discrete Geometry, Ed. G. Toussaint,
            School Computer Sci., McGill Univ., Montreal, TR SOCS-88.11,
            1988, pp. 181–188
[Sc86]   P. Schreiber: Zur Geschichte des sogenannten Steiner-Weber-Pro-
            blems. Wiss. Zeitschr. EMAU Greifswald, Math.-Naturw. Reihe 35
            (1986) 3, 53–58
[S86]   R. Seidel: Constructing higher-dimensional convex hulls at logarith-
            mic cost per face. Proc. 18th ACM Symp. Theory of Comput. 1986,
            pp. 404–413

[S90]   R. Seidel: Linear programming and convex hulls made easy. Proc.
            6th ACM Symp. Comput. Geom. 1990, pp. 211–215
[S78]   M.I. Shamos: Computational Geometry. Ph.D.Thes., Dept. Com-
            puter Sci., Yale Univ., New Haven 1978
[ST83]   E. Szemeredi and W. Trotter: Extremal problems in discrete geome-
            try. Combinatorica **3** (1983) 381–392
[T83]   G. Toussaint: Solving geometric problems with "rotating calipers".
            Proc. MELECON '83, IEE Medit. Electr. Engin. Conf. Athens,
            Greece, 1983, A10.02-A10.05
[T83]   G. Toussaint: Computing largest empty circles with location con-
            straints. J. Computer Syst. Sci. **12** (1983) 5, 347–358
[W75]   H.M. Wagner: Principles of Operations Research, Ed. 2. Prentice
            Hall, Englewood Cliffs, NJ, 1975
[W86]   A.D. Wainstein: A non-monotonous placement problem in the plane.
            Software Systems for Solving Optimal Planning Problems, 9th All-
            Union Sympos. Minsk, BSSR, USSR, 1986, Sympos Abstr. 70–71
[W88]   R. Wenger: Stabbing and separation. Thesis, McGill Univ., Montreal
            1988
[We75]   G. Wesolowsky: Location of the median line for weighted points.
            Environ. Planning **A7** (1975) 163–170
[YKII88]   P. Yamamoto, K. Kato, K. Imai and H. Imai: Algorithms for vertical
            and orthogonal L_1 linear approximation of points. Proc. 4th Ann.
            Sympos. Comput. Geom. 1988, pp. 352–361
[Z80]   E. Zemel: The linear multiple-choice knapsack problem. Oper. Res.
            28 (1980) 1412–1423
[Z84]   E. Zemel: An $O(n)$ algorithm for the linear multiple choice knapsack
            problem and related problems. Inform. Process. Lett. **18** (1984) 123–
            128

Notes Added in Proof. After the submission of this article, the authors
were informed by I. Bárány about recent results on halving hyperplanes in
d-dimensional space which might be useful for improvements of algorithmical
approaches to median hyperplanes (cf. Sect. 2.4.2 and [BFL90, ABFK91]). In
addition, the second named author collected a lot of further references with
respect to the Fermat-Torricelli problem (see Sect. 2.4.3). We can cite here
only a small selection which is mainly determined by the multitude of further
references given there and by the diversity of methods and kinds of extension.
In this sense we underline

(1) historical aspects [H69, W92],
(2) algorithmical aspects [ML87],
(3) geometrical and other generalizations [B49, E80, DM85, SM91],
(4) surveys in general [KP79, HT83, LLT89, KP90, FMW91].

Additional References

[ABFK91]   N. Alon, I. Bárány, Z. Füredi, D.J. Kleitman: Point selections and weak
           ε-nets for convex hulls. Submitted to: Combin. Probab. Comput., 1991
[BFL90]   I. Bárány, Z. Füredi and L. Lovász: On the number of halving planes.
           Combinatorica **10** (1990) 175–183
[B49]   H. Bieri: Ein geometrisches Minimumproblem. Comment. Math. Helvet.
           22 (1949) 103–114
[DM85]   R. Durier and C. Michelot: Geometrical properties of the Fermat-Weber
           problem. Europ. J. Oper. Res. **20** (1985) 332–343
[E80]   U. Eckhardt: Weber's problem and Weiszfeld's algorithm in general
           spaces. Math. Progr. **18** (1980) 186–196
[FMW91]   R.L. Francis, L.F. McGinnis and J.A. White: Facility Layout and Loca-
           tion: An Analytical Approach. Prentice Hall, N.J. 1991
[HT83]   P. Hansen and J.F. Thisse: Recent advances in continuous location the-
           ory. Sistemi Urbani **1** (1983) 33–54
[H69]   J.E. Hofmann: Über die geometrische Behandlung einer Extremwertauf-
           gabe durch Italiener des 17. Jahrhunderts. Sudhoffs Arch. **53** (1969)
           86–99
[KP79]   J. Krarup and P.M. Pruznan: Selected families of location problems.
           Ann. Discrete Math. **5** (1979) 327–387
[KP90]   J. Krarup and P.M. Pruznan: Ingredient of locational analysis. In: Dis-
           crete Location Theory. P-M. Nirchaudani and R.L. Francis (eds.). Wiley
           & Sons, Inc, 1990
[LLT89]   F.V. Louveaux, M. Labbe and J. Thisse (eds.): Facility Location Anal-
           ysis. Theory and Applications. Ann. Oper. Res. **18**, no. 1–4, Basel 1989
[ML87]   C. Michelot and O. Lefebvre: A primal-dual algorithm for the Fermat-
           Weber problem involving mixed gauges. Math. Progr. **39** (1987) 319–335
[SM91]   R.N.T. Sakai and M. Morimoto: Generalized Fermat's problem. Canad.
           Math. Bull. **34** (1) (1991) 96–104
[W92]   G.O. Wesolowsky: The Weber problem: History and perspectives. Sub-
           mitted to: J. Location Sci., 1992

Chapter VII. Geometric Transversal Theory

Jacob E. Goodman, Richard Pollack, and Rephael Wenger

1. Introduction

Geometric transversal theory has its origins in Helly's theorem:

Theorem 1.1 (Helly's Theorem) [49]. *Suppose $\mathcal{A}$ is a family of at least $d + 1$ convex sets in $\mathbb{R}^d$, and $\mathcal{A}$ is finite or each member of $\mathcal{A}$ is compact. Then if every $d + 1$ members of $\mathcal{A}$ have a common point, there is a point common to all the members of $\mathcal{A}$.*

This fundamental result can be generalized in many different directions (see [20] for an excellent survey, up to 1963, which in fact summarizes much of the early work on geometric transversals). The direction we shall be concerned with here comes from the observation of Vincensini [94] that a *point* lying in the intersection of a family of sets is the special case $k = 0$ of a *k-transversal* for the family, i.e., a k-flat which meets every member of the family. It is therefore natural to ask whether a similar result can be established for $k > 0$:

Vincensini's Problem [94]. *For $0 < k < d$, can we find a number $r(d, k)$ such that for any (sufficiently large) family $\mathcal{A}$ of convex sets in $\mathbb{R}^d$, if every r members have a common k-transversal, then all the members of $\mathcal{A}$ do?*

It is convenient, at this point, to avail ourselves of a bit of notation that has become fairly standard for this problem: Let $\mathbf{T}^k(r)$ denote the property that each subfamily of size $\leq r$ has a k-transversal, and $\mathbf{T}^k$ the same property for the entire family. (We abbreviate $\mathbf{T}^1$ as $\mathbf{T}$.) Then Helly's theorem asserts that $\mathbf{T}^0(d + 1) \Rightarrow \mathbf{T}^0$, and Vincensini asked whether there is an r such that $\mathbf{T}^k(r) \Rightarrow \mathbf{T}^k$ for $k > 0$ as well.

In the same paper in which he first announced this problem, Vincensini provided a proof that the answer is positive for $d = 2$ and $k = 1$, with $r = 6$, in other words that if any six (or fewer) members of a planar family of convex sets have a common line transversal, then all the sets do.

Unfortunately his proof was in error, and Santaló was able to show several years later [88] that $r(2,1)$ does not exist, in general, i.e., that there is no Helly-type theorem for transversals of positive dimension.[1]

In answer to a question of Klee, in fact, Hadwiger and Debrunner showed subsequently that $r(2,1)$ did not exist even if the convex sets in question were assumed to be disjoint. Figure 1.1 illustrates their example showing that $r(2,1) \neq 3$ for disjoint sets, as it appears on the cover of [47]; this can be generalized to show that *no* $r(2,1)$ exists for disjoint sets, as they indicated.

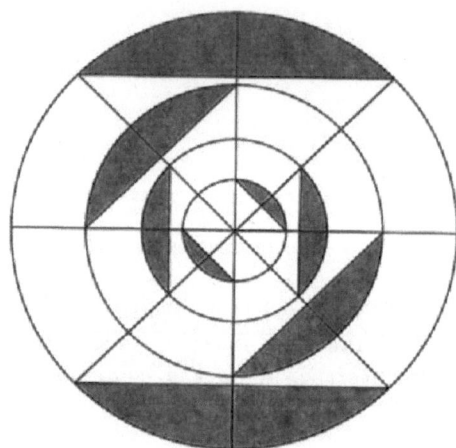

Fig. 1.1

It followed from these examples that in order for Vincensini's problem to make sense, it was necessary to impose additional restrictions on the sets in question, or on the way in which subfamilies were met by transversals. This was done by a number of mathematicians beginning with Santaló in 1940. Many of the results which were obtained are described in [20], especially in §5 of that survey. We shall therefore not go into detail on most of these theorems, nor include in our bibliography papers published before 1963, except for some that seem relevant to the more recent developments in the subject. Among these one can single out the following three results.

- Horn's 1949 theorem, which gives a sufficient (but far from necessary) condition for the existence of a transversal of a given dimension:

Theorem 1.2 [51]. *For a family of at least r compact convex sets in $\mathbb{R}^d$, the condition $\mathbf{T}^0(r)$ implies that each $(d-r)$-flat in $\mathbb{R}^d$ is contained in a $(d-r+1)$-transversal of the family.*

---

[1] This did not prevent Vincensini from repeating the problem in the same form in 1953 [95], even including a reference to Santaló's paper in another connection!

- The "special position" result proved independently by Grünbaum and Klee:

Theorem 1.3 [40,69]. *For a family of compact convex sets in* $\mathbb{R}^2$, *every two of which are separated by a translate of a fixed line,* $\mathbf{T}(3) \Rightarrow \mathbf{T}$.

- Valentine's generalization of Theorem 1.3 to three dimensions, which has a similar "special position" hypothesis, but was one of the first results on plane transversals for arbitrary compact convex sets in $\mathbb{R}^3$:

Theorem 1.4 [93]. *Suppose* P_i ($i = 1, 2, 3$) *are 3 distinct planes in* $\mathbb{R}^3$ *through a common line, and suppose* $\mathcal{A}$ *is a family of compact convex sets in* $\mathbb{R}^3$ *such that for every triple* $a_1, a_2, a_3 \in \mathcal{A}$, *each* a_j *is strictly separated from the remaining* a_k's *by a translate of some* P_i, *and the induced correspondence on the indices is cyclic. Then* $\mathbf{T}^2(4) \Rightarrow \mathbf{T}^2$ *for the family* $\mathcal{A}$.

We shall see below that some of these results follow from more recent work.

Much of the work to date in geometric transversal theory has been concerned with providing "local" conditions which are sufficient for the existence of a "global" transversal to a family of sets. Work of this sort is described in §2 (on results growing out of the Hadwiger theorem on ordered families of sets) and in §4 (on the special case of families of translates). In the past decade, however, several new areas of study have blossomed, and these we turn to in §3 and §5.

The first of these is the question of the combinatorial complexity of the space of transversals to a family of sets in $\mathbb{R}^d$. This question has its origin in the observation that different line transversals may meet the same family of sets in different orders. One question this raises is: how many different orders can there be? This has been answered precisely in the plane, and in higher dimensions various asymptotic bounds have been established. This and related questions are discussed in §3.

Another recent development in geometric transversal theory, which we describe in §5, is concerned with algorithms for computing the space of transversals to a family of suitably presented compact convex bodies in $\mathbb{R}^d$ (for example polytopes), a problem which belongs to the field of *computational geometry*. This is intimately linked with the combinatorial complexity results discussed in §3, and we detail a number of results on the complexity of such algorithms in various dimensions.

We conclude, in §6, by describing some additional results on transversals, as well as some open problems, which point the way toward further research.

Acknowledgements. The work of the first author was supported by NSA grant MDA904-89-H-2038 and PSC-CUNY grant 666426. The second author was supported by NSF grant CCR-8901484, NSA grant MDA904-89-H-2030, and the third one by NSA grant MDA904-89-H-2030 and DIMACS. We would like to express our appreciation to a number of people whose comments were very helpful to us in correcting a number of omissions and imprecise statements in an earlier version of this paper; they are Herbert Edelsbrunner, Ted Lewis, Joseph O'Rourke, Helge Tverberg, and especially Jürgen Eckhoff, who made a number of valuable contributions.

2. Hadwiger-Type Theorems

As we have seen, Santaló [88] showed that Vincensini's problem has a negative answer without additional hypotheses, and Hadwiger-Debrunner [47] showed that there was no Helly-type theorem for transversals even if the sets in question were pairwise disjoint. Thus it seemed that there was no way to pursue this problem other than by imposing additional conditions either of the sort that the sets are of some special type or that they lie in some special position (see §4 for an example of the former).

In 1957, however, the line of research proposed by Vincensini was revived when Hadwiger [45] realized that if a family of *disjoint* convex sets have a line transversal then this transversal meets the sets in a specific order (if we take the line to be directed). Thus, if the family $A = \{a_1, \ldots, a_n\}$ is met by a directed line, then the line meets the family in some order $a_{\pi(1)}, \ldots, a_{\pi(n)}$ which is described by the permutation π. Moreover, any three sets of the family are met by a directed line (namely by the given transversal) in an order consistent with the linear ordering given by π. This observation led to a proof of the fact that for pairwise disjoint compact convex sets in the plane, $T_{ord}(3) \Leftrightarrow T$:

Theorem 2.1 (Hadwiger's Transversal Theorem) [45]. *A family A of pairwise disjoint compact convex sets in the plane has a line transversal if and only if there is a linear ordering of A such that every three convex sets are met by a directed line consistently with that ordering.*

This theorem was proved by first shrinking the sets a_i (each one by a dilation about a point in the set) as far as possible while maintaining the hypothesis. At that point there must be a line which is "pinned" by three sets of the family. The proof is then completed by a case analysis which shows that this line must be a tranversal for the whole family. It should be noted that Hadwiger's theorem immediately implies Theorem 1.3, first proved by Grünbaum and Klee, since the hypothesis of the latter implies that $T(3) \Rightarrow T_{ord}(3)$.

In some sense this theorem gives a satisfactory answer to Vincensini's problem in dimension 2. (The only restriction here is that the family is required to be pairwise disjoint; we will see later that this condition was subsequently removed by Wenger [97].) A major problem is therefore how to generalize this theorem to k-transversals in dimensions $d > 2$. We shall see below a sequence of theorems generalizing the result to dimension d, but only for the case where $k = d - 1$, i.e., for hyperplane transversals. The case $0 < k < d - 1$ has seen no progress, and in fact we conclude the section with an example (Theorem 2.9) showing that there cannot be a theorem of exactly this type for line transversals in $\mathbb{R}^3$.

The first generalization of Hadwiger's transversal theorem to higher dimensions was found by Katchalski in 1980 when he gave sufficient conditions for a family of compact convex sets to have a hyperplane transversal. He proved that for pairwise disjoint compact convex sets in $\mathbb{R}^d$, $\mathbf{T}_{\mathrm{ord}}(3) \Rightarrow \mathbf{T}^{d-1}$:

Theorem 2.2 [57]. *A finite family $\mathcal{A}$ of pairwise disjoint compact convex sets in $\mathbb{R}^d$ has a hyperplane transversal if there is a linear ordering of $\mathcal{A}$ such that every three convex sets are met by a directed line consistently with that ordering.*

The proof of this theorem follows Hadwiger's idea of shrinking the sets in $\mathcal{A}$ until three sets, a_i, a_j, a_k are found, for which the hypothesis would fail if the sets were shrunk any further. At that point there is a hyperplane which separates two of the sets from the third. A case analysis shows that this hyperplane must be a transversal for the whole family.

This theorem is clearly a generalization of Hadwiger's theorem (it *is* Hadwiger's theorem when $d = 2$). But the assumption that any three of the sets have a line transversal means that the sets are in a very special position; it is somewhat akin to saying that a set of points lies in a hyperplane if every 3 of them lie on a line. (Notice the similarity to the case $r = 2$ of Theorem 1.2, which is also a special position result.) If all we are going to conclude about a set of points is that they lie in a hyperplane, then we would like to allow them to lie in general position on that hyperplane, i.e., no d of them should lie on a $(d-2)$-flat. Notice that in dimension $d = 2$, the requirement that no d sets of a family have a $(d-2)$-transversal simply means that the sets are disjoint. It is therefore natural to think of a family of (at least $k+1$) sets in $\mathbb{R}^d$ as being in "general position in dimension k" if no $k+1$ of the sets have a $(k-1)$-transversal, i.e., if the sets are "$(k-1)$-separated" in this sense. Thus a pairwise disjoint family is in general position in dimension 1, or 0-separated.

(It is easy to see that just as any two sets in a family of pairwise disjoint compact convex sets can be strictly separated by a hyperplane, so too a family of compact convex sets is $(k-1)$-separated if and only if for every $k+1$ sets of the family any j of them can be strictly separated from the remaining $k+1-j$ by a hyperplane.)

A natural extension of Hadwiger's theorem to the problem of hyperplane transversals in $\mathbb{R}^d$ should not assume any restriction on the position of the sets other than an analogue of their disjointness in the plane. A natural candidate is therefore that they be $(d-2)$-separated (or simply "separated", in the terminology of [39]). But what can be used to replace Hadwiger's hypothesis that every 3 of the sets are met by a line consistently with a certain ordering? What does it mean to talk about the "order" in which a hyperplane meets a family of sets, i.e., about $\mathbf{T}_{\mathrm{ord}}^{d-1}$?

The answer is provided by the concept of the *order type* of a set of points. By the *d-order type* of a numbered set of points $P = \{p_1, \ldots, p_n\} \subset \mathbb{R}^d$ we mean the family of orientations of its $(d+1)$-tuples, i.e., the family

$$\left(\operatorname{sgn}\ \det \begin{pmatrix} 1 & p_{i_0}^1 & \cdots & p_{i_0}^d \\ \vdots & \vdots & \ddots & \vdots \\ 1 & p_{i_d}^1 & \cdots & p_{i_d}^d \end{pmatrix} \right)_{1 \le i_0 < \cdots < i_d \le n}$$

(see [38] for details). This order type is *non-trivial* if the set P affinely spans $\mathbb{R}^d$, i.e., if at least one of these determinants is not zero. If the d-order type of P is non-trivial and $Q \subset P$ lies in an oriented k-flat, $k < d$, then the induced k-order type of Q is determined by the d-order type of P. A *k-ordering* of a family of sets, $\mathcal{A} = \{a_1, a_2, \ldots, a_n\}$, is defined via a corresponding set of points $P = \{p_1, p_2, \ldots, p_n\}$ in R^k. We assign as the orientation of a $(k+1)$-tuple of $\mathcal{A}$, say $(a_{i_0}, a_{i_1}, \ldots, a_{i_k})$, the orientation of the corresponding $(k+1)$-tuple of P, $(p_{i_0}, p_{i_1}, \ldots, p_{i_k})$. We say that an oriented k-flat F intersects a subfamily $\mathcal{B} \subset \mathcal{A}$ *consistently* with a given k-ordering of $\mathcal{A}$ if we can choose from each $a \in \mathcal{B}$ a point in $a \cap F$, so that the chosen points have the same order type as the corresponding points in R^k. If the family is $(k-1)$-separated, every choice of corresponding points generates the same order type. Thus consistency is determined by F alone.

With this concept Goodman and Pollack found the natural generalization of Hadwiger's theorem to higher dimensions, and proved that for separated compact convex sets in $\mathbb{R}^d$, $\mathbf{T}_{\mathrm{ord}}^{d-1}(d+1) \Leftrightarrow \mathbf{T}^{d-1}$:

Theorem 2.3 [39]. *A separated family $\mathcal{A}$ of compact convex sets in $\mathbb{R}^d$ has a hyperplane transversal if and only if there is a $(d-1)$-ordering of $\mathcal{A}$ such that every $d+1$ sets are met by an oriented hyperplane consistently with that $(d-1)$-ordering.*

The proof of this theorem again starts with Hadwiger's shrinking technique, but the case analysis is managed in a uniform way by a generalization of the Steinitz exchange theorem to Radon partitions.

We should note that an immediate consequence of this is Theorem 1.4, first proved by Valentine, since under his hypotheses $\mathbf{T}^2(4) \Rightarrow \mathbf{T}_{\mathrm{ord}}^2$, where the order type is obtained by projecting a point from each set along the line in which the three planes meet to an orthogonal plane.

Both Hadwiger's theorem and the Goodman-Pollack generalization might appear to imply, in a casual reading, that under the stated hypothesis the family will have a transveral which meets the sets consistently with the given ordering. This is unfortunately not the case. A simple example of G. Kertész [39] shows that this is not true (see Figure 2.1). The example consists of an ordered family of four compact convex sets in the plane such that every three of the sets can be met by a directed line consistently with the ordering, yet there is no directed line which meets all four consistently with the ordering. Wenger has proved, however, that for pairwise disjoint compact convex sets in $\mathbb{R}^2$, $\mathbf{T}_{\mathrm{ord}}(4) \Rightarrow \mathbf{T}_{\mathrm{ord}}$:

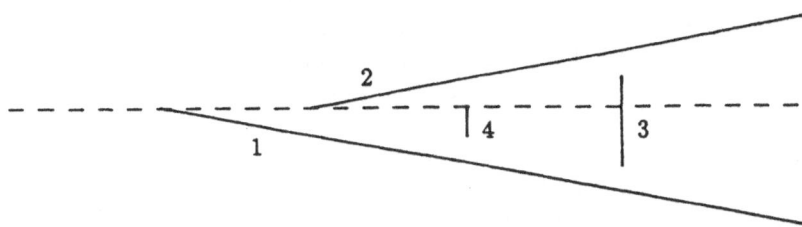

Fig. 2.1. $T_{ord}(3) \not\Rightarrow T_{ord}$

Theorem 2.4 [97]. *An ordered family $\mathcal{A}$ of pairwise disjoint compact convex sets in $\mathbb{R}^2$ has a line transversal meeting the sets consistently with the ordering of $\mathcal{A}$ if and only if every four of the sets are met by a directed line consistently with that ordering.*

This was also proved independently by H. Tverberg [92], who used it to prove that $T_{ord}(3) \Rightarrow T_{ord}$ for translates.

Theorem 2.4 can easily be generalized to hyperplane transversals in $\mathbb{R}^d$:

Theorem 2.5. *A $(d-1)$-ordered family $\mathcal{A}$ of pairwise disjoint compact convex sets in $\mathbb{R}^d$ has a hyperplane transversal meeting the sets consistently with the ordering of $\mathcal{A}$ if and only if every $2d+1$ of the sets are met by a hyperplane transversal consistently with that ordering.*

For $d = 2$, however, Theorem 2.5 requires that every five sets be met by a line, while we know from Theorem 2.4 that the weaker condition that every four sets are so met suffices. It seems probable that the $2d+1$ in Theorem 2.5 can be replaced by $2d$, although it may be possible that the correct constant is lower, perhaps even $d+2$.

For more than thirty years the only proof of Hadwiger's transversal theorem that was known required disjointness. It was an open question whether this hypothesis was really needed or was simply an artifact of the proof technique. Wenger settled this question in 1987 by giving a proof of the Hadwiger transversal theorem without this hypothesis. He proved that $T_{ord}(3) \Rightarrow T$ without any restriction on the position of the sets:

Theorem 2.6 [97]. *A family $\mathcal{A}$ of compact convex sets in the plane has a line transversal if and only if there is a linear ordering of $\mathcal{A}$ such that every three convex sets are met by a directed line consistently with that ordering.*

Of course we must now make clear what it means for a directed line to be transversal to three of the sets consistently with an ordering since the order in which the line meets the three sets is no longer uniquely determined. This is resolved by saying that a transversal is consistent with an ordering if we can choose points from each of the sets on the line so that the order of these points on the line is consistent with the given ordering. The proof of the theorem considers the set of directions (interpreted as an arc on the unit circle) in which each pair of sets have a directed line transversal which meets the pair consistently with the ordering, and then applies Helly's theorem for the circle to this family of arcs.

In an unpublished manuscript, Pollack and Wenger proved a family of theorems which had Katchalski's theorem at one end and the theorem of Goodman-Pollack at the other. They proved that for compact convex sets in general position in dimension k, $\mathbf{T}^k_{\text{ord}}(k+2) \Rightarrow \mathbf{T}^{d-1}$:

Theorem 2.7 [83]. *For $0 \leq k < d$, a $(k-1)$-separated family $\mathcal{A}$ of compact convex sets in $\mathbb{R}^d$ has a hyperplane transversal if there is a k-ordering of $\mathcal{A}$ such that every $k+2$ convex sets are met by an oriented k-flat consistently with that ordering.*

The proof of this theorem is a blend of the techniques used by Katchalski and by Goodman-Pollack. Notice that Hadwiger's transversal theorem is the case $k = 1$ and $d = 2$, Katchalski's theorem is the case $k = 1$, and the theorem of Goodman-Pollack is the case $k = d - 1$.

Just as Theorem 2.6 eliminated the hypothesis of general position from Hadwiger's transversal theorem, Pollack and Wenger removed the hypothesis of general position from Theorem 2.7 by proving that for compact convex sets in $\mathbb{R}^d$, $\mathbf{T}^k_{\text{ord}}(k+2) \Leftrightarrow \mathbf{T}^{d-1}$ for some k:

Theorem 2.8 [84]. *A family $\mathcal{A}$ of compact convex sets in $\mathbb{R}^d$ has a hyperplane transversal if and only if for some k, $0 \leq k < d$, there is a non-trivial k-ordering of $\mathcal{A}$ such that every $k+2$ of the sets are met by some oriented k-flat consistently with that k-ordering.*

Since this theorem subsumes all the theorems of Hadwiger type that we have discussed, we include its proof, taken from [84].

Proof. Let $P = \{p_1, \ldots, p_n\} \subset \mathbb{R}^k$ be the point set corresponding to $\mathcal{A}$ which defines a non-trivial k-ordering of $\mathcal{A}$ such that every $k+2$ sets of $\mathcal{A}$ are met by an oriented k-flat consistently with this k-ordering. For a hyperplane H orthogonal to a vector $v \in \mathbf{S}^{d-1}$, let H^+ denote the closed half-space bounded by H into which v points, with H^- the other half-space bounded by H. There is a unique hyperplane orthogonal to v, denoted $H_l(v)$, such that every $b \in \mathcal{A}$ meets $H_l^+(v)$ and some $b \in \mathcal{A}$ is a subset of $H_l^-(v)$. Define $H_r(v) = H_l(-v)$. Finally, let $H(v)$ be the hyperplane orthogonal to v which is half-way between $H_l(v)$ and $H_r(v)$.

With $d(b, H)$ denoting the (undirected) distance between the compact set b and the hyperplane H, the function $f : \mathbf{S}^{d-1} \to \mathbb{R}^k$ defined by

$$f(v) = \sum_{\substack{b_i \cap H^+(v) \neq \emptyset \\ b_j \cap H^-(v) \neq \emptyset}} (p_i - p_j) \min\{d(b_i, H(v)), d(b_j, H(v))\}$$

is continuous and $f(-v) = -f(v)$. Hence, by the Borsuk-Ulam theorem (see [77], for example), there is a v such that $f(v) = 0$. This means that either $H(v)$ is a transversal for $\mathcal{A}$ and $\min\{d(b_i, H(v)), d(b_j, H(v))\} = 0$ for all i, j or some convex combination of the $p_i - p_j$ equals zero. (Notice that if $H(v)$ missed some set in $\mathcal{A}$ it would have to separate two sets in $\mathcal{A}$ by construction.)

Suppose a convex combination of the vectors $p_i - p_j$ equalled zero. In that case, P would have two subsets S, T whose convex hulls met, where $S = \{p_i \mid b_i \in H^+(v)\}$ and $T = \{p_i \mid b_i \in H^-(v)\}$. By Kirchberger's theorem (see [20] or [68]), there would then be a $(k+2)$-element subset U of $S \cup T$ such that $\mathrm{conv}(S \cap U) \cap \mathrm{conv}(T \cap U) \neq \emptyset$. The corresponding subset of $\mathcal{A}$, say U', is met by a hyperplane H consistently with the order type of U. This implies that the points from the sets S' and T' corresponding to $S \cap U$ and $T \cap U$ form a Radon partition, and therefore that the convex hull of S' meets the convex hull of T'. But this is impossible since the sets S' and T' corresponding to S and T are separated by the hyperplane $H(v)$. We conclude that $H(v)$ is a hyperplane transversal for $\mathcal{A}$. $\qquad\qquad\qquad\qquad\qquad\qquad\qquad\qquad\qquad\qquad$ $\square$

It was observed by G. Kalai [personal communication] that the condition in Theorem 2.3 remains sufficient if the set P of points in $\mathbb{R}^d$ which defines the $(d-1)$-ordering of the hypothesis is generalized to an oriented matroid of rank d; the same holds for Theorem 2.7. (See [14,15,37] for definitions and discussions of oriented matroids.) Since the proof of Theorem 2.8 relies on this set P in an integral way, however, it is not clear that this observation can be extended to Theorem 2.8 as well.

Theorem 2.8 can be regarded as a complete solution to Vincensini's problem for the case of hyperplane transversals, but it leaves open the cases of k-transversals in $\mathbb{R}^d$ whenever k is positive but strictly less than $d-1$. The simplest such case is that of line transversals in $\mathbb{R}^3$. Here, Aronov, Goodman, Pollack, and Wenger have found the following example [3] which shows that there can be no pure Hadwiger-type theorem for this case.

Theorem 2.9 [3]. *For each $n > 2$ there is a pairwise disjoint family $\mathcal{A}$ of $n+2$ compact convex sets in $\mathbb{R}^3$ and a linear ordering of $\mathcal{A}$ such that any $n-2$ members of $\mathcal{A}$ are met by a directed line consistently with that ordering, yet $\mathcal{A}$ has no line transversal.*

Similar examples were used in [24] and [70] to establish (among others) the following Gallai-type analogue of Theorem 2.9:

Theorem 2.10. *For each positive integer n there is a pairwise disjoint family of compact convex sets in $\mathbb{R}^d$ and a linear ordering of the family such that any $2d-2$ members are met by a directed line consistently with the ordering, yet the family cannot be stabbed by n lines.*

The hyperbolic paraboloid $\mathcal{P}$ defined by $z = xy$ is used to visualize the construction, and its properties to justify the result. Let the sets a_i, $i = $

$1, 0, -1$, be the line segments $\{(x, y, z) \mid z = iy, \, x = i, \, 0 \leq y \leq n\}$. Now for $2 \leq i \leq n$ let a_i be the convex hull of the point set $C_i = \{(i/j, j, i) \mid 2 \leq j \leq n, \, j \neq i\}$. Notice that C_i lies in the intersection of the plane $z = i$ with the hyperbolic paraboloid $\mathcal{P}$, and that by the strict convexity of this intersection there are no other points of a_i, $i \geq 2$, which lie in $\mathcal{P}$.

Recall that the only lines which meet a_{-1}, a_0, a_1 are the rulings of $\mathcal{P}$ defined by $y = y_0, z = y_0 x$ with $0 \leq y_0 \leq n$. Thus for $i > 1$ the only lines which meet a_{-1}, a_0, a_1, a_i, are the rulings l_j defined by $y = j, z = jx$ with $j = 2, 3, \ldots, n$ and $j \neq i$. It follows that there is no line which meets all the sets in $\mathcal{A} = \{a_{-1}, \ldots, a_n\}$. On the other hand, any $n - 2$ members of $\mathcal{A}$ must omit some a_i, $i > 1$, and therefore the line l_i meets each of these $n - 2$ sets.

The 1940 example of Santaló was a barrier to pursuing Vincensini's problem in the plane (or for hyperplane transversals in higher dimensions, for that matter) until the observation was made that a line transversal meets a disjoint family in some particular order, or that a hyperplane transversal meets a separated family in some particular order type. Armed with this information, it then became natural to discover and to prove theorems of Hadwiger type rather than theorems of Helly type. Now, in view of the example of Theorem 2.9, the problem becomes that of finding a property of line transversals in space which is as natural as the fact that lines meet disjoint sets in a particular order, which will enable us to break through the barrier imposed by this example. It may then be possible to find conditions for the existence of k-transversals in $\mathbb{R}^d$ similar to the Helly-type and Hadwiger-type conditions which work for point and hyperplane transversals.

3. The Combinatorial Complexity of the Space of Transversals

The attempts described above to generalize Helly's theorem from point transversals to k-transversals have led, in recent years, to the study of the space of k-transversals, i.e., the topological space of all k-flats intersecting every member of a given family of convex sets. The space of *point* transversals for a family $\mathcal{A}$ of convex sets has a relatively simple structure, since the intersection of a family convex sets is just a convex set. For $k \geq 1$, however, the space of k-transversals of $\mathcal{A}$ need not be connected, much less convex.

A major breakthrough on an old conjecture of Grünbaum [41] came in 1986 when Katchalski, Lewis, and Liu proved that the space of line transversals for a pairwise disjoint family of translates of a compact convex set has at most three connected components [62]. Katchalski was then able to use this result to prove a Helly-type theorem for line transversals of translates, thereby proving a weak form of Grünbaum's conjecture [59]. The success of this approach prompted questions about the number of connected components of the space of k-transversals for more general families. (Katchalski's result, as well as others concerning translates, are discussed in §4.)

In the past decade computer scientists, particularly computational geometers, have become interested in algorithms to find k-transversals (see §5). Often the algorithms produce not just a single k-transversal, but a data structure representing the entire space of k-transversals. The time and space used by an algorithm of this nature are therefore bounded below by the size of such a data structure, which is in turn a function of the "combinatorial complexity" of the space of transversals. Thus computer scientists began asking: What is the combinatorial complexity of this space?

Given a family $\mathcal{A}$ of convex sets in $\mathbb{R}^d$, let $T_k^d(\mathcal{A})$ be the space of all k-transversals of $\mathcal{A}$. $T_k^d(\mathcal{A})$ is a subspace of the "noncompact Grassmannian" $\mathcal{G}_k^d$, the space of all k-flats in $\mathbb{R}^d$. If $\mathcal{A}$ is finite and the convex sets in $\mathcal{A}$ are compact, then $T_k^d(\mathcal{A})$ itself is compact. Its boundary consists of k-transversals each tangent to at least one of the sets of the family.

There are three different measures of the "combinatorial complexity" of $T_k^d(\mathcal{A})$ that are natural to consider. First, this complexity can be the number of distinct "orders" in which k-transversals intersect the convex sets in the family. These "orders" are related to the connected components of $T_k^d(\mathcal{A})$. Second, the complexity of $T_k^d(\mathcal{A})$ can be the number of "topological faces" on its boundary, where a *topological face* is a connected component of k-transversals all tangent to the same subfamily of $\mathcal{A}$. Finally, the convex sets may happen to be polytopes whose boundaries are themselves partitioned into polytopal faces. Faces of $T_k^d(\mathcal{A})$ would then be connected components of k-transversals all tangent to the same faces of the polytopes in $\mathcal{A}$, and its complexity would then mean the number of such faces. We will review each of these measures of complexity in turn. As in §2, most of the recent progress has been on hyperplane transversals.

With regard to our first measure of complexity, what specifically do we mean by the "order" in which a k-transversal meets a family $\mathcal{A}$ of convex sets?

A line transversal for a family $\mathcal{A}$ of pairwise disjoint convex sets induces a pair of orderings on $\mathcal{A}$, one for each orientation of the line. This pair of orderings is known as a *geometric permutation* [66]. Similarly, if an oriented k-flat intersects a $(k-1)$-separated family of compact convex sets, one can choose a point from the intersection of each of the sets and the k-flat and look at the order type of the resulting set of points. Since the family is $(k-1)$-separated, this order type will be independent of the choice of points. A k-transversal for a $(k-1)$-separated family of convex sets will therefore induce a pair of k-orderings on the sets, one for each orientation of the k-transversal. This pair of k-orderings is also known as a *geometric permutation* [16]. By definition, point transversals (0-transversals) all induce the same geometric permutation; when $k > 0$, however, a family may have many different geometric permutations induced by k-transversals.

Geometric permutations are reflected in the topology of the space of k-transversals. If two k-transversals for a $(k-1)$-separated family can be continously transformed into one another through the space of transversals, then they must induce the same geometric permutation, or the family would not be $(k-1)$-separated. In other words, if $\mathcal{A}$ is a $(k-1)$-separated family, all the k-transversals in the same connected component of $T_k^d(\mathcal{A})$ induce the same geometric permutation. The converse is also true, if $k = d - 1$:

Theorem 3.1 [99]. *Let $\mathcal{A}$ be a $(d-2)$-separated family of compact convex sets in $\mathbb{R}^d$. Two hyperplane transversals induce the same geometric permutation on $\mathcal{A}$ if and only if they lie in the same connected component of $T_{d-1}^d(\mathcal{A})$.*

In general, however, two k-transversals in $\mathbb{R}^d$ may induce the same geometric permutation without being isotopic. The example that shows this is similar to the construction in Theorem 2.9. For $i = -1, 0, 1$, let $s_i = \{(x, y, z) \mid z = iy, x = i, 0 \le y \le n\}$, forming three line segments lying on the hyperbolic paraboloid $z = xy$. Let a be the convex hull of the point set $\{(2/j, j, 2) \mid 2 \le j \le n\}$. Then the lines $\{(x, y, z) \mid z = jx, y = j\}$, $j = 2, \ldots, n$, constitute $n - 1$ distinct connected components of the transversal space, each line intersecting the sets in the same order s_{-1}, s_0, s_1, a.

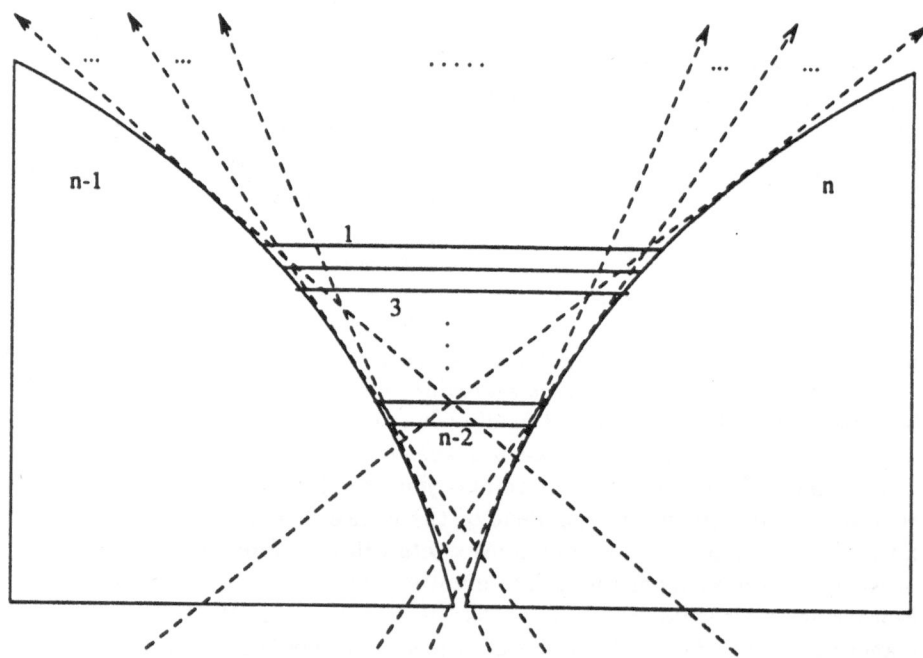

Fig. 3.1. n convex sets with $2n - 2$ geometric permutations

Given a $(k - 1)$-separated family of n compact convex sets in $\mathbb{R}^d$, what is the maximum number of geometric permutations induced by k-flats that this family can have? Let $g_k^d(n)$ represent this maximum, taken over all such families of size n. The first interesting case is when $k = 1$ and $d = 2$, i.e., the case of line transversals in the plane. There are many constructions showing that $g_1^2(n) = \Omega(n)$, even under the restriction that the convex sets in the family are all line segments or all circles [47,66,72]. Katchalski, Lewis, and Zaks showed that $g_1^2(n) \geq 2n - 2$ by constructing n pairwise disjoint convex sets with $2n - 2$ geometric permutations [66] (see Figure 3.1). Edelsbrunner and Sharir proved that this is the largest number of geometric permutations one can achieve in the plane [34]: Any line transversal in the plane can be moved to a line transversal with the same geometric permutation which is tangent to two convex sets and does not separate the sets; an argument based on *Davenport-Schinzel sequences*, sequences with forbidden $i \ldots j \ldots i \ldots j$ subsequences, then shows there are at most $2n - 2$ such line transversals. Thus,

Theorem 3.2 [34,66]. $g_1^2(n) = 2n - 2$.

These planar constructions generalize to constructions of n pairwise disjoint convex sets in $\mathbb{R}^d$ with $\Omega(n^{d-1})$ geometric permutations induced by line transversals, hence

Theorem 3.3 [64,66]. $g_1^d(n) = \Omega(n^{d-1})$.

An upper bound, on the other hand, comes from separating every pair of convex sets by a hyperplane. The arrangement of these $\binom{n}{2}$ hyperplanes defines $O(n^{2d-2})$ unbounded cells, or $O(n^{2d-2})$ directions, in $\mathbb{R}^d$. Two line transversals inducing different geometric permutations must point in different directions in this sense, so that

Theorem 3.4 [98]. $g_1^d(n) = O(n^{2d-2})$.

It is easy to show that $g_k^d(n) \geq g_1^{d-k+1}(n)$, but this bound seems very loose. No better lower bounds are known in general.

For geometric permutations induced by hyperplanes, we have

Theorem 3.5 [16]. $g_{d-1}^d(n) = O(n^{d-1})$.

This upper bound comes from a corresponding upper bound on the number of faces in the space of transversals which we discuss below (Theorem 3.6). Since this upper bound applies *a fortiori* to the number of connected components in that space, and since the transversals belonging to each connected component all induce the same geometric permutation, this upper bound applies as well to the number of geometric permutations.

The next measure of complexity of the space of transversals is the number of topological faces on its boundary, topological faces being connected components of k-transversals all "tangent" (i.e., supporting) to the same convex sets. Notice that we can have arbitrarily many lines tangent to two *intersecting* convex sets in the plane, hence arbitrarily many vertices (0-faces) in the space of transversals. To avoid this, therefore, we again assume when dealing with k-transversals that the family $\mathcal{A}$ of convex sets is $(k-1)$-separated.

Even if $\mathcal{A}$ is $(k-1)$-separated, however, the complexity of $T_k^d(\mathcal{A})$ cannot in general be bounded in terms of the size of $\mathcal{A}$. We refer to the example above of three skew lines on the hyperbolic paraboloid $z = xy$ and a polygon a touching the paraboloid at its $n-2$ vertices. Let $\mathcal{A}$ consist of these four objects. Each vertex of $\mathcal{A}$ generates one extreme line transversal for $\mathcal{A}$, i.e., one vertex of the boundary of $T_1^3(\mathcal{A})$. Since the number of vertices of the polygon can be arbitrarily large, so can the complexity of $T_1^3(\mathcal{A})$.

Nevertheless, the complexity of the space of *hyperplane* transversals can be bounded by a function of the size of $\mathcal{A}$, at least under certain circumstances. Cappell, Goodman, Pach, Pollack, Sharir, and Wenger have shown:

Theorem 3.6 [16]. *Let $\mathcal{A}$ be a $(d-2)$-separated family of n compact and strictly convex sets in $\mathbb{R}^d$. Then the number of topological faces of $T_{d-1}^d(\mathcal{A})$ is $O(n^{d-1})$.*

A convex set is *strictly* convex if every tangent hyperplane intersects the set at exactly one point.

In their proof, Cappell *et al.* first show that the space of oriented hyperplanes tangent to $k \leq d$ strictly convex sets in $\mathbb{R}^d$ is homeomorphic to 2^k copies of S^{d-k}. Thus, the hyperplanes tangent to a given convex set form a topological sphere and the intersection of several of these topological spheres is again a topological sphere. They next prove that under these conditions the boundary of the union of n topological spheres has $O(n^{d-1})$ complexity, generalizing a planar result in [65]. Applying these two results to the boundary of $T_{d-1}^d(\mathcal{A})$ gives the upper bound in Theorem 3.6.

Cappell *et al.* also construct a $(d-2)$-separated family of n compact convex sets which has $O(n^{\lfloor d/2 \rfloor})$ topological faces. The construction is a family $\mathcal{A}$ of n vertical line segments each with its "top" endpoint lying on a vertex of a cyclic polytope in $\mathbb{R}^d$. Each hyperplane containing a "lower" facet of this polytope is a transversal to the family of line segments which is tangent to d of them. Thus the $\Omega(n^{\lfloor d/2 \rfloor})$ lower facets determine as many vertices on the boundary of $T_{d-1}^d(\mathcal{A})$. The line segments can be made strictly convex by slightly "fattening" each of them.

It seems likely that Theorem 3.6 should apply to all convex sets, not just strictly convex ones. This would immediately follow if one could show that the space of oriented hyperplanes tangent to $k \leq d$ convex sets in $\mathbb{R}^d$ was homeomorphic to 2^k copies of S^{d-k} without assuming the sets were strictly convex. Using an approximation argument, Cappell *et al.* were able to generalize their upper bound to the number of connected components of the transversal space, without any assumption of strict convexity. As mentioned above, this bound holds also for the number of geometric permutations (Theorem 3.5).

Houle, Imai, Imai, Robert, and Yamamoto give a much better bound for $T_{d-1}^d(\mathcal{A})$ when $\mathcal{A}$ is a family of n $(d-1)$-spheres in $\mathbb{R}^d$ [53]. Under a suitable transformation, the hyperplanes tangent to n spheres in $\mathcal{A}$ form n (true) $(d-1)$-spheres in $\mathbb{R}^d$. These $(d-1)$-spheres can be mapped into n half-spaces in $\mathbb{R}^{d+1}$ [6]. The space $T_{d-1}^d(\mathcal{A})$ is represented by the intersection of these half-spaces and a suitable quadric surface. The convex polytope in $\mathbb{R}^{d+1}$ formed by intersecting these n half-spaces has $O(n^{\lceil d/2 \rceil})$ faces. Intersecting this convex polytope with the quadric surface only increases the number of faces by a factor of two.

Theorem 3.7 [53]. *Let $\mathcal{A}$ be a family of n $(d-1)$-spheres in $\mathbb{R}^d$. The number of topological faces of $T_{d-1}^d(\mathcal{A})$ is $O(n^{\lceil d/2 \rceil})$.*

No assumption is needed about the family of spheres being separated. Theorem 3.7 also bounds the number of geometric permutations generated by hyperplane transversals for a $(d-2)$-separated family of n spheres in $\mathbb{R}^d$. An open question is whether this $O(n^{\lceil d/2 \rceil})$ bound can be extended to families of convex sets, greatly improving upon Theorem 3.6.

It is possible, at least in $\mathbb{R}^2$, to loosen the restriction of pairwise disjointness and still bound the number of topological faces on the boundary of the space of line transversals. In place of pairwise disjointness, we will assume that the boundaries of any two sets meet transversally in at most s points, where s is an (even) integer at least two. Examples in $\mathbb{R}^2$ of families of convex sets with such a property are translates of a compact convex set ($s=2$), circular disks ($s=2$), or convex polygons with at most k vertices ($s=2k$). Bounds on the transversal space of such a family come from viewing the transversals as points in dual space.

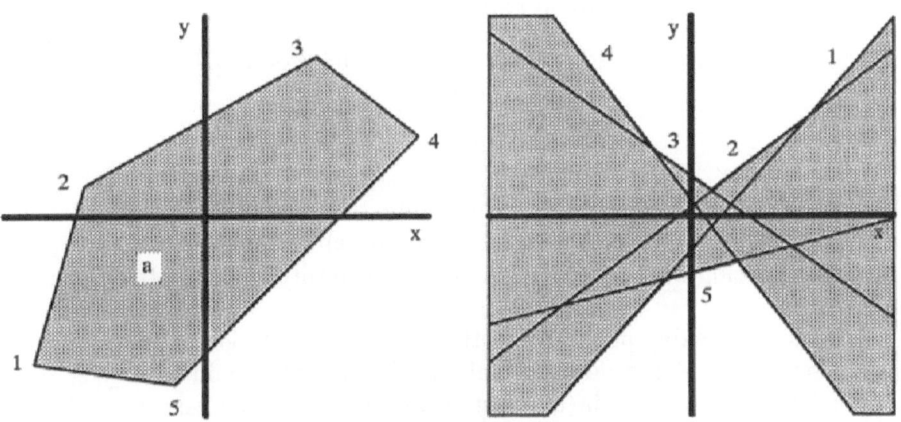

Fig. 3.2. Lines intersecting a in dual space

One way of transforming lines to points is by mapping the line $y = mx + b$ to the point (m, b) in dual space. This mapping takes the non-vertical lines intersecting a compact convex set a to a closed region sandwiched between the graphs of two real-valued functions, $\sigma^+(a)$ and $\sigma^-(a)$ (See Figure 3.2); the graph of $\sigma^+(a)$ (resp. $\sigma^-(a)$) corresponds to the set of supporting lines to a lying above a (resp. below a).

The *upper envelope* (resp. *lower envelope*) of a set of continuous functions on $\mathbb{R}$ is their pointwise maximum (resp. minimum). Given such an upper or lower envelope, $\mathbb{R}$ can be partitioned into maximal intervals, each associated with a function, so that the envelope is given by the corresponding function on the interval. The complexity of the envelope is then the number of such intervals and their endpoints.

In the dual space, the line transversals to a family $\{a_i\}_{1 \leq i \leq n}$ correspond to the points lying between the lower envelope of $\{\sigma^+(a_i)\}_{1 \leq i \leq n}$ and the upper envelope of $\{\sigma^-(a_i)\}_{1 \leq i \leq n}$. The partition of the boundary of the space of line transversals into topological faces matches the partition of the upper and lower envelopes into intervals, as described above. Thus, the complexity of the space of line transversals of the family $\{a_i\}$ is the joint complexity of these upper and lower envelopes.

Let $\lambda_s(n)$ be the maximum complexity of the upper or lower envelope of any set of n continuous functions on $\mathbb{R}$ which intersect pairwise in at most s points. The following bounds are known for $\lambda_s(n)$:

$$\lambda_1(n) = n \quad \text{(trivial)},$$
$$\lambda_2(n) = 2n - 1 \quad [4],$$
$$\lambda_3(n) = \Theta(n\,\alpha(n)) \quad [48],$$
$$\lambda_4(n) = \Theta(n\,2^{\alpha(n)}) \quad [2],$$
$$\lambda_s(n) \leq n\,2^{K_s\,\alpha(n)^{(s-2)/2}} \quad \text{for } s \text{ even} \quad [2],$$
$$\lambda_s(n) \leq n\,2^{K_s\,\log(\alpha(n))\,\alpha(n)^{(s-3)/2}} \quad \text{for } s \text{ odd} \quad [2],$$
$$\lambda_s(n) \geq n\,2^{K_s'\,\alpha(n)^{(s-2)/2}} \quad \text{for } s \text{ even} \quad [2].$$

Here $\alpha(n)$ is the inverse of the Ackermann function and grows very slowly with n. (For a definition of the Ackermann function and its relationship to upper and lower envelopes, see [48].) K_s and K_s' are positive constants depending on s.

Atallah and Bajaj pointed out that if the boundaries of two compact convex sets $a_1, a_2 \in \mathbb{R}^2$ intersect pairwise in at most s points, $s \geq 2$, then $\sigma^+(a_1)$ intersects $\sigma^+(a_2)$ in at most s points [5]. (If the sets are pairwise disjoint, $\sigma^+(a_1)$ may still intersect $\sigma^+(a_2)$ in two points.) The same holds true for the number of intersections between $\sigma^-(a_1)$ and $\sigma^-(a_2)$. Hence

Theorem 3.8 [5]. *Let A be a family of n compact convex sets in $\mathbb{R}^2$ whose boundaries intersect pairwise in at most s points, $s \geq 2$. Then the maximum number of topological faces on the boundary of $T_1^2(A)$ is $\lambda_s(n)$.*

For the final measure of complexity, consider a family $\mathcal{A}$ of polytopes whose boundaries are partitioned into polytopal faces. Each face of $T_k^d(\mathcal{A})$ is then a connected component of k-transversals all tangent to some particular set of faces of members of $\mathcal{A}$, and the complexity of $T_k^d(\mathcal{A})$ is the total number of its faces. This is a refinement of the previous measure of complexity, since two transversals may be tangent to the same polytopes yet tangent to them along different faces. The $(k-1)$-separatedness condition is no longer required, since the complexity can now be measured as a function of the total number of polytope faces, not just as a function of the number of polytopes.

Again we use a transformation to dual space, mapping the hyperplane $x_d = \mu_1 x_1 + \cdots + \mu_{d-1} x_{d-1} + \mu_d$ to the point $(\mu_1, \ldots, \mu_d)$ [29]. The set of vertices of the polytopes in $\mathcal{A}$ maps to an arrangement of hyperplanes in this dual space, and $T_{d-1}^d(\mathcal{A})$ maps to a union of faces of this arrangement. Notice also that the faces on the boundary of $T_{d-1}^d(\mathcal{A})$ are all linear, i.e., are k-dimensional open balls lying in a k-flat for some k, $0 < k < d$. Let n_0 be the total number of vertices over all the polytopes in $\mathcal{A}$. Since the full arrangement has $O(n_0^d)$ linear faces, so does $T_{d-1}^d(\mathcal{A})$. Better bounds are known, however, when $\mathcal{A}$ is a family of simplices or line segments.

The hyperplanes intersecting a compact convex set a map to a closed region between two surfaces defined by functions $\sigma^+(a)$ and $\sigma^-(a)$ from $\mathbb{R}^{d-1}$ to $\mathbb{R}$. For a family $\mathcal{A} = \{a_i\}_{1 \le i \le n}$ of compact convex sets, we can again take the lower (resp. upper) envelope of $\{\sigma^+(a_i)\}$ (resp. $\{\sigma^-(a_i)\}$), i.e., the pointwise minimum (resp. maximum) of the functions $\sigma^+(a_i)$ (resp. $\sigma^-(a_i)$). When a is a convex polytope, these upper and lower envelopes are piecewise linear functions on $\mathbb{R}^{d-1}$ In this context, we can define the complexity of these envelopes as the number of their connected linear components.

The hyperplane transversals to a family $\{a_i\}$ of convex polytopes correspond, in the dual, to the points lying between the lower envelope of $\{\sigma^+(a_i)\}$ and the upper envelope of $\{\sigma^-(a_i)\}$. The number of linear faces on the boundary of $T_{d-1}^d(\mathcal{A})$ equals the number of linear components of these envelopes.

Hart and Sharir proved that the upper and lower envelopes of n line segments in $\mathbb{R}^d$ have at most $O(n\,\alpha(n))$ faces, where $\alpha(n)$ is the inverse of the Ackermann function [48]. Generalizing this result, Pach and Sharir proved that the upper and lower envelopes of n $(d-1)$-simplices in $\mathbb{R}^d$ have at most $O(n^{d-1}\,\alpha(n))$ facets [79], and Edelsbrunner extended this bound to the number of faces of all dimensions [30]. Edelsbrunner, Guibas and Sharir gave an even better bound of $O(n^{d-1})$ for the number of faces in the upper and lower envelopes of n half-hyperplanes $((d-1)$-dimensional half-spaces) [31].

For a family $\{a_i\}$ of n d-simplices in $\mathbb{R}^d$, the lower envelope of $\{\sigma^+(a_i)\}$ is a lower envelope of $O(n)$ $(d-1)$-simplices, and therefore has complexity $O(n^{d-1}\,\alpha(n))$. This bound on the complexity applies, in fact, whenever the number of vertices in each of the polytopes is bounded by a constant. For a family $\{a_i\}$ of n line segments in $\mathbb{R}^d$, on the other hand, the lower envelope of $\{\sigma^+(a_i)\}$ is a lower envelope of $2n$ half-hyperplanes, hence has complexity $O(n^{d-1})$. The same bounds hold of course for the upper envelopes of $\{\sigma^-(a_i)\}$.

These bounds on the complexity of the upper and lower envelopes for simplices and for half-hyperplanes are tight, with examples provided of n $(d-1)$-simplices and of n half-hyperplanes which achieve their respective bounds. These examples can be "reverse engineered" to give examples of a family $\mathcal{A}$ of n d-simplices (resp. n line segments) in $\mathbb{R}^d$ such that $\mathcal{T}_{d-1}^d(A)$ has complexity $\Omega(n^{d-1}\alpha(n))$ (resp. $\Omega(n^{d-1})$). Thus we have

Theorem 3.9 [30,31,48,79]. *The maximum number of linear faces on the boundary of the space of hyperplane transversals to n d-simplices in $\mathbb{R}^d$ is $\Theta(n^{d-1}\alpha(n))$. The maximum number of linear faces on the boundary of the space of hyperplane transversals to n line segments in $\mathbb{R}^d$ is $\Theta(n^{d-1})$.*

In $\mathbb{R}^2$ and $\mathbb{R}^3$, these bounds on the complexity of the transversal space can be given as functions solely of the total number of vertices over all convex polygons or polytopes in the family. Let n_0 be the total number of such vertices. Each polygon vertex in $\mathbb{R}^2$ maps to a line segment on the upper or lower envelope. Each polytope vertex in $\mathbb{R}^3$, similarly, maps to a polygon on the upper or lower envelope. These polygons can be triangulated to form a total of $O(n_0)$ triangles. From the results cited above on the upper and lower envelopes of line segments and triangles, we therefore have

Theorem 3.10 [48,79]. *The maximum number of linear faces on the boundary of the space of line transversals to convex polygons in $\mathbb{R}^2$ with a total of n_0 vertices is $\Theta(n_0\alpha(n_0))$. The maximum number of linear faces on the boundary of the space of plane transversals to convex polytopes in $\mathbb{R}^3$ with a total of n_0 vertices is $\Theta(n_0^2\alpha(n_0))$.*

In dimension > 3, unfortunately, the mapping of polytope vertices to simplices in the upper or lower envelopes can generate a superlinear number of such simplices. Nevertheless, we suspect that the number of linear faces should still be $\Theta(n_0^{d-1}\alpha(n_0))$ for families of convex polytopes in $\mathbb{R}^d$ with a total of n_0 vertices.

Very recently, Pellegrini and Shor gave almost tight bounds on the complexity of the space of line transversals of convex polytopes in $\mathbb{R}^3$ [81]. The space $\mathcal{G}_1^3$ of all lines in $\mathbb{R}^3$ can be represented as a set of points lying on a suitable quadric surface in $\mathbb{R}^5$. Lines which intersect a given convex polytope in $\mathbb{R}^3$ are then the intersection of this quadric surface with some convex polytopes in $\mathbb{R}^5$. A face in this representation is the intersection of of a polytope face with the quadric surface. Using random sampling and the fact that convex polytopes in $\mathbb{R}^5$ with n facets have $O(n^2)$ complexity, they prove:

Theorem 3.11 [81]. *The number of faces on the boundary of the space of line transversals to convex polytopes in $\mathbb{R}^3$ with a total of n_2 facets is bounded by $c_\epsilon n_2^{3+\epsilon}$ for any $\epsilon > 0$, the constant c_ϵ depending on ϵ.*

This upper bound almost meets an $\Omega(n_2^3)$ lower bound by Pellegrini [80].

4. Translates of a Convex Set

One course that research on common transversals has followed is concerned with transversals to a family of translates of a compact convex set, a chapter that has an interesting history in itself. We will go into some detail on this question, since significant progress has recently taken place here.

The story begins in 1940 with Santaló's observation that a pure Helly-type theorem could not exist for line transversals in the plane, and his realization that additional conditions, beyond the convexity of the sets in the family, would be necessary. We have already seen one direction that such investigation took, in §2. Santaló came up with an entirely different condition, and in fact proved, using Radon's theorem, that a generalization of it was sufficient in all dimensions for the existence of a hyperplane transversal:

Theorem 4.1 [88]. *For a family $\mathcal{A}$ of parallelotopes in $\mathbb{R}^d$ with edges parallel to the coordinate axes, $\mathbf{T}^{d-1}(2^{d-1}(d+1))$ implies $\mathbf{T}^{d-1}$.*

In particular, this says that $\mathbf{T}(6) \Rightarrow \mathbf{T}$ for translates of a parallelogram in the plane.

He also proved the corresponding statement for line transversals:

Theorem 4.2 [88]. *For a family of parallelotopes in $\mathbb{R}^d$ with edges parallel to the coordinate axes, $\mathbf{T}(2^{d-1}(2d-1))$ implies $\mathbf{T}$.*

The number $2^{d-1}(d+1)$ in Theorem 4.1 has been shown to be sharp by Kramer [70] (see Theorem 4.6 below), but whether $2^{d-1}(2d-1)$ is equally sharp in Theorem 4.2 seems to be unknown.

In 1955 Hadwiger [44] asked the corresponding question about circles: find the smallest r such that $\mathbf{T}(r) \Rightarrow \mathbf{T}$ for every family of disjoint unit disks, and remarked that if such a r exists, it must be at least 5.

This problem was solved by Danzer, who proved that 5 would indeed suffice:

Theorem 4.3 [19]. $\mathbf{T}(5) \Rightarrow \mathbf{T}$ *for a family of mutually disjoint unit disks in $\mathbb{R}^2$.*

Danzer also pointed out that if the disjointness hypothesis or the assumption that the disks are congruent is dropped, then the conclusion becomes invalid for *any* r in place of 5. In addition, he conjectured the existence of a number k_d such that $\mathbf{T}(k_d) \Rightarrow \mathbf{T}$ for any family of mutually disjoint unit balls in $\mathbb{R}^d$; $k_2 = 5$ by Theorem 4.3. Finally, he remarked that the minimal such k_d for each d is easily seen to satisfy $k_{d+1} \geq k_d$. He asked whether this can be replaced by a strict inequality, for example for $d = 2$.

In 1958, Grünbaum showed that Santaló's result $\mathbf{T}(6) \Rightarrow \mathbf{T}$ could be improved for disjoint translates of a single parallelogram:

Theorem 4.4 [41]. $T(5) \Rightarrow T$ *for a family of disjoint translates of a parallelogram in* $\mathbb{R}^2$.

His proof used several results of Hadwiger-Debrunner, among them

Theorem 4.5 [46]. *Given any family of parallelograms with parallel edges such that any three can be intersected by an ascending line, there exists an ascending line intersecting all the parallelograms.*

Grünbaum pointed out, in addition, that $T(4)$ would not suffice for the result in Theorem 4.4 (see Figure 4.1 for his example), nor could disjointness be dropped from the hypothesis, and that "translates" could not be weakened to "homothets". He also strengthened Danzer's theorem on mutually disjoint unit disks by showing that $T(4) \Rightarrow T$ for families of size at least 6.

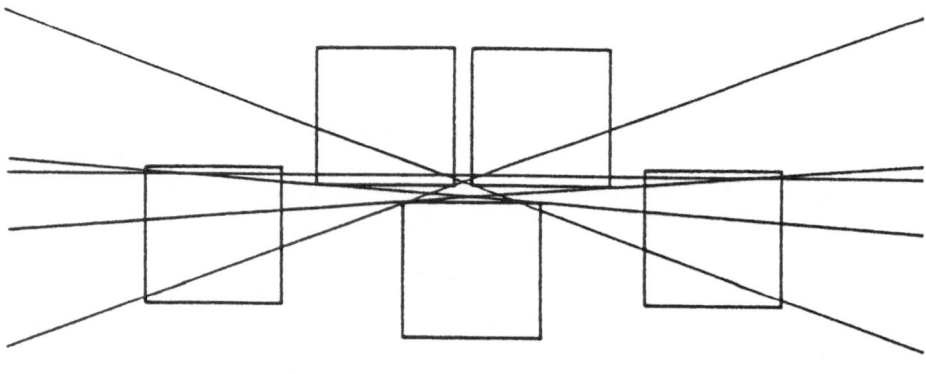

Fig. 4.1. $T(4) \not\Rightarrow T$

More significantly, it was in this paper that Grünbaum conjectured that $T(5) \Rightarrow T$ probably holds as well for a family of disjoint translates of *any* convex set, a conjecture which would take more than 30 years to establish:

Grünbaum's Conjecture [41]. *For a family A of disjoint translates of a compact convex set in* $\mathbb{R}^2$, *if any five members of A have a line transversal then A has a line transversal.*

There were several more developments on the "translate" problem before those which took place within the past decade and which culminated in a proof of Grünbaum's conjecture. In 1964 Grünbaum used Helly's theorem to generalize Santaló's 1940 theorem about parallelotopes in $\mathbb{R}^d$ to the case of hyperplane transversals to more general families of polytopes, whose vertex cones were "related" to those of a fixed polytope. We state his result just for translates, even though it holds in greater generality:

Theorem 4.6 [42]. *For a family of translates of a convex polytope P in $\mathbb{R}^d$ with n vertices, $\mathbf{T}^{d-1}(\binom{n}{2}(d+1)) \Rightarrow \mathbf{T}^{d-1}$. If P is centrally symmetric, moreover, we have $\mathbf{T}^{d-1}(\frac{n}{2}(d+1)) \Rightarrow \mathbf{T}^{d-1}$.*

(Notice that the second statement gives exactly the special case of Theorem 4.1 in which $\mathcal{A}$ consists of translates of a parallelotope.)

The number $\frac{n}{2}(d+1)$ was subsequently proved to be sharp by Kramer [70], for *any* centrally symmetric polytope in $\mathbb{R}^d$ with n vertices.

Grünbaum also proved that for a centrally symmetric convex body a in $\mathbb{R}^2$, if there exists a t (depending only on a) such that $\mathbf{T}(t) \Rightarrow \mathbf{T}$ for all families of homothets of a, then a must be a polygon, and asked whether this remained true if "homothet" is replaced by "translate", or if the assumption of central symmetry is dropped.

This was answered affirmatively in the following stronger form by Eckhoff [24]:

Theorem 4.7. *For a compact convex set a in $\mathbb{R}^d$, there exists a number t (depending only on a) such that $\mathbf{T}^{d-1}(t) \Rightarrow \mathbf{T}^{d-1}$ for all families of translates of a, if and only if a is a polytope.*

In his 1969 dissertation, Eckhoff considered a variant of the transversal problem for translates: If a family of translates of a compact convex set in $\mathbb{R}^d$ has a k-dimensional transversal for every subset of size r, can it be split into a finite number (m) of subfamilies, each having a k-transversal? Hadwiger-Debrunner [46] had already shown this to be true for a family of homothets in the case $d = 2$, $k = 1$, $r = 4$, for $m = 4$, i.e., if every four members have a line transversal in $\mathbb{R}^2$, then the family can be split into four subfamilies, each having a line transversal, and for $m = 3$ in the special case of translates [47] Translating the problem to a covering problem (much as Danzer had done in his 1957 paper), Eckhoff was able to prove the following "Gallai-type" results:

Theorem 4.8 [24, 25]. *If every r translates in a family $\mathcal{A}$ of translates in $\mathbb{R}^d$ have a k-transversal (i.e., $\mathbf{T}^k(r)$ holds for $\mathcal{A}$), $r \geq k + 2$, then $\mathcal{A}$ can be partitioned into a finite number, $\delta^k(r)$, of subfamilies, each of which have a k-transversal (i.e., $\mathbf{T}^k$ holds for each subfamily).*

Moreover, if every three translates in a family of translates in $\mathbb{R}^2$ have a line transversal, then there exist two parallel lines which together meet all the members of $\mathcal{A}$.

He also gave bounds on $\delta^k(r)$.

Kramer [70] was subsequently able to generalize the first statement of Theorem 4.8 to families of homothets, and to give corresponding upper bounds.

The next decade saw no progress on the transversal problem for translates. Then, in 1980, Katchalski and Lewis published a result related to the Grünbaum conjecture.

They had already, shortly before, succeeded in proving the

Theorem 4.9 [61]. *If A is a family of pairwise disjoint translates of a parallelogram, and A has property $T(4)$, then some subfamily $B \subset A$ with $\operatorname{card}(A \setminus B) \leq 2$ has property T.*

Their proof used the Hadwiger-Debrunner result (Theorem 4.5) mentioned above which Grünbaum had also used in his 1958 paper.

In their 1980 paper, they showed that a similar conclusion (but with 2 replaced by a much larger constant) could be established for much more general families, even under the weaker assumption $T(3)$:

Theorem 4.10 [60]. *For any family A of pairwise disjoint translates of a compact convex set, if A has property $T(3)$ then some subfamily B with $\operatorname{card}(A \setminus B) \leq q$, where q is some constant independent of A, has property T.*

They showed that q could be taken to be 192π, and conjectured that $q = 2$ was sufficient:

Katchalski-Lewis Conjecture [61]. *For a family A of pairwise disjoint translates of a compact convex set in $\mathbb{R}^2$, if A has property $T(3)$ then some subfamily B has property T, with $\operatorname{card}(A \setminus B) = 2$.*

(In 1987 A. Bezdek [personal communication] showed, by a construction, that there is an arrangement of n disjoint unit disks, any 3 of which have a common transversal, but no $n - 1$ of which do. Hence q cannot be smaller than 2.)

The proof of Theorem 4.10 involved showing that for any family A satisfying the hypothesis of the theorem, each of whose members had diameter $\leq r$, one could find 3 disks of radius $3r$ such that all the members of A not meeting these disks had a common transversal. This, in turn, was proved by using a contraction argument, and with the help of the planar case of Jung's theorem [55], which asserts that a subset of $\mathbb{R}^d$ of diameter $\leq r$ is contained in a ball of radius $(d/(2r + 2))^{1/2}r$. (As is pointed out in [20, p. 112], Jung's theorem itself can be proved using Helly's theorem.)

In his 1986 paper [59], Katchalski continued the progress toward establishing Grünbaum's conjecture by proving the following "weak Grünbaum conjecture" (which actually constituted the first Helly-type theorem for common transversals of translates):

Theorem 4.11 [59]. *For a family A of pairwise disjoint translates of a compact convex set in $\mathbb{R}^2$, $T(r_0) \Rightarrow T$, where r_0 is some constant less than 128.*

To prove this, he used Hadwiger's planar transversal theorem ($T_{ord}(3) \Rightarrow T$), together with the following basic facts about geometric permutations (see §3) for translates, the first of which follows by refining a result of [62]:

Theorem 4.12. *If A is a family of pairwise disjoint translates of a compact convex set in $\mathbb{R}^2$, then any two geometric permutations induced by common transversals on the members of A differ by at most 5 consecutive indices.*

Theorem 4.13 [63,64]. *If A is a family of pairwise disjoint translates of a compact convex set in $\mathbb{R}^2$, then A has at most three geometric permutations.*[2]

What these theorems say is that certain compatibility relations must hold between distinct geometric permutations arising from the same family of disjoint translates of a compact convex set in the plane. This idea was exploited in Tverberg's brilliant proof of the Grünbaum conjecture, which appeared in 1989:

Theorem 4.14 [91]. *For a family A of disjoint translates of a compact convex set in $\mathbb{R}^2$, $\mathbf{T}(5) \Rightarrow \mathbf{T}$.*

The proof of Tverberg's theorem proceeds by a sequence of reductions. Assuming a counterexample, he first easily reduces to the situation where A is finite, the convex set is a centrally symmetric polygon, and the translation vectors are convexly independent points in general position.

Next, shrinking each set around an interior point, just as Hadwiger had done in [45] (and as Klee had done even earlier in [69]), he obtains a counterexample with only six members. Moreover, the centers of these six translates must be in convex position, forming a hexagon.

A case analysis ensues, which proceeds by considering the shape of the hexagon of centers, where "shape" means equivalence class under the mapping which assigns to each edge of a hexagon its furthest vertex. It turns out that there are only four possible shapes in this sense, and this knowledge enables Tverberg to limit the possibilities for the geometric permutations arising in each case.

He then extends results on incompatible geometric permutations (in the sense described above) and shows that certain geometric permutations cannot coexist in his presumptive counterexample.

Finally, a computer check (carried out independently by hand) reduces the number of possible sextuples of permutations (one for each subset of cardinality 5) to 2. A careful geometric argument is then used to rule these out as well, and the proof is complete.

The main problem which remains, in the area of common transversals to translates of planar sets, concerns the Katchalski-Lewis conjecture. Using the last result of Eckhoff's mentioned above, Tverberg [92] can reduce the Katchalski-Lewis constant of 192π to 108. On the other hand, as Tverberg has observed, it follows from recent results of Tuza [89] that if there is a

---

[2] The original proof in [63] did not apply to families of five or six pairwise disjoint translates, but Katchalski, Lewis, and Liu extended the theorem to these special cases in [64].

counterexample to the conjecture, there must be one with at most 49 sets. It may then be possible, he believes, to use similar methods to those used in [91] to reduce this still further, and ultimately to prove the full conjecture.

In higher dimensions, finally, there is the problem of bounding the number of geometric permutations induced by hyperplane transversals of separated families of n compact convex translates. Just as in the plane the number of geometric permutations for a family of compact convex sets drops considerably when they form a set of translates, a similar result should hold in higher dimensions: there one should be able to reduce the n^{d-1} upper bound proved in [16] significantly in the case of translates.

In addition, there is the problem of finding a Helly-type theorem for hyperplane transversals of separated families of compact convex translates, which may well exist in higher dimensions just as it does in the plane.

5. Transversal Algorithms

Computer scientists have recently begun designing algorithms for finding k-transversals and for constructing the entire space of k-transversals, $\mathcal{T}_k^d(\mathcal{A})$, for a family $\mathcal{A}$ of objects in $\mathbb{R}^d$. To *construct* the space of transversals means to build a data structure that represents the simplicial complex forming the boundary of the space of transversals (see §3). Many different families $\mathcal{A}$ have been considered, including families of line segments, simplices, spheres, convex polytopes, and translates of a compact convex set.

The problem of finding point transversals for families of half-spaces lies at the center of the field of linear programming. We will not venture to discuss, even briefly, such a vast and deep area. It will suffice to mention that Megiddo showed how to solve linear programming problems with d variables and n constraints in $O(2^{2^d} n)$ time [76]. Thus point transversals for n half-spaces in $\mathbb{R}^d$ can be found in $O(2^{2^d} n)$ time which, for fixed d, is linear in n. (See [1,17,18,23] for subsequent improvements.)

After problems concerning point transversals, the problem of finding hyperplane transversals seems most tractable. Many of the algorithms for finding hyperplane transversals have been obtained with the help of the dualizing process described in §3. Given a family $\mathcal{A}$ of convex polytopes, for example, map the vertices of the polytopes to hyperplanes in dual space [29]. These hyperplanes form an arrangement, partitioning dual space into faces of various dimensions. The space of hyperplane transversals, $\mathcal{T}_{d-1}^d(\mathcal{A})$, maps to a union of faces in this arrangement. As mentioned in §3, if n_0 is the total number of vertices over all the polytopes in $\mathcal{A}$, then this arrangment has $O(n_0^d)$ complexity. An algorithm of Edelsbrunner, O'Rourke, and Seidel constructs the complete arrangement in optimal $O(n_0^d)$ time [29,33]. It is easy to modify this algorithm to determine which cells correspond to hyperplane transversals.

Theorem 5.1 [29,33]. *Let $\mathcal{A}$ be a family of convex polytopes having a total of n_0 vertices. Then $\mathcal{T}_{d-1}^d(\mathcal{A})$ can be constructed in $O(n_0^d)$ time.*

A different algorithm, using the linear time algorithm for linear programming in fixed dimension, was proposed by Avis and Doskas [7]. An *antipodal pair* of vertices on a convex polytope is a pair of vertices which lie on two parallel hyperplanes supporting the polytope. Each direction gives rise to an antipodal pair of vertices lying on supporting hyperplanes having the given direction as normal. A hyperplane intersects the convex polytope if and only if it separates the antipodal pair associated to its normal direction.

Instead of partitioning the dual space of hyperplanes, Avis and Doskas partition the space of normals to hyperplanes. Each edge in a convex polytope with endpoints $(\mu_1, \ldots \mu_d)$ and $(\eta_1, \ldots, \eta_d)$ is mapped to a hyperplane passing through the origin, $(\mu_1 - \eta_1)x_1 + \cdots + (\mu_d - \eta_d)x_d = 0$. The resulting arrangement of hyperplanes through the origin partitions the space of normal directions into $(d-1)$-dimensional cells in such a way that two normal directions in the same cell generate the same antipodal pair for each convex polytope in the family. Finding a hyperplane transversal with normal in a given cell is then equivalent to separating each of the antipodal pairs for the cell.

Finding a hyperplane which separates two sets of n points can be easily formulated as a problem in linear programming with d variables and $2n$ constraints. For fixed d this linear programming problem can be solved in linear time. Thus for each cell in the partition of normal directions, one can determine in linear time whether there exists a hyperplane transversal with normal in the given cell. Finally, we need only note that an arrangement of n_1 hyperplanes through the origin contains $O(n_1^{d-1})$ faces and can be constructed in $O(n_1^{d-1})$ time [29,33]. The result is

Theorem 5.2 [7]. *Let $\mathcal{A}$ be a family of n convex polytopes with a total of n_1 edges. A hyperplane transversal for $\mathcal{A}$, if one exists, can be found in $O(n\,n_1^{d-1})$ time.*

The *directions* determined by an edge are the directions of the two oriented lines containing the edge. The complexity of the algorithm by Avis and Doskas really depends on the number of different directions determined by the edges, not on the number of edges themselves. Even before this algorithm was found, Edelsbrunner noticed that if the number of such edge directions is bounded by a constant, then a hyperplane transversal for n convex polytopes can be found in $O(n)$ time. In particular, a hyperplane transversal for a family of n isothetic rectangles in $\mathbb{R}^d$, each the Cartesian product of d intervals on the d coordinate axes, can be found in linear time:

Theorem 5.3 [27, 28]. *Let $\mathcal{A}$ be a family of n convex polytopes with a total of $\hat{n}_1$ directions determined by polytope edges. A hyperplane transversal for $\mathcal{A}$, if one exists, can be found in $O(n\,\hat{n}_1^{d-1})$ time.*

The space of hyperplane transversals for a family of n simplices in $\mathbb{R}^d$ has at most $O(n^{d-1}\alpha(n))$ faces (Theorem 3.9). The algorithms of Edelsbrunner,

O'Rourke, and Seidel and of Avis and Doskas run in $O(n^d)$ time on simplices, off by a factor of roughly n. Better algorithms are known in lower dimensions.

One of the first papers on line transversal algorithms contained a simple "divide-and-conquer" algorithm, published in 1982 by Edelsbrunner, Maurer, Preparata, Rosenberg, Welzl, and Wood, for constructing the space of line transversals to n line segments in $\mathbb{R}^2$ [32]. This algorithm runs in $O(n \log n)$ time. Again, the line transversals are viewed as points in a dual space. Partition the n line segments into two groups of $n/2$ segments each. Recursively, run the algorithm on each group to calculate the space of transversals for each group. Finally, intersect the two spaces to form the space of transversals for the original n line segments.

As noted in § 3, the space of transversals corresponds, in the dual, to the set of points lying between the upper and lower envelopes of $\sigma^-(a_i)_{1 \le i \le n}$ and $\sigma^+(a_i)_{1 \le i \le n}$, respectively. The algorithm of Edelsbrunner *et al.* intersects two transversal spaces by merging their upper and lower envelopes. Since these upper and lower envelopes are graphs of continuous functions on $\mathbb{R}$, a planar sweep merges the envelopes in time proportional to the complexity of the envelopes.

The steps outlined here have become part of the general approach for many of the successful attacks on constructing the transversal space in two or three dimensions (as well as for many other problems related to upper and lower envelopes.) In $\mathbb{R}^2$, the main problem was to calculate the complexity of the envelopes: either the number of topological faces or, for convex polygons, the number of linear faces. The running time of the algorithm above is $\log n$ times that complexity. In $\mathbb{R}^3$, on the other hand, new techniques were needed for merging the upper and lower envelopes. Unfortunately, the techniques that were developed did not generalize to higher dimensions.

Using the complexity bounds in Theorem 3.10, the divide-and-conquer algorithm applied to a family of convex polygons with a total of n_0 vertices runs in $O(n_0 \, \alpha(n_0) \log n_0)$ time [31]. Using some additional data structures, Hershberger was able to improve this by a factor of $\alpha(n_0)$, giving an $O(n_0 \log n_0)$ time algorithm [50].

Atallah and Bajaj extended the divide-and-conquer algorithm of [32] to a family of simple convex sets whose boundaries intersect pairwise at most s times [5]. (A family of *simple* convex sets consists of sets with a constant size description which can be used to compute the intersection and common tangents of any pair in constant time.) Using Theorem 3.8, they showed that the algorithm runs in $O(\lambda_s(n) \log n)$ time on these families. In particular, since the boundaries of two circles in $\mathbb{R}^2$, the boundaries of two convex translates in $\mathbb{R}^2$, and the boundaries of two convex homothets in $\mathbb{R}^2$ can intersect in at most two components, the corresponding transversal spaces can be constructed in $O(\lambda_2(n) \log n) = O(n \log n)$ time. (A set b is a *homothet* of a set a if $b = ma + v$, $m \in \mathbb{R}$, $m > 0$, $v \in \mathbb{R}^d$.) In fact, well before the paper by Atallah and Bajaj, Edelsbrunner gave specialized algorithms running in $O(n \log n)$ time for finding line transversals for translates and homothets [27,28]. Bajaj and Li also gave a specialized algorithm running in $O(n \log n)$ time for finding line transversals for circles of equal radii [11].

By showing how to merge upper and lower envelopes in $\mathbb{R}^3$, Edelsbrunner, Guibas, and Sharir extended the divide-and-conquer algorithm of [32] to plane transversals of convex polytopes in $\mathbb{R}^3$ [31]. Their algorithm runs in time proportional to the worst-case complexity of the space of plane transversals (see Theorem 3.10). They were unable to improve upon the $O(n^d)$ time algorithms in higher dimensions, however.

These results are summarized below.

Theorem 5.4 [50]. *Let $\mathcal{A}$ be a family of convex polygons in $\mathbb{R}^2$ with a total of n_0 vertices. Then $T_1^2(\mathcal{A})$ can be constructed in $O(n_0 \log n_0)$ time.*

Theorem 5.5 [5]. *Let $\mathcal{A}$ be a family of n simple convex sets in $\mathbb{R}^2$ whose boundaries intersect pairwise in at most s components. Then $T_1^2(\mathcal{A})$ can be constructed in $O(\lambda_s(n) \log n)$ time. In particular, if $\mathcal{A}$ is a family of simple convex translates or of simple convex homothets in $\mathbb{R}^2$, then $T_1^2(\mathcal{A})$ can be constructed in $O(n \log n)$ time.*

Theorem 5.6 [31]. *Let $\mathcal{A}$ be a family of convex polytopes in $\mathbb{R}^3$ with a total of n_0 vertices. Then $T_2^3(\mathcal{A})$ can be constructed in $O(n_0^2 \alpha(n_0))$ time. If $\mathcal{A}$ consists of n line segments in $\mathbb{R}^3$, then $T_2^3(\mathcal{A})$ can be constructed in $O(n^2)$ time.*

Suppose $\mathcal{A}$ is a family of n convex polygons in $\mathbb{R}^2$. Since the full space $T_1^2(\mathcal{A})$ can have $\Omega(n)$ lines on its boundary and these lines appear in sorted order, it is simple to show that it takes $\Omega(n \log n)$ time to construct $T_1^2(\mathcal{A})$. On the other hand, it is conceivable that one element of $T_1^2(\mathcal{A})$ might be found in linear time. After all, while it takes $\Omega(n \log n)$ time to construct the convex polygon in which n half-planes meet, a single point of that polygon can be found in $O(n)$ time [22,75,76]. Avis, Robert, and Wenger, however, proved a lower time bound of $\Omega(n \log n)$ on an algebraic decision tree for finding a line transversal for a family of line segments in $\mathbb{R}^2$. They also proved the same lower bound for the problem of finding a line transversal for a family of circles in $\mathbb{R}^2$:

Theorem 5.7 [8]. *Finding a line transversal for a family of n line segments or a family of n circles in $\mathbb{R}^2$ takes $\Omega(n \log n)$ time on an algebraic decision tree.*

This $\Omega(n \log n)$ lower bound is perhaps not so surprising, since there is no Helly-type theorem for line transversals in the plane (see § 1). There is, however, a Helly-type theorem (Theorem 4.11) for line transversals to a planar family of pairwise disjoint compact convex translates. In addition, such a family has at most three geometric permutations independently of the size of the family (Theorem 4.13). Accordingly, Egyed and Wenger gave an $O(n)$ algorithm for finding a line transversal to a family of n disjoint convex translates. The only assumption about the translates is that they have an $O(n)$

size storage description, and that the common tangents to two translates can be found in $O(n^{1-(\log 3)/2})$ time. While the algorithm cannot construct a representation of the entire space of line transversals, it does find all of their slopes.

Theorem 5.8 [35]. *A line transversal for a family of n pairwise disjoint convex translates in $\mathbb{R}^2$, if one exists, can be found in $O(n)$ time.*

There is also a Helly-type theorem (Theorem 2.4) for the existence of a line transversal consistent with a given ordering for a family of pairwise disjoint convex sets in the plane. Accordingly, Egyed and Wenger gave a linear time algorithm for finding a line transversal consistent with a given ordering for a family of simple convex sets. This algorithm also finds all the slopes of such line transversals.

Theorem 5.9 [36]. *Let $\mathcal{A}$ be an ordered family of n pairwise disjoint simple convex sets in $\mathbb{R}^2$. A line transversal which meets the members of $\mathcal{A}$ in the given order, if one exists, can be found in $O(n)$ time.*

Faster algorithms are available in higher dimensions, as well, for finding hyperplane transversals to special families of sets. Houle, Imai, Imai, Robert, and Yamamoto give an algorithm for constructing the space of hyperplane transversals to a family of spheres in $\mathbb{R}^d$ as the intersection of a certain polytope and cylinder in $\mathbb{R}^{d+1}$ [53]. Constructing the polytope takes $O(n^{\lfloor d/2 \rfloor + 1})$ time, while intersecting the polytope and the cylinder increases the time complexity by a constant factor.

Theorem 5.10 [53]. *Let $\mathcal{A}$ be a family of n spheres in $\mathbb{R}^d$. Then $T_{d-1}^d(\mathcal{A})$ can be constructed in $O(n^{\lfloor d/2 \rfloor + 1})$ time.*

Cappell *et al.* prove that the space of hyperplane transversals to a $(d-2)$-separated family of n compact and strictly convex sets has $O(n^{d-1})$ topological faces (Theorem 3.6). If the sets are simple (see above), their proof extends to an $O(n^{d-1} \log^2 n)$ time algorithm for constructing the transversal space. If Theorem 3.6 could be proved for all $(d-2)$-separated families of compact convex sets, not just for strictly convex ones, then this algorithm would run in the same time on all such families.

Theorem 5.11 [16]. *Let $\mathcal{A}$ be a $(d-2)$-separated family of n simple, compact, and strictly convex sets in $\mathbb{R}^d$. Then $T_{d-1}^d(\mathcal{A})$ can be constructed in $O(n^{d-1} \log^2 n)$ time.*

Much progress has been made on algorithms for finding line transversals to convex polytopes in $\mathbb{R}^3$. Avis and Wenger gave the first $O(n^4 \log n)$ algorithm [10] which was improved upon to $O(n^4 \alpha(n))$ by McKenna and O'Rourke [74]. Very recently, Pellegrini and Shor gave an algorithm which runs in nearly $O(n^3)$ time.

Theorem 5.12 [81]. *Let $\mathcal{A}$ be a family of convex polytopes in $\mathbb{R}^3$ with a total of n_2 facets. Then $T_1^3(\mathcal{A})$ can be constructed in $c_\epsilon n_2^{3+\epsilon}$ time for any $\epsilon > 0$, where the constant c_ϵ depends on ϵ.*

Since the complexity of $T_1^3(\mathcal{A})$ is $\Omega(n^3)$ [80], their algorithm is nearly optimal.

The general problem of finding all k-transversals to a family of n polytopes in $\mathbb{R}^d$ can, of course, be formulated as the problem of solving a certain system of algebraic inequalities. The space of transversals can be constructed by general techniques in computer algebra for solving such systems. These techniques, however, while quite powerful, are in principle inefficient because of their generality, ignoring, as they do, the underlying geometric nature of the problem. Nevertheless, we do not know of algorithms to find k-transversals in $\mathbb{R}^d$ for arbitrary k and d which improve on these general techniques.

6. Other Directions

We have indicated above some problems arising in the areas covered by each of the preceding sections. In this section we will touch on some additional results on transversals which point to still other directions in which fruitful research may be pursued.

For $\rho \geq 1$, we say that a set in $\mathbb{R}^d$ is *ρ-stubby* if it is contained in a ball of radius ρ and contains a ball of radius one. Families of ρ-stubby sets behave in some respects like (solid) ellipsoids of bounded aspect, and reduce to spheres when $\rho = 1$. More to the point, any family of d-dimensional translates of a convex body can be mapped, via a linear transformation, into a family of d-stubby sets. (If E is the ellipsoid of smallest volume containing a d-dimensional convex body a, then the ellipsoid E' which is concentric and homothetic to E at the ratio $1/d$ is contained in a, by a theorem of John [54]; a linear transformation then sends these ellipsoids to spheres of radius d and 1, respectively.) Thus we should expect families of ρ-stubby sets to have many of the same properties that families of translates have. In fact, Katchalski and Lewis proved Theorem 4.10 by proving the more general

Theorem 6.1 [60]. *For any family $\mathcal{A}$ of pairwise disjoint ρ-stubby sets in $\mathbb{R}^2$, if $\mathcal{A}$ has property $\mathbf{T}(3)$, then $\mathcal{B}$ has property $\mathbf{T}$ for some subfamily $\mathcal{B}$ with $\mathrm{card}(\mathcal{A} \setminus \mathcal{B}) \leq q_\rho$, where q_ρ is a constant dependent only on ρ.*

In addition, Katchalski's first proof [58] of a Helly-type theorem for line transversals of translates (cf. Theorem 4.9) was accomplished by proving such a theorem for ρ-stubby sets. The proof depended also on bounding the number of geometric permutations for ρ-stubby sets:

Theorem 6.2 [58]. *A family of pairwise disjoint ρ-stubby sets in $\mathbb{R}^2$ has at most m_ρ geometric permutations, where m_ρ is a constant dependent only on ρ.*

Theorem 6.3 [58]. *For any family* $\mathcal{A}$ *of pairwise disjoint* ρ-*stubby sets in* $\mathbb{R}^2$, $\mathbf{T}(r_\rho) \Rightarrow \mathbf{T}$, *where* r_ρ *is a constant dependent only on* ρ.

Families of ρ-stubby sets also behave well with respect to transversals in higher dimensions. M. Perles [82], has proved the following Ramsey-type theorem for k-transversals of ρ-stubby sets in $\mathbb{R}^d$:

Theorem 6.4. *If* $\mathcal{A}$ *is a family of* ρ-*stubby sets in* $\mathbb{R}^d$, *then either* $\mathcal{A}$ *has an infinite subfamily which is* k-*separated or there is a finite number of* k-*flats whose union meets every set in* $\mathcal{A}$.

Since ρ-stubby sets seem to have transversal properties very similar to those of translates, all of the open questions about translates can reasonably be posed about families of ρ-stubby sets. In particular, is the number of geometric permutations for a $(d-2)$-separated family of ρ-stubby sets in $\mathbb{R}^d$ bounded, and is there a Helly-type theorem for ρ-stubby sets in $\mathbb{R}^d$?

As noted in §1, from the property $\mathbf{T}(r)$ we can not, in general, conclude the existence of a transversal, no matter how large r may be. In his dissertation, however, Eckhoff proved the following Gallai-type result, similar to Theorems 4.8 and 6.4:

Theorem 6.5 [24]. *If* $\mathcal{A}$ *is a family of compact convex sets in the plane satisfying* $\mathbf{T}(4)$, *then there are two lines whose union meets every member of* $\mathcal{A}$.

For a variant, see [96].

What if the family satisfies $\mathbf{T}(3)$? In this case, Eckhoff has shown [25], by means of an example consisting of 14 sets, that at least 3 lines are then required to cover $\mathcal{A}$, while Kramer [70] showed that 5 lines will always do; the latter has recently been improved to 4 by Eckhoff [26]. Are 4 necessary, in the worst case?

It would also be interesting to know to what extent these results can be generalized to higher dimensions.

Recently, Dol'nikov has generalized the concepts of transversals and disjointness in such a way that he was able to prove a theorem whose corollaries, for suitable values of the parameters, include Helly's theorem, the Borsuk-Lyusternik-Shnirel'man theorem on covering a sphere by sets of smaller diameter, the Borsuk-Ulam theorem, and in addition some new results about transversals (in the classical sense) as well as some purely combinatorial theorems.

In order to state his main result we need the following terminology. Let $\mathcal{P}(d, n)$ denote the space of real polynomials in d variables of degree at most n, whose linear dimension is $N = \binom{d+n}{d}$. We say that a polynomial p *separates* $V \subset \mathbb{R}^d$ if there are points $x, y \in V$ such that $p(x)p(y) \leq 0$. For a family

$\mathcal{A}$ of sets in $\mathbb{R}^d$, a subspace $L \subset \mathcal{P}(d,n)$ of dimension $N - k - 1$ is called a (k,n)-*transversal* of $\mathcal{A}$ if every polynomial $p \in L$ separates every member of $\mathcal{A}$. The family $\mathcal{A}$ is said to be *n-inseparable* if it has a $(0,n)$-transversal; otherwise $\mathcal{A}$ is said to *n-separable*.

It is easy to see that these concepts generalize those of transversals and disjointness by observing that

- A family of compact convex sets in $\mathbb{R}^d$ has a k-transversal if and only if it has a $(k,1)$-transversal.
- If a family of sets has nonempty intersection, then it is n-inseparable for every n.
- If a family $\mathcal{A}$ of closed sets in $\mathbb{R}^d$ is n-inseparable for $n \geq 2$ and all the polynomials belonging to its $(0,n)$ transversal have a common zero, then $\mathcal{A}$ has nonempty intersection.

Dol'nikov's main result is

Theorem 6.6 [21]. *Let $\mathcal{A}_i$, $1 \leq i \leq k + 1$, $0 \leq k \leq N - 2$ be families of sets in $\mathbb{R}^d$, and suppose that in each family $\mathcal{A}_i$, every subfamily of size $l \leq N - k$ is n-inseparable. Then the family $\mathcal{A} = \cup \mathcal{A}_i$ has a (k,n)-transversal.*

As a special case of Theorem 6.6 relating to "ordinary" transversals, we mention:

Theorem 6.7. *Given families $\mathcal{A}_i$ of convex sets in $\mathbb{R}^d$, where $1 \leq i \leq k + 1$ and $0 \leq k \leq d - 1$, suppose that in each family any $d - k + 1$ sets have a common point. Then the family $\mathcal{A} = \cup_{i=1}^{k+1} \mathcal{A}_i$ has a k-transversal.*

In view of the wide-ranging scope of Theorem 6.6, we believe that these ideas should certainly be pursued. It would be interesting, in particular, to know what additional results can be obtained if we adjoin the concept of order type to Dol'nikov's ideas.

Finally, in [12] and [13], Bárány, Katchalski, and Pach have introduced various quantitative versions of Helly's theorem. One example of a theorem of this type is their

Theorem 6.8 [13]. *There is a constant $c(d)$ such that a family of convex sets in $\mathbb{R}^d$ has intersection with volume at least 1 if every subfamily of size $2d$ has intersection with volume at least $c(d)$.*

There are similar results for surface area, diameter, and so on.

Pach has asked [personal communication] whether one can prove quantitative theorems of this type for transversals. For example, might it be true that there are constants $k(d), c(d)$ such that a family of convex sets has a hyperplane transversal which intersects each member of the family in a set of $(d - 1)$-volume at least 1 if every $k(d)$-element subfamily has a transversal which intersects each of the $k(d)$ members in a set of $(d-1)$-volume at least $c(d)$, and all of these transversals meet the corresponding subfamilies consistently with a fixed order type?

References

[1] I. Adler and R. Shamir: A randomization scheme for speeding up algorithms for linear and convex programming problems with high constraints-to-variables ratio. Tech. Rep. TR-89-7, DIMACS, 1989

[2] P. Agarwal, M. Sharir and P. Shor: Sharp upper and lower bounds on the length of general Davenport-Schinzel sequences. J. Comb. Theory, Ser. A **52** (1989) 228–274

[3] B. Aronov, J. E. Goodman, R. Pollack and R. Wenger: Hadwiger's theorem for line transversals does not generalize to higher dimensions. Unpublished manuscript

[4] M. Atallah: Dynamic computational geometry. Comput. Math. Appl. **11** (1985) 1171–1181

[5] M. Atallah and C. Bajaj: Efficient algorithms for common transversals. Inform. Process. Lett. **25** (1987) 87–91

[6] F. Aurenhammer: Power diagrams: properties, algorithms and applications. SIAM J. Comput. **16** (1987) 78–96

[7] D. Avis and M. Doskas: Algorithms for high dimensional stabbing problems. Discrete Appl. Math. **27** (1990) 39–48

[8] D. Avis, J. Robert and R. Wenger: Lower bounds for line stabbing. Inform. Process. Lett. **33** (1989) 59–62

[9] D. Avis and R. Wenger: Algorithms for line stabbers in space. In: Proc. Third ACM Symp. on Computational Geometry, 1987, pp. 300–307

[10] D. Avis and R. Wenger: Polyhedral line transversals in space. Discrete Comput. Geom. **3** (1988) 257–265

[11] C. Bajaj and M. Li: On the duality of intersection and closest points. In: Proc. 21st Annual Allerton Conf. 1983, pp. 459–461

[12] I. Bárány, M. Katchalski and J. Pach: Quantitative Helly-type theorems. Proc. Amer. Math. Soc. **86** (1982) 109–114

[13] I. Bárány, M. Katchalski and J. Pach: Helly's theorem with volumes. Amer. Math. Monthly **91** (1984) 362–365

[14] A. Björner, M. Las Vergnas, B. Sturmfels, N. White and G. Ziegler: Oriented matroids. Cambridge University Press, Cambridge 1992

[15] R.G. Bland and M. Las Vergnas: Orientability of matroids. J. Combin. Theory, Ser. B **23** (1978) 94–123

[16] S. E. Cappell, J. E. Goodman, J. Pach, R. Pollack, M. Sharir and R. Wenger: The combinatorial complexity of hyperplane transversals. In: Proc. Sixth Annual ACM Symp. on Computational Geometry, 1990, pp. 83–91

[17] K. L. Clarkson: Linear programming in $O(n3^{d^2})$ time. Inform. Process. Lett. **22** (1986) 21–24

[18] K. L. Clarkson: A Las Vegas algorithm for linear programming when the dimension is small. In: Proc. 29th Annual Symp. on the Foundations of Computer Science, 1989, pp. 452–456

[19] L. Danzer: Über ein Problem aus der kombinatorischen Geometrie. Arch. Math. **8** (1957) 347–351

[20] L. Danzer, B. Grünbaum and V. Klee: Helly's theorem and its relatives. In: Convexity, Proc. Symp. Pure Math., vol. 7. Amer. Math. Soc., Providence 1963, pp. 100–181

[21] V. L. Dol'nikov: Generalized transversals of families of sets in $\mathbb{R}^n$ and connections between the Helly and Borsuk theorems. Soviet Math. Dokl. **36** (1988) 519–522

[22] M.E. Dyer: Linear time algorithms for two- and three-variable linear programs. SIAM J. Comput. **13** (1984) 31–45

[23] M.E. Dyer: On a multidimensional search technique and its application to the Euclidean one-centre problem. SIAM J. Comput. **15** (1986) 725–738

[24] J. Eckhoff: Transversalenprobleme vom Gallai'schen Typ. Ph.D. dissertation, Georg-August-Universität, Göttingen, 1969

[25] J. Eckhoff: Transversalenprobleme in der Ebene. Arch. Math. **24** (1973) 195–202

[26] J. Eckhoff: A transversal problem in the plane. Discrete Comput. Geom. (To appear)

[27] H. Edelsbrunner: Intersection problems in computational geometry. Ph.D. dissertation, Tech. Univ. Graz, 1982

[28] H. Edelsbrunner: Finding transversals for sets of simple geometric figures. Theor. Comput. Science **35** (1985) 55–69

[29] H. Edelsbrunner: Algorithms in Combinatorial Geometry. Springer, Berlin 1987

[30] H. Edelsbrunner: The upper envelope of piecewise linear functions: tight bounds on the number of faces. Discrete Comput. Geom. **4** (1989) 337–343

[31] H. Edelsbrunner, L. J. Guibas and M. Sharir: The upper envelope of piecewise linear functions: algorithms and Applications. Discrete Comput. Geom. **4** (1989) 311–336

[32] H. Edelsbrunner, H. A. Maurer, F. P. Preparata, A. L. Rosenberg, E. Welzl and D. Wood: Stabbing line segments. BIT **22** (1982) 274–281

[33] H. Edelsbrunner, J. O'Rourke and R. Seidel: Constructing arrangements of lines and hyperplanes with applications. SIAM J. Comput. **15** (1986) 341–363

[34] H. Edelsbrunner and M. Sharir: The maximum number of ways to stab n convex non-intersecting objects in the plane is $2n-2$. Discrete Comput. Geom. **5** (1990) 35–42

[35] P. Egyed and R. Wenger: Stabbing pairwise disjoint translates in linear time. Proc. Fifth Annual ACM Symp. on Comput. Geometry 1989, pp. 364–369

[36] P. Egyed and R. Wenger: Ordered stabbing of pairwise disjoint convex sets in linear time. Discrete Appl. Math. (To appear)

[37] J. Folkman and J. Lawrence: Oriented matroids. J. Combin. Theory, Ser. B **25** (1978) 199–236

[38] J. E. Goodman and R. Pollack: Multidimensional sorting. SIAM J. Comput. **12** (1983) 484–507

[39] J. E. Goodman and R. Pollack: Hadwiger's transversal theorem in higher dimensions. J. Amer. Math. Soc. **1** (1988) 301–309

[40] B. Grünbaum: On a theorem of L. Santaló. Pacific J. Math. **5** (1955) 351–359

[41] B. Grünbaum: On common transversals. Arch. Math. **9** (1958) 465–469

[42] B. Grünbaum: Common secants for families of polyhedra. Arch. Math. **15** (1964) 76–80

[43] R. H. Güting: Stabbing c-oriented polygons. Inform. Process. Lett. **16** (1983) 35–40

[44] H. Hadwiger: Ungelöste Probleme, No. 7. Elem. Math. **10** (1955) 111

[45] H. Hadwiger: Über Eibereiche mit gemeinsamer Treffgeraden. Portugal. Math. **16** (1957) 23–29

[46] H. Hadwiger and H. Debrunner: Ausgewählte Einzelprobleme der kombinatorischen Geometrie in der Ebene. L'Enseignement Math. **1** (1955) 56–89

[47] H. Hadwiger, H. Debrunner and V. Klee: Combinatorial geometry in the plane. Holt, Rinehart and Winston, New York, 1964

[48] S. Hart and M. Sharir: Nonlinearity of Davenport-Schinzel sequences and of generalized path compression schemes. Combinatorica **6** (1986) 151–177

[49] E. Helly: Über Mengen konvexer Körper mit gemeinschaftlichen Punkten. Jber. Deutsch. Math. Verein, **32** (1923) 175–176

[50] J. Hershberger: Finding the upper envelope of n line segments in $O(n \log n)$ time. Inform. Process. Lett. (To appear)

[51] A. Horn: Some generalizations of Helly's theorem on convex sets. Bull. Amer. Math. Soc. **55** (1949) 923–929

[52] M. E. Houle, H. Imai, K. Imai and J.-M. Robert: Weighted orthogonal linear L_∞-approximation and applications. In: Algorithms and data structures. Springer Lecture Notes in Computer Science, vol. 382. Springer, Berlin Heidelberg 1989, pp. 183–191

[53] M. E. Houle, H. Imai, K. Imai, J. Robert and P. Yamamoto: Orthogonal Weighted Linear L_1 and L_∞-Approximation and Applications. Discrete Appl. Math. to appear

[54] F. John: Extremum problems with inequalities as subsidiary conditions. In: Studies and Essays Presented to R. Courant on His 60th Birthday, Interscience, New York, 1948, pp. 187–204

[55] H. W. E. Jung: Über die kleinste Kugel, die eine räumliche Figur einschließt. J. Reine Angew. Math. **123** (1901) 241–257

[56] M. Katchalski: A Helly type theorem for convex sets. Canad. Math. Bull. **21** (1978) 121–123

[57] M. Katchalski: Thin sets and common transversals. J. Geometry **14** (1980) 103–107

[58] M. Katchalski: On a conjecture of Grünbaum on common transversals. Tech. Rep. MT-620, Technion, 1983

[59] M. Katchalski: A conjecture of Grünbaum on common transversals. Math. Scand. **59** (1986) 192–198

[60] M. Katchalski and T. Lewis: Cutting families of convex sets. Proc. Amer. Math. Soc. **79** (1980) 457–461

[61] M. Katchalski and T. Lewis: Cutting rectangles in the plane. Discrete Math. **42** (1982) 67–71

[62] M. Katchalski, T. Lewis and A. Liu: Geometric permutations and common transversals. Discrete Comput. Geom. **1** (1986) 371–377

[63] M. Katchalski, T. Lewis and A. Liu: Geometric permutations of disjoint translates of convex sets. Discrete Math. **65** (1987) 249–259

[64] M. Katchalski, T. Lewis and A. Liu: The different ways of stabbing disjoint convex sets. Discrete Comput. Geom. **7** (1992) 197–206

[65] K. Kedem, R. Livne, J. Pach and M. Sharir: On the Union of Jordan Regions and Collision-Free Translational Motion Amidst Polygonal Obstacles. Discrete Comput. Geom. **1** (1986) 59–71

[66] M. Katchalski, T. Lewis and J. Zaks: Geometric permutations for convex sets. Discrete Math. **54** (1985) 271–284

[67] M. Katchalski and A. Liu: Symmetric twins and common transversals. Pacific J. Math. **86** (1980) 513–515

[68] P. Kirchberger: Über Tschebyschefsche Annäherungsmethoden. Math. Ann. **57** (1903) 509–540

[69] V. Klee: Common secants for plane convex sets. Proc. Amer. Math. Soc. **5** (1954) 639–641

[70] D. Kramer: Transversalenprobleme vom Helly'schen und Gallai'schen Typ. Ph.D. dissertation, Universität Dortmund, 1974

[71] S. Lay: Convex Sets. John Wiley & Sons, New York, 1982

[72] T. Lewis: Two counterexamples concerning transversals for convex subsets of the plane. Geom. Dedicata **9** (1980) 461–465

[73] A. Mandel: Topology of Oriented Matroids. Ph.D. dissertation, University of Waterloo, 1982

[74] M. McKenna and J. O'Rourke: Arrangements of lines in 3-space: a data structure with applications. Proc. Fourth ACM Symp. on Computational Geometry, (1988) 371–380

[75] N. Megiddo: Linear-time algorithms for linear programming in $\mathbb{R}^3$ and related problems. SIAM J. Comput. **12** (1983) 759–776

[76] N. Megiddo: Linear programming in linear time when the dimension is fixed. J. Assoc. Comput. Mach. **31** (1984) 114–127

[77] J.R. Munkres: Elements of Algebraic Topology. Addison Wesley, Menlo Park, 1984

[78] J. O'Rourke: An on-line algorithm for fitting straight lines between data ranges. Comm. Assoc. Comput. Mach. **24** (1981) 574–578

[79] J. Pach and M. Sharir: The upper envelope of piecewise linear functions and the boundary of a region enclosed by convex plates: combinatorial analysis. Discrete Comput. Geom. **4** (1989) 291–309

[80] M. Pellegrini: Stabbing and ray shooting in 3 dimensional space. In: Proc. Sixth Annual ACM Symp. on Computational Geometry, 1990, 177–186

[81] M. Pellegrini and P. Shor: Finding Stabbing Lines in 3-Dimensional Space. Discrete Comput. Geom. **8** (1992) 191–208

[82] M. Perles: Personal communication

[83] R. Pollack and R. Wenger: A family of generalizations of Hadwiger's theorem to hyperplane transversals. Unpublished manuscript

[84] R. Pollack and R. Wenger: Necessary and sufficient conditions for hyperplane transversals. Combinatorica **10** (1990) 307–311

[85] F. P. Preparata and M. I. Shamos: Computational Geometry. Springer New York, 1985

[86] P. Ramanan: Obtaining lower bounds using artificial components. Inform. Process. Lett. **24** (1987) 243–246

[87] P. Ramanan: $\Omega(n \log n)$ lower bound for computing the width of a planar point set. Tech. Rep. TRCS-88-14, University of California, Santa Barbara, 1988

[88] L. Santaló: Un teorema sóbre conjuntos de paralelepipedos de aristas paralelas. Publ. Inst. Mat. Univ. Nac. Litoral **2** (1940) 49–60

[89] Z. Tuza: The minimum number of elements representing a set system of given rank. J. Combin. Theory, Ser. A **52** (1989) 84–89

[90] H. Tverberg: A separation property of plane convex sets. Math. Scand. **45** (1979) 255–260

[91] H. Tverberg: Proof of Grünbaum's conjecture on common transversals for translates. Discrete Comput. Geom. **4** (1989) 191–203

[92] H. Tverberg: On geometric permutations and the Katchalski-Lewis conjecture on partial transversals for translates. Discrete and Computational Geometry: Papers from the DIMACS Special Year, Amer. Math. Soc., Providence, 1991

[93] F.A. Valentine: The dual cone and Helly type theorems. In: Convexity, Proc. Symp. Pure Math. Amer. Math. Soc., vol. 7, Providence, (1963) 473–493

[94] P. Vincensini: Figures convexes et variétés linéaires de l'espace euclidien à n dimensions. Bull. Sci. Math. **59** (1935) 163–174

[95] P. Vincensini: Les ensembles d'arcs d'un même cercle dans leurs relations avec les ensembles de corps convexes du plan euclidien. Atti IV Congr. Un. Mat. Ital. **2** (1953) 456–464

[96] B. Wegner: Ein ebenes Transversalenproblem. Monatsh. Math. **77** (1973) 72–81

[97] R. Wenger: A generalization of Hadwiger's theorem to intersecting sets. Discrete Comput. Geom. **5** (1990) 383–388

[98] R. Wenger: Upper bounds on geometric permutations for convex sets. Discrete Comput. Geom. **5** (1990) 27–33

[99] R. Wenger: Geometric permutations and connected components. Tech.Rep. TR-90-50, DIMACS, 1990

Chapter VIII. Hadwiger-Levi's Covering Problem Revisited

Károly Bezdek

0. Introduction

A well-known problem of discrete geometry is due to Hadwiger (1957), (1960) and Levi (1955). The following conjecture concerning this problem was published by Hadwiger (1957), (1960) and also by Gohberg and Markus (1960): Any convex body of $\mathbf{E}^d$, $d \geq 1$ (i.e. any compact convex subset of the d-dimensional Euclidean space $\mathbf{E}^d$ with non-empty interior) can be covered by 2^d smaller homothetic bodies and equality is attained only for d-dimensional parallelotopes. This conjecture has stimulated a lot of research in geometry. To survey the basic results we need some simple definitions.

Let $\mathbf{K} \subset \mathbf{E}^d$, $d \geq 1$ be a d-dimensional closed convex set. We say that a non-zero vector i.e. a direction $\underline{v} \in \mathbf{E}^d$ (a point i.e. a light-source $L \in \mathbf{E}^d \setminus \mathbf{K}$, resp.) illuminates the boundary point P of $\mathbf{K}$ ($P \in$ bd $\mathbf{K}$, resp.) if and only if the open ray $r_{\underline{v}}^P$ ($r_{\overrightarrow{LP}}^P$, resp.) emanating from P having direction vector $\underline{v}$ ($\overrightarrow{LP}$, resp.) has a non-empty intersection with the interior of $\mathbf{K}$ (int $\mathbf{K}$, resp.). Furthermore, we say that the directions $\underline{v}_1, \underline{v}_2, \ldots, \underline{v}_n$ (the light-sources $\{L_1, L_2, \ldots, L_n\} \subset \mathbf{E}^d \setminus \mathbf{K}$, resp.) illuminate bd $\mathbf{K}$ if and only if every boundary point of $\mathbf{K}$ is illuminated by at least one of the directions $\underline{v}_1, \underline{v}_2, \ldots, \underline{v}_n$ (the light-sources $L_1, L_2, \ldots, L_n$, resp.). Let

(1) $\mathcal{H}(\mathbf{K})$ = the smallest number of smaller homothetic copies of $\mathbf{K}$ with which it is possible to cover $\mathbf{K}$;

(2) $I_0(\mathbf{K})$ = the smallest number of light-sources lying outside $\mathbf{K}$ which illuminate the boundary of $\mathbf{K}$;

(3) $I_\infty(\mathbf{K})$ = the smallest number of directions which illuminate the boundary of $\mathbf{K}$.

The planar case of Hadwiger-Levi's covering problem was solved by Levi (1955). He showed that if $\mathbf{B}$ is a planar convex domain other than a parallelogram, then $\mathcal{H}(\mathbf{B}) = 3$; if $\mathbf{B}$ is a parallelogram, then $\mathcal{H}(\mathbf{B}) = 4$. The same result was proved independently by Gohberg and Markus (1960) and also by Boltjansky (1960) and it is reproved by a new method in Lemma 7. Boltjansky (1960) has proved that $\mathcal{H}(\mathbf{B}) = I_\infty(\mathbf{B})$ for any convex body $\mathbf{B}$ of $\mathbf{E}^d$, $d \geq 1$. Soltan (1963) has shown the inequalities $I_\infty(\mathbf{K}) \leq I_0(\mathbf{K}) \leq \mathcal{H}(\mathbf{K})$ for any closed convex set $\mathbf{K} \subset \mathbf{E}^d$, $d \geq 1$ with non-empty interior. He has observed that $I_\infty(\mathbf{B}) = I_0(\mathbf{B}) = \mathcal{H}(\mathbf{B})$ for every convex body $\mathbf{B}$ of $\mathbf{E}^d$, $d \geq 1$. Furthermore, Soltan (1963) has proved that $\mathcal{H}(\mathbf{K})$ is finite if and only if $\mathbf{K}$ is almost conic (see Definition 3). It is a surprising result of Boltjansky and Soltan (1978) that $I_0(\mathbf{K})$ is finite for a d-dimensional closed convex set $\mathbf{K} \subset \mathbf{E}^d$, $d \geq 1$ if and only if $\mathbf{K}$ is almost conic. We reprove this result (see Theorem 4 and Lemma 5) introducing the notion of almost bounded sets (see Definition 2) and using a new method which we call separation by hyperplanes. Section 1 discusses the method in detail (see Lemma 3, Corollary 1, Theorem 1, Lemma 4, Corollary 2, Theorem 2 and Theorem 3). Throughout this paper we use this method to prove new results and also to reprove some old ones.

Section 2 deals with the following general illumination problem. Let $\mathbf{K} \subset \mathbf{E}^d$, $d \geq 1$ be a d-dimensional closed convex set. We say that the affine subspace $L \subset \mathbf{E}^d \setminus \mathbf{K}$ of dimension $0 \leq \dim L \leq d-1$ illuminates the boundary point P of $\mathbf{K}$ if and only if there exists a point of L which illuminates P. Furthermore, we say that the affine subspaces $L_1, L_2, \ldots, L_n \subset \mathbf{E}^d \setminus \mathbf{K}$ illuminate bd $\mathbf{K}$ if and only if every boundary point of $\mathbf{K}$ is illuminated by at least one of the affine subspaces $L_1, L_2, \ldots, L_n$. Finally, let

(4) $I_l(\mathbf{K})$ = the smallest number of affine subspaces of dimension l, $0 \leq l \leq d-1$ lying in $\mathbf{E}^d \setminus \mathbf{K}$ which illuminate the boundary of $\mathbf{K}$.

Obviously, $I_{d-1}(\mathbf{K}) \leq I_{d-2}(\mathbf{K}) \leq \ldots \leq I_0(\mathbf{K}) \leq \mathcal{H}(\mathbf{K})$. The problem of finding upper bounds for $I_l(\mathbf{K})$ is discussed in Theorem 6 (which is a generalization of Theorem 4) and in Theorem 7. (In the proofs we apply a generalized version (Theorem 5) of our separation technique.) Theorem 8 (b) and (c) show that it is "hard" to find a "good" upper bound on $I_\infty(\mathbf{B}) = I_0(\mathbf{B}) = \mathcal{H}(\mathbf{B})$ for convex bodies $\mathbf{B}$ of $\mathbf{E}^d$. Namely, it was conjectured in Grünbaum (1967), pp. 422–423 that if the directions $\underline{v}_1, \underline{v}_2, \ldots, \underline{v}_n \in \mathbf{E}^d$ illuminate bd $\mathbf{B}$ such that removing any of them the remaining directions do not illuminate bd $\mathbf{B}$, then $n \leq 2(2^d - 1)$. Theorem 8 (b) gives a counter-example to that conjecture. Theorem 8 (c) was already stated in Grünbaum (1967), pp. 423. Theorem 8 (a) about k−fold illumination can also be formulated as a Helly-type statement on the sphere.

One of the most important corollaries of the theorems of Section 1 is Lemma 6 of Section 3. We say that an interior point O of the convex body $\mathbf{B}$ of $\mathbf{E}^d$, $d \geq 1$ is strictly separated by the hyperplanes $H_1, H_2, \ldots, H_n$ from the faces of $\mathbf{B}$ ($F \neq \emptyset$ is a face of the convex body $\mathbf{B} \subset \mathbf{E}^d$ if and only if there exist a hyperplane H which is called a supporting hyperplane and a closed half-space H^+ which is called a supporting half- space bounded by H such that $F = H \cap \mathbf{B}$ and $H^+ \supset \mathbf{B}$) if and only if every face of $\mathbf{B}$ is strictly separated by at least one of the hyperplanes $H_1, H_2, \ldots, H_n$ from the point O. (Two sets of $\mathbf{E}^d$ are strictly separated by a hyperplane H if and only if they are disjoint from H and lie on different sides of H). If $O(\in \text{int } \mathbf{B})$ is the origin of $\mathbf{E}^d$, $d \geq 1$, then the set $\mathbf{B}^* = \{X \in \mathbf{E}^d | \langle \overrightarrow{OX}, \overrightarrow{OY} \rangle \leq 1 \text{ for all } Y \in \mathbf{B}\}$ is called the polar body of $\mathbf{B}$. (Here $\langle ., . \rangle$ denotes the usual dot product of $\mathbf{E}^d$.) Lemma 6 of Section 3 claims that if $\mathbf{B}$ is a convex body of $\mathbf{E}^d$, $d \geq 1$ that contains the origin O as an interior point, then $I_\infty(\mathbf{B}) = I_0(\mathbf{B}) = \mathcal{H}(\mathbf{B})$ is the smallest number of hyperplanes of $\mathbf{E}^d$ which strictly separate O from the faces of the polar body $\mathbf{B}^*$. (I thank M. Lassak and V. P. Soltan for calling my attention to the paper P. S. Soltan and V. P. Soltan (1986). It contains the idea of Corollary 2 and Theorem 2 and so it can lead to a statement similar to Lemma 6.) Lemma 6 has many applications. Lemma 7 mentiones some. It is worth mentioning that other proofs of Lemma 7 can be found in Boltjansky and Gohberg (1985), Boltjansky and Soltan (1987) and Martini (1985).

Borsuk's conjecture (Borsuk (1933), Grünbaum (1963) and Hadwiger (1945/46)) that any set of $\mathbf{E}^d$ of diameter 1 can be partitioned into $d+1$ subsets of diameter < 1 also gave some attention to Hadwiger–Levi's covering problem. For example Boltjansky (1960) proved that if the convex body $\mathbf{B}$ of $\mathbf{E}^d$, $d \geq 3$ has no more than d corners (i.e. boundary points with non-uniquely determined supporting planes), then $\mathcal{H}(\mathbf{B}) = d+1$. (See Weissbach (1981) for a related result). Thus Borsuk's conjecture is true for this particular $\mathbf{B}$. (Grünbaum (1967), pp. 420-421 and Schramm (1988) give more details of this relation. However, the very recent paper of Kahn and Kalai (1992) gives a counter-example to Borsuk's conjecture.) Theorem 9 of Section 4 was first proved by Charazishvili (1973) but also it is in Boltjansky and Soltan (1978). This result is an extension of the above theorem of Boltjansky and shows the surprising feature of convex bodies with finitely many corner points. Our proof is based on the separation technique of Section 1.

There are upper bounds on $\mathcal{H}(\mathbf{B})$

- for convex bodies of $\mathbf{E}^d$ (see Lassak (1988) and Lassak (Tagungsbericht Oberwolfach 32/1984)),
- for centrally symmetric convex bodies of $\mathbf{E}^d$ (see Rogers, Levin and Petunin cited in Boltjansky and Gohberg (1985), pp. 45),
- for convex bodies with constant width of $\mathbf{E}^d$ (see Schramm (1988)) and
- for zonoids of $\mathbf{E}^d$ (Boltjansky and Soltan (1990)).

However, Hadwiger-Levi's covering problem remains unsolved for $d \geq 3$. Lassak (1984) (P. S. Soltan and V. P. Soltan (1986), resp.) proved that if $\mathbf{B}$ is a centrally symmetric convex body (centrally symmetric convex polyhedron, resp.) of $\mathbf{E}^3$, then $\mathcal{H}(\mathbf{B}) \leq 8$. The main result of Section 5 is Theorem 11. It claims that if $\mathbf{P}$ is a convex polyhedron of $\mathbf{E}^3$ with affine symmetry (i.e. the affine symmetry group of $\mathbf{P}$ consists of the identity and at least one other affinity of $\mathbf{E}^3$), then $\mathcal{H}(\mathbf{P}) \leq 8$. We give an outline of the proof. For the complete proof see Bezdek (1991/a).

Finally, we mention that there are further problems related to Hadwiger-Levi's covering problem which we did not survey. For the terminology and details, the reader should consult Bezdek (1983), Bezdek (1991/e), Buchman and Valentine (1970), (1976), Fejes Tóth (1977), Lassak (1985), (1986) and Martini (1985).

1. On $I_0(K)$ and $I_\infty(K)$

The following lemma is a more general version of Lemma 6, 7 and 8 of Mc-Mullen and Shephard (1971), pp. 64-65.

Lemma 1. *Let* $K \subset E^d, d \geq 1$ *be a closed convex set that contains the origin* O *and let* $\mathcal{F} \neq \emptyset$ *be the set of all faces of* K *which do not contain* O. *Then the polar set* $K^* = \{X \in E^d | \langle \overrightarrow{OX}, \overrightarrow{OY} \rangle \leq 1 \text{ for all } Y \in K\}$ *is a closed convex set of* E^d *with* $O \in K^*$. *If* $\mathcal{F}^*$ *denotes the set of all faces of* K^* *which are disjoint from* O, *then the map*

$$* : \mathcal{F} \to \mathcal{F}^*$$

$$F \mapsto F^* = \{X \in K^* | \langle \overrightarrow{OX}, \overrightarrow{OY} \rangle = 1 \text{ for all } Y \in F\}$$

is a one-to-one map between $\mathcal{F}$ *and* $\mathcal{F}^*$ *and it is inclusion reversing.*

Proof. First, we prove that F^* is a face of K^* with $O \notin F^*$. Since $F \in \mathcal{F}$ is a face of K with $O \notin F$ therefore there exists a supporting hyperplane $H = \{Y \in E^d | \langle \overrightarrow{OY}, \overrightarrow{OX_o} \rangle = 1\}$ of K with $H \cap K = F$ and $O \in K \subset H^+ = \{Y \in E^d | \langle \overrightarrow{OY}, \overrightarrow{OX_o} \rangle \leq 1\}$. Consequently, $X_o \in F^*$, i.e. $F^* \neq \emptyset$. Now let Y_o be a relative interior point of F i.e. $Y_o \in \text{relint } F$. Then $\mathbf{H} = \{X \in E^d | \langle \overrightarrow{OX}, \overrightarrow{OY_o} \rangle = 1\}$ is a supporting hyperplane of K^* because $K^* \subset \mathbf{H}^+ = \{X \in E^d | \langle \overrightarrow{OX}, \overrightarrow{OY_o} \rangle \leq 1\}$ and $(\emptyset \neq) F^* \subset F' = \mathbf{H} \cap K^*$. We prove that also $F^* \supset F'$ which then implies $F^* = F'$ (finishing the proof of the fact that F^* is a face of K^* with $O \notin F^*$.) Suppose that there exists $X_o \in F' \setminus F^*$. Then we have a point $Y_1 \in F$ such that $\langle \overrightarrow{OX_o}, \overrightarrow{OY_1} \rangle < 1$. Since $Y_1 \neq Y_o$ and $Y_o \in \text{relint } F$ therefore there exists a point $Y_2 \in F$ with $\overrightarrow{OY_o} = \lambda \cdot \overrightarrow{OY_1} + (1 - \lambda) \cdot \overrightarrow{OY_2}$, $0 < \lambda < 1$. But $\langle \overrightarrow{OX_o}, \overrightarrow{OY_2} \rangle \leq 1$ consequently, $\langle \overrightarrow{OX_o}, \overrightarrow{OY_o} \rangle = \lambda \cdot \langle \overrightarrow{OX_o}, \overrightarrow{OY_1} \rangle + (1 - \lambda) \cdot \langle \overrightarrow{OX_o}, \overrightarrow{OY_2} \rangle < \lambda + 1 - \lambda = 1$, a contradiction.

Secondly, we observe that $(K^*)^* = K$. If Y is an arbitrary point of K, then $\langle \overrightarrow{OX}, \overrightarrow{OY} \rangle \leq 1$ for all $X \in K^*$. Hence $K \subset (K^*)^*$. We prove that $K \supset (K^*)^*$. Let $Y_o \in E^d \setminus K$. So there exsists a hyperplane $H = \{Y \in E^d | \langle \overrightarrow{OY}, \overrightarrow{OX_o} \rangle = 1\}$ which strictly separates Y_o from $(O \in)K$, i.e. $\langle \overrightarrow{OY_o}, \overrightarrow{OX_o} \rangle > 1$ and $\langle \overrightarrow{OY}, \overrightarrow{OX_o} \rangle < 1$ for all $Y \in K$. But then $X_o \in K^*$ and so $Y_o \in E^d \setminus (K^*)^*$.

We finish the proof of Lemma 1 showing that $(F^*)^* = F$ for any face $F \in \mathcal{F}$. We know that $(F^*)^* = \{Y \in (\mathbf{K}^*)^* | \langle \overrightarrow{OY}, \overrightarrow{OX} \rangle = 1$ for all $X \in F^*\} = \{Y \in \mathbf{K} | \langle \overrightarrow{OY}, \overrightarrow{OX} \rangle = 1$ for all $X \in F^*\} \supset F$. We have to show that $(F^*)^* \subset F$. We have seen above that $F = H \cap \mathbf{K}$ with $H = \{Y \in \mathbf{E}^d | \langle \overrightarrow{OY}, \overrightarrow{OX_o} \rangle = 1\}$ and $\mathbf{K} \subset H^+ = \{Y \in \mathbf{E}^d | \langle \overrightarrow{OY}, \overrightarrow{OX_o} \rangle \leq 1\}$. Hence $X_o \in F^*$. So if $Y_o \in \mathbf{K} \setminus F$, then $\langle \overrightarrow{OY_o}, \overrightarrow{OX_o} \rangle < 1$ i.e. $Y_o \in \mathbf{K} \setminus (F^*)^*$. $\qquad \square$

Then we need

Definition 1. Let $\mathbf{K} \subset \mathbf{E}^d, d \geq 1$ be a closed convex set and $P(F,$ resp.$)$ be a boundary point (face, resp.) of $\mathbf{K}$. If O denotes the origin of $\mathbf{E}^d$, then we define the following closed convex cones:

$\overline{C}_P = \cap\{H^+ | H^+$ is a supporting half-space to $\mathbf{K}$ bounded by H with $P \in H\}$,

$C_P = \overrightarrow{PO} + \overline{C}_P$ the translate of $\overline{C}_P$ by the vector $\overrightarrow{PO}$),

$\overline{C}_F = \cap\{H^+ | H^+$ is a supporting half-space to $\mathbf{K}$ bounded by H with $F \subset H\}$,

$C_F = \overrightarrow{PO} + \overline{C}_F$ with any $P \in \mathrm{aff}\, F$ (here aff(.) denotes the affine hull of the corresponding set) and

$C_F^* = \{X \in \mathbf{E}^d | \langle \overrightarrow{OX}, \overrightarrow{OY} \rangle \leq 0$ for all $Y \in C_F\}$(called polar cone of C_F). Finally, if $O \in \mathbf{K} \setminus F$, then we define the convex cone $\tilde{C}_F^* = \{X \in \mathbf{E}^d | \overrightarrow{OX} = \lambda \cdot \overrightarrow{OY}$ with $\lambda \geq 0$ and $\langle \overrightarrow{OY}, \overrightarrow{OZ} \rangle \leq 1$ for all $Z \in \mathbf{K}$ and $\langle \overrightarrow{OY}, \overrightarrow{OZ_o} \rangle = 1$ for all $Z_o \in F\}$.

Lemma 2. *If F is a face of the closed convex set $\mathbf{K} \subset \mathbf{E}^d, d \geq 1$, with $O \in \mathbf{K} \setminus F$, then $\tilde{C}_F^* = \mathrm{pos}\, F^*$, $C_F^* = \mathrm{cl}\, \tilde{C}_F^*$ (where pos(.) and cl(.) denote the positive hull and the closure of the corresponding set) and $F^* = (\cup \mathcal{F}^*) \cap \tilde{C}_F^* = (\cup \mathcal{F}^*) \cap C_F^*$.*

Proof. Since F^* is convex therefore $\tilde{C}_F^* = \mathrm{pos}\, F^*$. Now let $Y \in F^*$. Then $Y \in (\cup \mathcal{F}^*) \cap \tilde{C}_F^*$. If $\overline{r}^O_{\overrightarrow{OY}}$ is the closed ray emanating from O and having direction vector $\overrightarrow{OY}$, then it is easy to see that $\overline{r}^O_{\overrightarrow{OY}} \cap (\cup \mathcal{F}^*) = \{Y\}$. Thus $F^* = (\cup \mathcal{F}^*) \cap \tilde{C}_F^*$. The claims left are trivial. $\qquad \square$

Remark 1. Lemma 1 and 2 help to determine the shape of $\mathbf{K}^*$ (Fig. 1).

We are now in a position to prove the key-lemma of this section.

Lemma 3. *Let $\mathbf{K} \subset \mathbf{E}^d, d \geq 1$ be a closed convex set with $O \in \mathbf{K}$ and let $P \in \mathrm{bd}\, \mathbf{K}$. Assume that the smallest dimensional face F_m of $\mathbf{K}$ which contains P is disjoint from O (i.e. $F_m \in \mathcal{F}$). Then the light-source $L \in \mathbf{E}^d \setminus \mathbf{K}$ illuminates P if and only if there exists $\varepsilon > 0$ such that the hyperplane $H_{L'} =$*

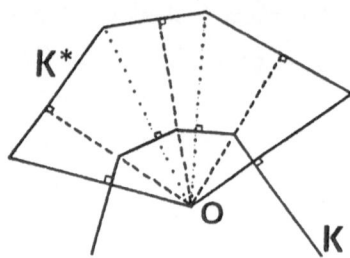

Fig. 1

$\{X \in \mathbf{E}^d | \langle \overrightarrow{OX}, \overrightarrow{OL'} \rangle = 1\}$ *strictly separates O from the polar face $F_m^* \in \mathcal{F}^*$ for any $L' \in \mathbf{E}^d$ with $d(L, L') \leq \varepsilon$ where $d(.,.)$ denotes the distance between two points.*

Proof. L illuminates P if and only if $r_{\overrightarrow{LP}}^P \subset \text{int } \overline{\mathbf{C}}_P$ i. e. $r_{\overrightarrow{LP}}^P \subset \text{int } \overline{\mathbf{C}}_{F_m}$. Thus for any $Y(\neq O) \in \mathbf{C}_{F_m}^*$ we have to have $\langle \overrightarrow{OY}, \overrightarrow{PL} \rangle > 0$ i.e. for any $Y(\neq O) \in \text{cl } \tilde{\mathbf{C}}_{F_m}^*$ it has to be true that $\langle \overrightarrow{OY}, \overrightarrow{PL} \rangle > 0$. We know that $\langle \overrightarrow{OY}, \overrightarrow{PL} \rangle > 0$ for any $Y(\neq O) \in \text{cl } [\text{pos } F_m^*]$ if and only if there exists $\varepsilon > 0$ such that $\langle \overrightarrow{OY}, \overrightarrow{PL'} \rangle > 0$ for any $Y \in F_m^*$ with any L' for which $d(L, L') \leq \varepsilon$. Thus we have obtained that L illuminates P if and only if there exists $\varepsilon > 0$ such that $\langle \overrightarrow{OY}, \overrightarrow{OL'} \rangle > \langle \overrightarrow{OY}, \overrightarrow{OP} \rangle = 1$ for any $Y \in F_m^*$ with any L' for which $d(L, L') \leq \varepsilon$. $\qquad \square$

Corollary 1. *Let $\mathbf{K} \subset \mathbf{E}^d, d \geq 1$ be a closed convex set with $O \in \text{int } \mathbf{K}$. If F_m is the smallest dimensional face of $\mathbf{K}$ which contains the boundary point P of $\mathbf{K}$, then the light-source $L \in \mathbf{E}^d \setminus \mathbf{K}$ illuminates P if and only if the hyperplane $H_L = \{X \in \mathbf{E}^d | \langle \overrightarrow{OX}, \overrightarrow{OL} \rangle = 1\}$ strictly separates O from the polar face $F_m^* \in \mathcal{F}^*$.*

Proof. F_m^* is bounded now. Thus $\langle \overrightarrow{OY}, \overrightarrow{PL} \rangle > 0$ for any $Y(\neq O) \in \text{cl } [\text{pos } F_m^*]$ if and only if $\langle \overrightarrow{OY}, \overrightarrow{PL} \rangle > 0$ for any $Y \in F_m^*$ i. e. $\langle \overrightarrow{OY}, \overrightarrow{OL} \rangle > \langle \overrightarrow{OY}, \overrightarrow{OP} \rangle = 1$ for any $Y \in F_m^*$. $\qquad \square$

Theorem 1. *The light-sources $\{L_1, L_2, \ldots, L_n\} \subset \mathbf{E}^d \setminus \mathbf{K}$ illuminate the boundary of the closed convex set $\mathbf{K} \subset \mathbf{E}^d, d \geq 1$, $O \in \text{int } \mathbf{K}$ if and only if the hyperplanes $H_{L_i} = \{X \in \mathbf{E}^d | \langle \overrightarrow{OX}, \overrightarrow{OL_i} \rangle = 1\}$, $1 \leq i \leq n$ strictly separate O from the faces $\mathcal{F}^* = \{F^* | F^* \text{ is a face of } \mathbf{K}^* \text{ with } O \notin F^*\}$ of the polar set $\mathbf{K}^* = \{X \in \mathbf{E}^d | \langle \overrightarrow{OX}, \overrightarrow{OY} \rangle \leq 1 \text{ for all } Y \in \mathbf{K}\}$. Hence $I_0(\mathbf{K})$ is the smallest number of the hyperplanes of $\mathbf{E}^d$ which strictly separate O from the faces $\mathcal{F}^*$ of $\mathbf{K}^*$.*

Proof. Use Corollary 1 and Lemma 1. $\qquad \square$

Lemma 4. Let $\mathbf{K} \subset \mathbf{E}^d, d \geq 1$ be a closed convex set with $O \in \mathbf{K}$ and let $P \in \text{bd } \mathbf{K}$. Assume that the smallest dimensional face F_m of $\mathbf{K}$ which contains P is disjoint from O (i.e. $F_m \in \mathcal{F}$). Then the direction $\underline{o} \neq \underline{v} \in \mathbf{E}^d$ illuminates P if and only if there exists $\varepsilon > 0$ such that the open half-space $\{X \in \mathbf{E}^d | \langle \overrightarrow{OX}, -\underline{v}' \rangle > 0\}$ contains the polar face $F_m^* \in \mathcal{F}^*$ for any $\underline{v}' \in \mathbf{E}^d$ with $\sphericalangle (\underline{v}, \underline{v}') \leq \varepsilon$.

Proof. Use the method of the proof of Lemma 3.   □

Corollary 2 (P. S. Soltan and V. P. Soltan (1986)). *Let $\mathbf{K} \subset \mathbf{E}^d, d \geq 1$ be a closed convex set with $O \in \text{int } \mathbf{K}$. If F_m is the smallest dimensional face of $\mathbf{K}$ which contains the boundary point P of $\mathbf{K}$, then the direction $\underline{o} \neq \underline{v} \in \mathbf{E}^d$ illuminates P if and only if the open half-space $\{X \in \mathbf{E}^d | \langle \overrightarrow{OX}, -\underline{v} \rangle > 0\}$ contains the polar face $F_m^* \in \mathcal{F}^*$.*

Proof. Apply the same argument as in the proof of Corollary 1.   □

Theorem 2 (P. S. Soltan and V. P. Soltan (1986)). *The directions $\underline{o} \notin \{\underline{v}_1, \underline{v}_2, \ldots, \underline{v}_n\} \subset \mathbf{E}^d, d \geq 1$ illuminate the boundary of the closed convex set $\mathbf{K} \subset \mathbf{E}^d$, $O \in \text{int } \mathbf{K}$ if and only if every face of $\mathcal{F}^* = \{F^* | F^* \text{ is a face of } \mathbf{K}^*$ with $O \notin F^*\}$ of the polar set $\mathbf{K}^* = \{X \in \mathbf{E}^d | \langle \overrightarrow{OX}, \overrightarrow{OY} \rangle \leq 1 \text{ for all } Y \in \mathbf{K}\}$ lies in at least one of the open half-spaces $\{X \in \mathbf{E}^d | \langle \overrightarrow{OX}, -\underline{v}_i \rangle > 0\}$, $1 \leq i \leq n$. Thus $I_\infty(\mathbf{K})$ is the smallest number of open half-spaces of $\mathbf{E}^d$ the boundaries of which pass through O such that every face of $\mathcal{F}^*$ lies in at least one of the given open half-spaces.*

Proof. Use Corollary 2 and Lemma 1.   □

Remark 2. It is obvious that $I_\infty(\mathbf{K}) \leq I_0(\mathbf{K})$ for any closed convex set $\mathbf{K} \subset \mathbf{E}^d, d \geq 1$.

Theorem 3. *Let $\mathbf{P} \subset \mathbf{E}^d, d \geq 1$ be a polyhedral convex set with $O \in \mathbf{P} \setminus \text{vert } \mathbf{P}$. Then $I_\infty(\mathbf{P}) = I_0(\mathbf{P}) < +\infty$ is the smallest number of polyhedral convex cones each of which possesses one apex namely, O such that every face of $\mathcal{F}^* = \{F^* | F^* \text{ is a face of the polar set } \mathbf{P}^* \text{ with } O \notin F^*\}$ lies in at least one of the interiors of the cones. If O is a vertex of the polyhedral convex set $\mathbf{P} \subset \mathbf{E}^d, d \geq 1$, then the smallest number of polyhedral convex cones each of which possesses O as the only apex such that every face of $\mathcal{F}^*$ lies in at least one of the interiors of the cones is at least $I_\infty(\mathbf{P}) - 1 = I_0(\mathbf{P}) - 1$.*

Proof. Use Lemma 3 and Lemma 4.   □

Definition 2. A d-dimensional closed convex set $\mathbf{K} \subset \mathbf{E}^d, d \geq 1$ is called almost bounded if and only if there exists a d-dimensional ball of $\mathbf{E}^d$ which intersects every supporting hyperplane of $\mathbf{K}$. ($\mathbf{E}^d$ is considered to be an almost bounded set.) Furthermore, if $\mathbf{K}$ is almost bounded and $\mathbf{C}$ denotes the closed convex cone with apex $O \in \text{int } \mathbf{K}$ which is the union of closed half-lines emanating from O and lying in $\mathbf{K}$, then $Pr_L : \mathbf{E}^d \to L$ denotes the orthogonal projection of $\mathbf{E}^d$ onto the affine subspace $O \in L$ which is the orthogonal complement of aff $\mathbf{C}$ in $\mathbf{E}^d$ and $I_0[\mathrm{cl}\ (Pr_L(\mathbf{K}))]$ denotes the corresponding illumination number of cl $(Pr_L(\mathbf{K}))$ in L.

We prove

Theorem 4. *Let* $\mathbf{K} \subset \mathbf{E}^d, d \geq 1$ *be a d-dimensional closed convex set. Then* $I_0(\mathbf{K})$ *is finite if and only if* $\mathbf{K}$ *is almost bounded. Furthermore, if* $\mathbf{K} \subset \mathbf{E}^d, d \geq 1$ *is a d-dimensional almost bounded closed convex set, then* $I_0(\mathbf{K}) \leq I_0[\mathrm{cl}\ (Pr_L(\mathbf{K}))] < +\infty.$

Proof. If $I_0(\mathbf{K})$ $(\mathbf{K} \subsetneqq \mathbf{E}^d, d \geq 1)$ is finite, then because of Theorem 1 dist $(\cup \mathcal{F}^*, O) > 0$ (we have choosen O in int $\mathbf{K}$) consequently, there exists a d-dimensional ball of $\mathbf{E}^d$ centered at O which intersects every supporting hyperplane H of $\mathbf{K}$ i.e. $\mathbf{K}$ is almost bounded.

Assume now that $\mathbf{K} \subsetneqq \mathbf{E}^d, d \geq 1$ is an almost bounded d-dimensional closed convex set with $O \in$ int $\mathbf{K}$ and prove that $I_0(\mathbf{K}) \leq I_0[\mathrm{cl}\ (Pr_L(\mathbf{K}))] < +\infty$. We know that $\mathbf{K}^* \subset \mathbf{E}^d$ is a compact convex set with $O \in \mathbf{K}^*$ and dist $(\cup \mathcal{F}^*, O) > 0$. Let F_m^* be the smallest dimensional face of $\mathbf{K}^*$ which contains O. (F_m^* can be identical to the improper face $\mathbf{K}^*$.)

We prove that $O \in$ relint F_m^* using induction on the dimension $d^*(\geq 1)$ of $\mathbf{K}^*$. If $d^* = 1$ or $d^* = 2$, then it is easy to see that $O \in$ relint F_m^*. So suppose that the claim is true for any $(2 \leq) d'(< d^*)$-dimensional compact convex set and take a d^*- dimensional compact convex set $\mathbf{K}^* \subset \mathbf{E}^{d^*}$ with $O \in$ bd $\mathbf{K}^*$ and dist$(\cup \mathcal{F}^*, O) > 0$. Let $\mathbf{C}^* = \cup\{\overline{r^O_{OX}}|\overline{r^O_{OX}}$ denotes the closed ray emanating

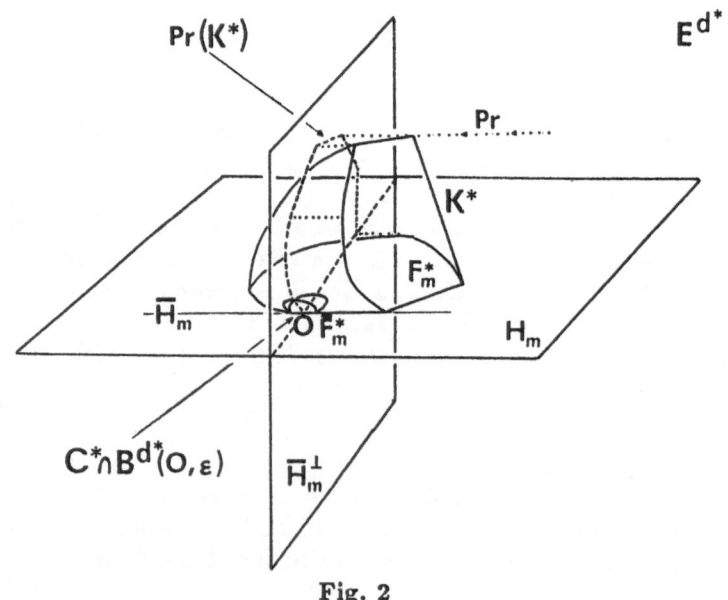

Fig. 2

from O having direction vector $\overrightarrow{OX}$ with $(O \neq)X \in \mathbf{K}^*\}$. It is obvious that dist $(\cup \mathcal{F}^*, O) > 0$ if and only if there exists a d^*-dimensional closed ball $B^{d^*}(O, \varepsilon) \subset \mathbf{E}^{d^*}$ centered at O with radius $\varepsilon > 0$ such that $\mathbf{K}^* \cap B^{d^*}(O, \varepsilon) = \mathbf{C}^* \cap B^{d^*}(O, \varepsilon)$. Let $H_m \subset \mathbf{E}^{d^*}$ be the supporting hyperplane of $\mathbf{K}^*$ for which $H_m \cap \mathbf{K}^* = F_m^*$. Since in case of $O \in$ relint F_m^* we are done we suppose that $O \in$ relbd $F_m^* = F_m^* \setminus$ relint F_m^*. (Consequently, dim $F_m^* \geq 1$.) By induction (dim $F_m^* <$ dim $\mathbf{K}^* = d^*$ and $F_m^* \cap B^{d^*}(O, \varepsilon) = (\mathbf{C}^* \cap H_m) \cap B^{d^*}(O, \varepsilon)$ i.e. the union of the faces of F_m^* which are disjoint from O lies at distance $\geq \varepsilon$ from O) we know that F_m^* possesses a face $\overline{F}_m$ with $O \in$ relint $\overline{F}_m$ (Fig. 2). Hence there exists a $(d^* - 2)$-dimensional affine subspace $\overline{H}_m$ which supports F_m^* in H_m such that $\overline{H}_m \cap F_m^* = \overline{H}_m \cap \mathbf{K}^* = \overline{F}_m^*$. Let $(O \in)\overline{H}_m^\perp$ be the 2-dimensional affine subspace of $\mathbf{E}^{d^*}$ which is totally orthogonal to $\overline{H}_m$ and let $Pr(\mathbf{K}^*)$ be the orthogonal projection of $\mathbf{K}^*$ onto $\overline{H}_m^\perp$ parallel to $\overline{H}_m$. Obviously, $Pr(\mathbf{K}^*)$ is a convex domain whose boundary contains O and $Pr(\mathbf{K}^*) \cap B^{d^*}(O, \varepsilon) = Pr(\mathbf{C}^*) \cap B^{d^*}(O, \varepsilon)$ i.e. every face of $Pr(\mathbf{K}^*)$ which is disjoint from O lies at distance $\geq \varepsilon$ from O. Consequently, by induction there exists a face $Pr(F_m^\perp)$ of $Pr(\mathbf{K}^*)$ with $O \in$ relint $Pr(F_m^\perp)$. Hence for the face $F_m^\perp$ of $\mathbf{K}^*$ whose orthogonal projection onto $\overline{H}_m^\perp$ is $Pr(F_m^\perp)$ we have that $O \in$ relint $F_m^\perp$ and $F_m^* \neq F_m^\perp$, a contradiction. Thus $O \in$ relint F_m^*.

It is easy to show that the cone $\mathbf{C}^*$ defined above is the polar cone of C. Thus we have $\mathbf{K}^* = \{X \in \mathbf{E}^d | \langle \overrightarrow{OX}, \overrightarrow{OY} \rangle \leq 1$ for all $Y \in \mathbf{K}\} = \{X \in \mathbf{E}^d | \langle \overrightarrow{OX}, \overrightarrow{OY} \rangle \leq 1$ for all $Y \in$ bd $\mathbf{K}\} \cap \{X \in \mathbf{E}^d | \langle \overrightarrow{OX}, \overrightarrow{OZ} \rangle \leq 0$ for all $Z \in$ C$\} = \{X \in \mathbf{E}^d | \langle \overrightarrow{OX}, \overrightarrow{OY} \rangle \leq 1$ for all $Y \in$ bd $\mathbf{K}\} \cap \mathbf{C}^*$ and $[Pr_L(\mathbf{K})]^{*L} = \{X \in L | \langle \overrightarrow{OX}, \overrightarrow{OPr_L(Y)} \rangle \leq 1$ for all $Y \in \mathbf{K}\} = \{X \in L | \langle \overrightarrow{OX}, \overrightarrow{OPr_L(Y)} \rangle \leq 1$ for all $Y \in$ bd $\mathbf{K}\} = \{X \in L | \langle \overrightarrow{OX}, \overrightarrow{OY} \rangle \leq 1$ for all $Y \in$ bd $\mathbf{K}\} = L \cap \{X \in \mathbf{E}^d | \langle \overrightarrow{OX}, \overrightarrow{OY} \rangle \leq 1$ for all $Y \in$ bd $\mathbf{K}\}$. Then $[Pr_L(\mathbf{K})]^{*L} \cap \mathbf{C}^* = L \cap (\{X \in \mathbf{E}^d | \langle \overrightarrow{OX}, \overrightarrow{OY} \rangle \leq 1$ for all $Y \in$ bd $\mathbf{K}\} \cap \mathbf{C}^*) = L \cap \mathbf{K}^*$. Since L is totally orthogonal to aff C therefore $L \subset \mathbf{C}^*$ and so $[Pr_L(\mathbf{K})]^{*L} = [Pr_L(\mathbf{K})]^{*L} \cap \mathbf{C}^* = L \cap \mathbf{K}^*$. We have proved that $O \in$ relint F_m^*. This implies that aff $F_m^* = L$ from which we get that $[Pr_L(\mathbf{K})]^{*L} = L \cap \mathbf{K}^* = F_m^*$. Thus by Theorem 1 $I_0[\text{cl } (Pr_L(\mathbf{K}))]$ is the smallest number of hyperplanes of aff F_m^* which strictly separate $O \in$ relint F_m^* from the faces of F_m^* in aff F_m^*. We distinguish Case 1: $F_m^* = \mathbf{K}^*$ and Case 2: F_m^* is a face of dimension $\leq d^* - 1$ of $\mathbf{K}^*$ (dim $\mathbf{K}^* = d^* \geq 1$).

Case 1. $\mathbf{K}^* \subset \mathbf{E}^{d^*}$ is a d^*-dimensional $(d^* \geq 1)$ compact convex set with $O \in$ int $\mathbf{K}^*$. Since $[Pr_L(\mathbf{K})]^{*L} = \mathbf{K}^*$ therefore by Theorem 1 $I_0(\mathbf{K}) = I_0[\text{cl } (Pr_L(\mathbf{K}))]$ and it is sufficient to show that there are finitely many hyperplanes of $\mathbf{E}^{d^*}$ which strictly separate O from the faces of $\mathbf{K}^*$. Case $d^* = 1$ is trivial and in case $d^* = 2$ we can prove a sharp upper bound on the number of hyperplanes (see Chapter 3). We assume now that $d^* \geq 3$. Let $B^{d^*}(O, r)(B^{d^*}(O, R)$, resp.) be a d^*- dimensional closed ball centered at O with radius $r > 0$ ($R > 0$, resp.) such that $B^{d^*}(O, r) \subset \text{int } \mathbf{K}^*$ (int $B^{d^*}(O, R) \supset \mathbf{K}^*$, resp.). Each boundary point of $\mathbf{K}^*$ is the center of an open ball of $\mathbf{E}^{d^*}$ with positive radius $< r$ lying in $B^{d^*}(O, R)$ and being disjoint from $B^{d^*}(O, r)$. Thus by Heine-Borel's theorem finitely many of these open balls cover bd $\mathbf{K}^*$. Let $B^{d^*}(P, \varepsilon)$ be the closure of one of these balls ($r > \varepsilon > 0, B^{d^*}(P, \varepsilon) \subset B^{d^*}(O, R)$ and int $B^{d^*}(P, \varepsilon) \cap$ int $B^{d^*}(O, r) = \emptyset$). Take a hyperplane H_1 of $\mathbf{E}^{d^*}$ passing through the centers P and O and let H_2 (H_3 resp.) be the hyperplane of $\mathbf{E}^{d^*}$ which is orthogonal to $\overrightarrow{OP}$ and which passes through O (O_ε where O_ε denotes the point of the segment OP with $d(O, O_\varepsilon) = \varepsilon$, resp.) (Fig. 3). Let $D_i = H_i \cap B^{d^*}(O, R)$, $1 \le i \le 3$ and if ($O_\varepsilon \ne$)$X \in H_3$, then let $S(X)$ denote the ($d^* - 2$)-dimensional affine subspace of H_3 which is orthogonal to $\overrightarrow{O_\varepsilon X}$ and passes through X. (Obviously, D_1, D_2 and D_3 are ($d^* - 1$)-dimensional closed balls.)

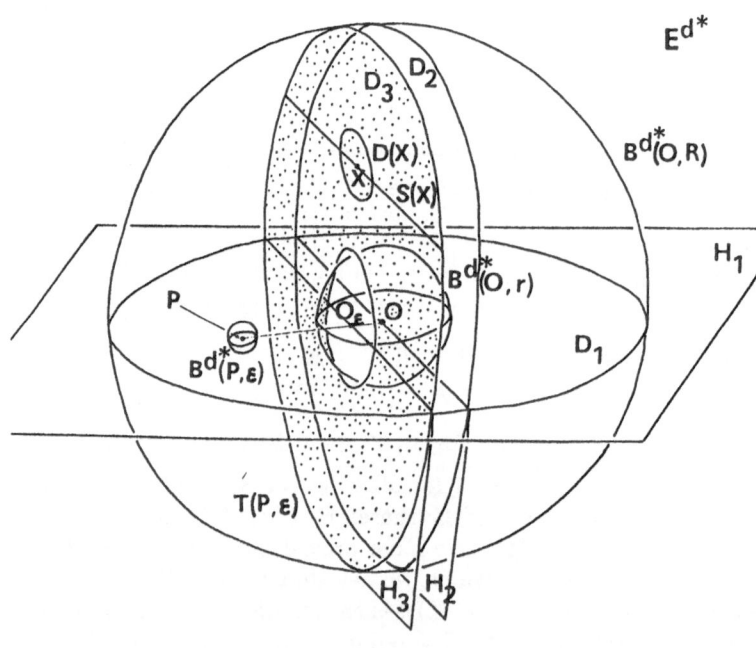

Fig. 3

Finally, let $T(P, \varepsilon) = \{X \in D_3 |$ there exists $Y \in B^{d^*}(P, \varepsilon)$ such that the hyperplane of $\mathbf{E}^{d^*}$ spanned by Y and $S(X)$ is disjoint from int $B^{d^*}(O, r)\}$.

It is sufficient to prove that the faces of $\mathcal{F}^*(P, \varepsilon) = \{F^* \in \mathcal{F}^* | F^* \cap B^{d^*}(P, \varepsilon) \ne \emptyset\}$ can be strictly separated by finitely many hyperplanes of $\mathbf{E}^{d^*}$ from O. The faces $F^* \in \mathcal{F}^*(P, \varepsilon)$ with $F^* \cap H_3 = \emptyset$ are strictly separated by H_3 from O. Thus we have to deal with the faces F^* of $\mathcal{F}^*(P, \varepsilon)$ which intersect H_3. If F^* is one of these faces, then there exist $Y \in B^{d^*}(P, \varepsilon)$

and $X \in T(P, \varepsilon)$ such that the hyperplane H of $\mathbf{E}^{d^*}$ spanned by Y and $S(X)$ satisfies $H \cap \mathbf{K}^* = F^*$. Thus it is sufficient to prove that the sets $D = H \cap B^{d^*}(O, R)$ with $H = \mathrm{span}\,(Y, S(X))$, $Y \in B^{d^*}(P, \varepsilon)$ and $X \in T(P, \varepsilon)$ can be strictly separated by finitely many hyperplanes of $\mathbf{E}^{d^*}$ from O. First, we observe that if $Y \in B^{d^*}(P, \varepsilon)$ and $X \in T(P, \varepsilon)$, then there are (small) open balls $B(Y) \subset \mathbf{E}^{d^*}$ and $D(X) \subset H_3$ of dimension d^* and $(d^* - 1)$ centered at Y and X such that the sets $D' = H' \cap B^{d^*}(O, R)$ with $H' = \mathrm{span}\,(Y', S(X'))$, $Y' \in B(Y)$ and $X' \in D(X)$ are strictly separated by a hyperplane of $\mathbf{E}^{d^*}$ from O. Secondly, we observe that in this way we get an open covering of the compact set $B^{d^*}(P, \varepsilon) \times T(P, \varepsilon)$ in $\mathbf{E}^{d^*} \times H_3$. Thus Heine-Borel's theorem implies the claim.

Case 2. F_m^* is a face of dimension $\leq (d^* - 1)$ of the d^*- dimensional compact convex set $\mathbf{K}^* \subset \mathbf{E}^{d^*}, d^* \geq 1$ with $O \in \mathrm{relint}\, F_m^*$ and dist $(\cup \mathcal{F}^*, O) > 0$. We have seen that $[Pr_L(\mathbf{K})]^{*L} = F_m^*$ where $O \in L = \mathrm{aff}\, F_m^*$ is the orthogonal complement of aff $\mathbf{C}$ in $\mathbf{E}^d$. Since F_m^* is bounded therefore Case 1 implies that there are finitely many say, n - in fact we can take $n = I_0[\mathrm{cl}\,(Pr_L(\mathbf{K}))]$ (by Theorem 1)-hyperplanes $H_1(m), H_2(m), \ldots, H_n(m)$ of aff F_m^* which strictly separate $O \in \mathrm{relint}\, F_m^*$ from the faces of F_m^* in aff F_m^*. Let $H(H^+, \mathrm{resp.}\,)$ be the supporting hyperplane (supporting half-space bounded by H, resp.) of $\mathbf{K}^*$ in $\mathbf{E}^{d^*}$ with $H \cap \mathbf{K}^* = F_m^*$. Let $H_i, 1 \leq i \leq n$ be the hyperplane of $\mathbf{E}^{d^*}$ orthogonal to H with $H_i \cap \mathrm{aff}\, F_m^* = H_i(m)$ and let $H_i^+, 1 \leq i \leq n$ denote the closed half-space of $\mathbf{E}^{d^*}$ bounded by H_i with $O \in H_i^+$. Finally, let $\mathcal{F}_i^*(m), 1 \leq i \leq n$ be the collection of the faces of F_m^* which are strictly separated by the hyperplane $H_i(m)$ of aff F_m^* from O. Without loss of generality we may suppose the existence of $\varepsilon > 0$ such that the ε-parallel body $[\cup \mathcal{F}_i^*(m)]_\varepsilon = (\cup \mathcal{F}_i^*(m)) + B^{d^*}(O, \varepsilon)$ is disjoint from H_i^+ for any $i \in \{1, 2, \ldots, n\}$. Let $B^{d^*}(O, R)$ be a d^*-dimensional closed ball of $\mathbf{E}^{d^*}$ centered at O with radius $R > O$ such that int $B^{d^*}(O, R) \supset \mathbf{K}^*$. We have to show that the faces $\mathcal{F}^*$ of $\mathbf{K}^*$ can be strictly separated by n hyperplanes of $\mathbf{E}^{d^*}$ from O. In order to prove this take the hyperplane $H_\varepsilon \subset H^+$ of $\mathbf{E}^{d^*}$ parallel to H at distance $\varepsilon > 0$ from H and choose the closed half-space H_ε^+ of $\mathbf{E}^{d^*}$ bounded by H_ε which contains O (Fig. 4). Since relbd $F_m^* \subset \mathrm{int}\, \bigcup_{i=1}^n [\cup \mathcal{F}_i^*(m)]_\varepsilon$ and dist $(\cup \mathcal{F}^*, O) > 0$ therefore it is easy to see the existence of $(\varepsilon \geq) \delta(\varepsilon) > 0$ such that dist $(F^*, H) > \delta(\varepsilon)$ for each $F^* \in \mathcal{F}^*$ with $F^* \cap \bigcup_{i=1}^n [\cup \mathcal{F}_i^*(m)]_\varepsilon = \emptyset$. Then we claim that there exists $\varepsilon_o > 0$ $(\varepsilon \geq \varepsilon_o)$ such that for any face $F^* \in \mathcal{F}^*$ with $F^* \cap \bigcup_{i=1}^n [\cup \mathcal{F}_i^*(m)]_{\varepsilon_o} \neq \emptyset$ we have $i(F^*) \in \{1, 2, \ldots, n\}$ with $F^* \cap (H^+ \cap H_{\varepsilon_o}^+ \cap H_{i(F^*)}^+) = \emptyset$. Else, there are sequences $(\varepsilon \geq) \varepsilon_1 > \varepsilon_2 > \ldots > \varepsilon_k \ldots$, $F_1^*, F_2^*, \ldots, F_k^*, \ldots$ with $\lim_{k \to +\infty} \varepsilon_k = 0$, $F_k^* \in \mathcal{F}^*$, $F_k^* \cap \bigcup_{i=1}^n [\cup \mathcal{F}_i^*(m)]_{\varepsilon_k} \neq \emptyset$ and $F_k^* \cap (H^+ \cap H_{\varepsilon_k}^+ \cap H_j^+) \neq \emptyset$ for any $j \in \{1, 2, \ldots, n\}$. Then consider the sequence of the non-empty compact convex sets $(F_k^* \cap H^+ \cap H_{\varepsilon_k}^+)$, $k = 1, 2, \ldots$ lying in $B^{d^*}(O, R)$. Blaschke selection theorem (Lay (1982), pp. 98) asserts that the sequence in question contains a convergent subsequence. (We are using the usual Hausdorff metric of all non-empty compact convex subset of $\mathbf{E}^{d^*}$). Without loss of generality we may assume that the sets $(F_k^* \cap H^+ \cap H_{\varepsilon_k}^+)$ $k = 1, 2, \ldots$ converge to the compact convex set S of $\mathbf{E}^{d^*}$. It is easy to prove that there exists a face $F^*(m)$ of F_m^* with $F^*(m) \supset S$. Let $F^*(m) \in \mathcal{F}_j^*(m)$. Thus for $[\cup \mathcal{F}_j^*(m)]_\varepsilon$ there exists $N > 0$ such that for any $k \geq N$ $(F_k^* \cap H^+ \cap H_{\varepsilon_k}^+) \subset [\cup \mathcal{F}_j^*(m)]_\varepsilon$. Since $[\cup \mathcal{F}_j^*(m)]_\varepsilon \cap H_j^+ = \emptyset$ therefore $F_k^* \cap (H^+ \cap H_{\varepsilon_k}^+ \cap H_j^+) \subset [\cup \mathcal{F}_j^*(m)]_\varepsilon \cap H_j^+ = \emptyset$, $k \geq N$, a contradiction. Thus the hyperplanes $\mathbf{H}_i$, $1 \leq i \leq n$ of $\mathbf{E}^{d^*}$ with $\mathbf{H}_i \cap H = H_i \cap H$ and $\mathbf{H}_i \cap (H^+ \cap B^{d^*}(O, R)) \subset H^+ \cap H_{\delta(\varepsilon_o)}^+ \cap H_i^+$ strictly separate O from the faces $\mathcal{F}^*$ of $\mathbf{K}^*$. This completes the proof of the theorem.   □

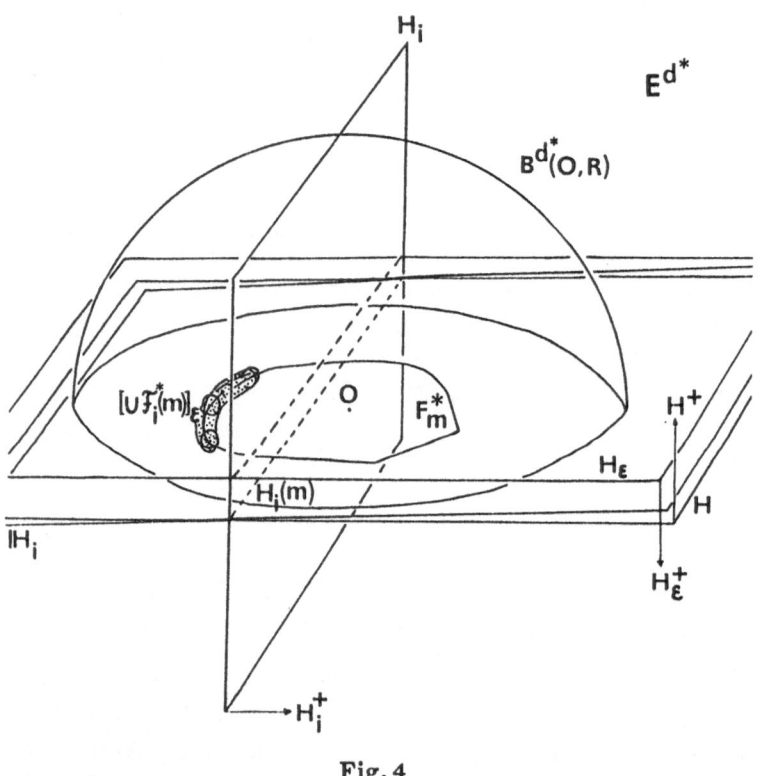

Fig. 4

The following definition is due to P. S. Soltan (1963).

Definition 3. The d-dimensional closed convex set $\mathbf{K} \subset \mathbf{E}^d, d \geq 1$ is called almost conic if and only if it possesses a point Q and a distance $0 \leq s < +\infty$ such that every point of $\mathbf{K}$ lies at distance $\leq s$ from the closed convex cone with apex Q which is the union of closed rays emanating from Q and lying in $\mathbf{K}$. ($\mathbf{E}^d$ is considered to be almost conic.)

Remark 3. It was shown by Boltjansky and Soltan (1978) that $I_0(\mathbf{K})$ is finite for a d-dimensional closed convex set $\mathbf{K} \subset \mathbf{E}^d, d \geq 1$ if and only if $\mathbf{K}$ is almost conic.

Thus we complete this chapter with a direct proof of

Lemma 5. *Let $\mathbf{K} \subset \mathbf{E}^d, d \geq 1$ be a d-dimensional closed convex set. Then $\mathbf{K}$ is almost bounded if and only if it is almost conic.*

Proof. Let $O \in \text{int } \mathbf{K}$ and let $\mathbf{C}$ be the inscribed cone of $\mathbf{K}$ with apex O. First, assume that $\mathbf{K}$ is almost conic. That means that we have a distance $0 \leq s < +\infty$ such that every point of $\mathbf{K}$ lies at distance $\leq s$ from $\mathbf{C}$. Thus the d-dimensional closed ball centered at O having radius s intersects every supporting hyperplane of $\mathbf{K}$ i.e. $\mathbf{K}$ is almost bounded. Secondly, suppose that $\mathbf{K}$ is almost bounded. Then $\mathbf{K}^* \subset \mathbf{E}^d$ is a compact convex set with $O \in \mathbf{K}^*$ and dist $(\cup \mathcal{F}^*, O) > 0$. We have seen above that if F_m^* is the smallest dimensional face of $\mathbf{K}^*$ which contains O, then $O \in \text{relint } F_m^*$ and we have proved that $\mathbf{C}^* = \cup \{\overrightarrow{r_{\overrightarrow{OX}}^O} | \overrightarrow{r_{\overrightarrow{OX}}^O}$ denotes the closed ray emanating from O having direction vector $\overrightarrow{OX}$ with $(O \neq)X \in \mathbf{K}^*\}$ is the polar cone of $\mathbf{C}$ with $\mathbf{K}^* \cap B^{d^*}(O, \varepsilon) = \mathbf{C}^* \cap B^{d^*}(O, \varepsilon)$ for an $\varepsilon > 0$. Let $X \in \mathbf{K} \setminus \mathbf{C}$ be an arbitrary point and let $Y \in \mathbf{C}$ be the closest point of $\mathbf{C}$ to X. It is clear that for $\overrightarrow{OZ} = \overrightarrow{YX}$ we have $Z \in \mathbf{C}^*$. Finally, let $Z^* = r_{\overrightarrow{OZ}}^O \cap (\cup \mathcal{F}^*) \in \mathbf{K}^*$. Since $\|\overrightarrow{OZ^*}\| \geq \varepsilon$ and

$$1 \geq \langle \overrightarrow{OZ^*}, \overrightarrow{OX} \rangle = \langle \overrightarrow{OZ^*}, \overrightarrow{OY} + \overrightarrow{YX} \rangle = \langle \overrightarrow{OZ^*}, \overrightarrow{YX} \rangle = \|\overrightarrow{OZ^*}\| \cdot \|\overrightarrow{YX}\| \text{ therefore}$$

$\|\overrightarrow{YX}\| \leq \frac{1}{\varepsilon}$. Thus every point of $\mathbf{K}$ lies at distance $\leq \frac{1}{\varepsilon}$ from $\mathbf{C}$. This completes the proof of the lemma. □

2. On $I_l(\mathbf{K})$ and k-fold Illumination

Definition 4. If $(O \notin)L$ is an affine subspace of $\mathbf{E}^d$, $d \geq 1$, $0 \leq \dim L \leq d - 1$, then let $\hat{L} = \cap \{H_Q | H_Q = \{X \in \mathbf{E}^d | \langle \overrightarrow{OX}, \overrightarrow{OQ} \rangle = 1\}, Q \in L\}$.

The following two theorems are generalizations of Theorem 1 and Theorem 4.

Theorem 5. *The light-sources (i.e. affine subspaces) $L_1, L_2, \ldots, L_n \subset \mathbf{E}^d \setminus \mathbf{K}$ of dimension $\dim L_1 = \ldots = \dim L_n = l$, $0 \leq l \leq d - 1, d \geq 1$ illuminate the boundary of the closed convex set $\mathbf{K} \subset \mathbf{E}^d$, $O \in \text{int } \mathbf{K}$ if and only if any face of $\mathcal{F}^* = \{F^* | F^*$ is a face of $\mathbf{K}^*$ with $O \notin F^*\}$ of the polar set $\mathbf{K}^* = \{X \in \mathbf{E}^d | \langle \overrightarrow{OX}, \overrightarrow{OY} \rangle \leq 1$ for all $Y \in \mathbf{K}\}$ can be strictly separated from O by a hyperplane of $\mathbf{E}^d$ which contains at least one of the affine subspaces $\hat{L}_1, \hat{L}_2, \ldots, \hat{L}_n$ of dimension $d - l - 1$. Thus $I_l(\mathbf{K})$ is the smallest number of $(d-l-1)$-dimensional affine subspaces $\hat{L}_1, \hat{L}_2, \ldots, \hat{L}_n$ of $\mathbf{E}^d$ such that any face* of $\mathcal{F}^*$ can be strictly separated from O by a hyperplane of $\mathbf{E}^d$ which contains at least one of the affine subspaces $\hat{L}_1, \hat{L}_2, \ldots, \hat{L}_n$.

Proof. Use the method of the proof of Theorem 1.   □

Theorem 6. *Let* $\mathbf{K} \subset \mathbf{E}^d$, $d \geq 1$ *be a d-dimensional almost bounded closed convex set and let* $0 \leq l \leq d-1$. *Then* $Pr_L(\mathbf{K})$ *is bounded and* $I_l(\mathbf{K}) \leq I_l[\mathrm{cl}\ (Pr_L(\mathbf{K}))] < +\infty$.

Proof. One can prove this theorem by combining Theorem 5 with the proof of Theorem 4. For more details see Bezdek (1991/c).   □

Thus it is sufficient to inevestigate bounded closed convex sets only.

Definition 5. Let $d \geq 1$, $0 \leq l \leq d-1$ and $f(d,l) = \max\{I_l(\mathbf{K}) | \mathbf{K}$ is a convex body of $\mathbf{E}^d\}$.

Theorem 7. (a) *If* $d \geq 1$, $0 \leq l \leq d-1$, *then* $f(d,l) \leq f(d-l,0)$. *Furthermore,* $f(3,2) = 2$ *and* $2 \leq f(3,1) \leq 3$.
(b) *If* $\mathbf{P}$ *is a convex d-polytope of* $\mathbf{E}^d$, $d \geq 3$, *with affine symmetry, then* $I_{d-3}(\mathbf{P}) \leq 8$ *and* $I_{d-2}(\mathbf{P}) = 2$.
(c) *Let* $\mathbf{B}$ *be a smooth convex body of* $\mathbf{E}^d$ *and let* $1 \leq l \leq d-1$. *Then* $I_l(\mathbf{B}) = \left\lfloor \dfrac{d - \left\lceil \frac{d}{l+1} \right\rceil}{l} \right\rfloor + 1 = \left\lceil \dfrac{d+1}{l+1} \right\rceil$.
(d) *Let* $\mathbf{B}$ *be a d-dimensional zonoid of* $\mathbf{E}^d$ *and let* $d+1 = 2^p, p \geq 2$. *Then* $\mathbf{I}_1(\mathbf{B}) \leq 2^{2^p - p - 1} = \dfrac{2^d}{d+1}$.

Proof. (a) Take a convex body $\mathbf{K}$ of $\mathbf{E}^d$ such that $f(d,l) = I_l(\mathbf{K})$. Without loss of generality we may suppose that $O \in \mathrm{int}\ \mathbf{K}$ and $l \geq 1$. Obviously, the polar set $\mathbf{K}^*$ is a convex body of $\mathbf{E}^d$ with $O \in \mathrm{int}\ \mathbf{K}^*$. If H is a hyperplane of $\mathbf{E}^d$ passing through the origin O, then $d - l - 1 = (d-1) - (l-1) - 1$, Theorem 5 and Theorem 15 of McMullen and Shephard (1971), pp. 70 imply that $I_l(\mathbf{K}) \leq I_{l-1}(\mathbf{K}^\perp)$ where $\mathbf{K}^\perp$ is the orthogonal projection of $\mathbf{K}$ onto H. Thus $f(d,l) \leq f(d-1, l-1)$. Then apply this step l times until getting $f(d,l) \leq f(d-l,0)$. We leave the easy details of the proof of $f(3,2) = 2$ and $2 \leq f(3,1) \leq 3$ to the reader.
  (b) See Bezdek (1991/a).
  (c) For the proof the reader should consult Bezdek (1991/b) and (1991/d).
  (d) See Bezdek, Kiss and Mollard (1991).   □

Theorem 7 implies that any upper bound for $f(d-l,0)$ is an upper bound for $f(d,l)$. However, it looks hard to achieve non-trivial upper bounds (see Theorem 8 (b) and (c)). Finally, we complete this chapter with the statement Theorem 8 (a) about k-fold illumination. This problem arises naturally from the original illumination problem.

Definition 6. Let **B** be a convex body of $\mathbf{E}^d$, $d \geq 2$. We say that the directions $\underline{o} \notin \{\underline{v}_i | i \in I\} \subset \mathbf{E}^d$ form a k-fold illumination of **B** if and only if every boundary point of **B** is illuminated by at least k directions, $k \geq 1$.

Theorem 8. (a) *Let $\underline{o} \notin \{\underline{v}_i | i \in I\} \subset \mathbf{E}^d, d \geq 2$ be a k-fold, $k \geq 1$ illumination of the smooth convex body **B** of $\mathbf{E}^d$. Then there are $\lceil \frac{k}{2(d-1)} \rceil$ pairwise disjoint subsets of $\{\underline{v}_i | i \in I\}$ each of which illuminates **B** and each of which consists of at most 2d directions.*

(b) *For any $d \geq 3$ and $n \geq d+1$ there exist a convex d-polytope $\mathbf{P} \subset \mathbf{E}^d$ and directions $\underline{o} \notin \{\underline{v}_1, \underline{v}_2, \ldots, \underline{v}_n\} \subset \mathbf{E}^d$ such that $\{\underline{v}_1, \underline{v}_2, \ldots, \underline{v}_n\}$ illuminates **P** but no real subset of $\{\underline{v}_1, \underline{v}_2, \ldots, \underline{v}_n\}$ illuminates **P**.*

(c) *(Grünbaum (1967)) If the directions $\underline{o} \notin \{\underline{v}_i | i \in I\} \subset \mathbf{E}^2$ illuminate the convex domain $\mathbf{B} \subset \mathbf{E}^2$, then also at most 6 of them illuminate **B**.*

Proof. (a) Let $O \in \text{int } \mathbf{B}$. As **B** is smooth the faces of the convex body $\mathbf{B}^*$ of $\mathbf{E}^d$ ($O \in \text{int } \mathbf{B}^*$) are points (i.e. 0-dimensional faces) (see also Lemma 1). Thus Corollary 2 implies that $\{\underline{v}_i | i \in I\}$ forms a k-fold illumination of **B** if and only if each point of bd $\mathbf{B}^*$ lies in at least k of the open half-spaces $\{X \in \mathbf{E}^d | \langle \overrightarrow{OX}, -\underline{v}_i \rangle > 0\}$, $i \in I$. Let $\mathbf{S}^{d-1}$ be the sphere of $\mathbf{E}^d$ centered at O with radius $\varepsilon > 0$. Obviously, the open hemispheres $\mathbf{HS}_i^{d-1} = \{X \in \mathbf{E}^d | \langle \overrightarrow{OX}, -\underline{v}_i \rangle > 0\} \cap \mathbf{S}^{d-1}$, $i \in I$ form a k-fold covering of $\mathbf{S}^{d-1}$ i.e. each point of $\mathbf{S}^{d-1}$ belongs to at least k open hemispheres. It is sufficient to prove that any k-fold, $k \geq 1$ covering of $\mathbf{S}^{d-1}, d \geq 2$ with open hemispheres $\mathbf{HS}_i^{d-1}$, $i \in I$ contains $\lceil \frac{k}{2(d-1)} \rceil$ simple coverings (i.e. 1-fold coverings) of $\mathbf{S}^{d-1}$ each of which consists of at most 2d open hemispheres. (Here $\lceil x \rceil$ denotes the upper integer part of x.) We recall the following statement: Let $\mathbf{P} \subset \mathbf{E}^d, d \geq 2$ be a convex polytope with $O \in \text{int } \mathbf{P}$. Then either there are at most $(2d-1)$ vertices of **P** the interior of the convex hull of which contains O or there exist diagonals of **P** intersecting in O the interior of the convex hull of which contains O. (A slightly weaker dual version of this statement can be found in Grünbaum (1967), pp. 67-68.) Observe now that a collection of open hemispheres covers $\mathbf{S}^{d-1}$ if and only if the interior of the convex hull of the centers of a finite subcollection of the given open hemispheres contains O. Thus there exists at most 2d open hemispheres of $\{\mathbf{HS}_i^{d-1} | i \in I\}$ which cover $\mathbf{S}^{d-1}$. As each point of $\mathbf{S}^{d-1}$ belongs to at most $2(d-1)$ of these hemispheres and $\{\mathbf{HS}_i^{d-1} | i \in I\}$ is a k-fold covering of $\mathbf{S}^{d-1}$ removing the open hemispheres we have found (the number of which is at most 2d) we get a $(k - 2(d-1))$-fold covering of $\mathbf{S}^{d-1}$. Apply this step a total of $\lceil \frac{k}{2(d-1)} \rceil$ times until we get $\lceil \frac{k}{2(d-1)} \rceil$ simple coverings of $\mathbf{S}^{d-1}$ each of which consists of at most 2d open hemispheres.

(b) Let H be a hyperplane of $\mathbf{E}^d, d \geq 3$ and take an arbitrary convex $(d-1)$-polytope of H with $(n-1)$ vertices $V_1, V_2, \ldots, V_{n-1}$. Then let $V_n \in \mathbf{E}^d \setminus H$. We construct the directions $\underline{v}_1, \underline{v}_2, \ldots, \underline{v}_n$ for the convex d-polytope $\mathbf{P} = \text{conv } \{V_1, V_2, \ldots, V_n\}$ in the following way (conv (.) stands for the convex hull). Let $\underline{v}_n = \overrightarrow{V_n X}$ with $X \in \text{relint } (\text{ conv } \{V_1, V_2, \ldots, V_{n-1}\})$. Then it is

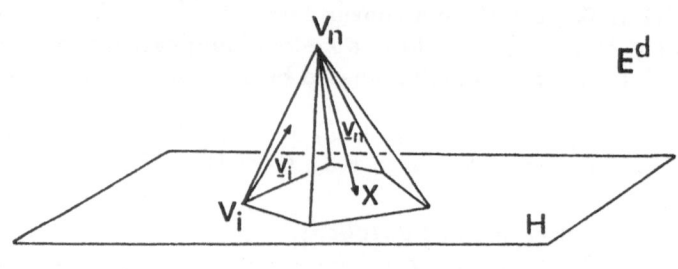

Fig. 5

easy to see that $\underline{v}_n$ illuminates V_n but does not illuminate any other vertex of $\mathbf{P}$. Finally, let $V_i \in \{V_1, V_2, \ldots, V_{n-1}\}$ be an arbitrary vertex. It is easy to prove that there exists a vector $\underline{v}_i$ close to $\overrightarrow{V_i V_n}$ which illuminates V_i but does not illuminate any other vertex of $\mathbf{P}$ (see Fig. 5). Thus the directions $\{\underline{v}_1, \underline{v}_2, \ldots, \underline{v}_n\}$ illuminate $\mathbf{P}$ but no real subset of $\{\underline{v}_1, \underline{v}_2, \ldots, \underline{v}_n\}$ illuminates $\mathbf{P}$.

(c) Let $O \in \text{int } \mathbf{B}$. Then $\mathbf{B}^* \subset \mathbf{E}^2$ is a convex domain with $O \in \text{int } \mathbf{B}^*$. According to Theorem 2 the directions $\{\underline{v}_i | i \in I\}$ illuminate $\mathbf{B}$ if and only if every face of $\mathbf{B}^*$ lies in at least one of the open half-planes $\{X \in \mathbf{E}^2 | \langle \overrightarrow{OX}, -\underline{v}_i \rangle > 0\}$, $i \in I$. Let $\mathcal{C} \subset \mathbf{E}^2$ be a circle centered at O with positive radius. If $\{\hat{F}_j^* | j \in J\}$ is the set of all faces of $\mathbf{B}^*$, then let $\{\hat{F}_j^* | j \in J\}$ be the set of the central projections of the faces of $\mathbf{B}^*$ from O onto $\mathcal{C}$. (Every $\hat{F}_j^*$ is either a point of $\mathcal{C}$ or a closed arc the central angle of which is smaller than π.) Take the open hemicircles $HC_i = \{X \in \mathbf{E}^2 | \langle \overrightarrow{OX}, -\underline{v}_i \rangle > 0\} \cap \mathcal{C}$, $i \in I$. They cover $\mathcal{C}$ thus by a lemma used in (a) there are either 3 say, HC_1, HC_2, HC_3 or 4 say, HC_1, HC_2, HC_3, HC_4 of them which also cover $\mathcal{C}$. In addition in the latter case we may assume that $HC_1 \cap HC_2 = \emptyset$ and $HC_3 \cap HC_4 = \emptyset$. First, assume that $HC_1 \cup HC_2 \cup HC_3 = \mathcal{C}$. Then the sets $c_1 = HC_2 \cap HC_3, c_2 = HC_1 \cap HC_3$ and $c_3 = HC_1 \cap HC_2$ are non-empty pairwise disjoint open arcs of $\mathcal{C}$. If $\hat{F}_j^*$ $(j \in J)$ is not contained by any of the open hemicircles HC_1, HC_2 and HC_3, then $\hat{F}_j^*$ contains at least one of the open arcs c_1, c_2 and c_3. So the number of the closed arcs $\hat{F}_j^*$ which are not contained by any of HC_1, HC_2 and HC_3 is at most 3. Finally, for each of these arcs take an open hemicircle $HC_i (i \in I)$ which contains that. Thus we have found at most 6 open hemicircles of $\{HC_i | i \in I\}$ such that every closed arc $\hat{F}_j^*$ is contained by at least one of them. In other words we have proved that at most 6 of the directions $\{\underline{v}_i | i \in I\}$ illuminate $\mathbf{B}$ provided that $HC_1 \cup HC_2 \cup HC_3 = \mathcal{C}$. Secondly, one can apply a similar argument to get the desired claim in the case left. □

Corollary 3. *Any $(2d-1)-$fold $, d \geq 2$ illumination of the smooth convex body $\mathbf{B}$ of $\mathbf{E}^d$ can be decomposed into two simple (i.e. $1-$fold) illuminations of $\mathbf{B}$.*

3. Some Simple Remarks on $\mathcal{H}(\mathbf{B})$

The following lemma describes a separation problem which is equivalent to Hadwiger's covering problem.

Lemma 6. *If $\mathbf{B}$ is a convex body of $\mathbf{E}^d, d \geq 1$ that contains the origin O as an interior point, then $\mathcal{H}(\mathbf{B}) = I_0(\mathbf{B})$ is the smallest number of hyperplanes of $\mathbf{E}^d$ which strictly separate O from the faces of the polar body $\mathbf{B}^* = \{X \in \mathbf{E}^d | \langle \overrightarrow{OX}, \overrightarrow{OY} \rangle \leq 1$ for all $Y \in \mathbf{B}\}$.*

Proof. It was shown by P. S. Soltan (1963) that $\mathcal{H}(\mathbf{B}) = I_0(\mathbf{B})$ for every convex body $\mathbf{B}$ of $\mathbf{E}^d, d \geq 1$. Thus it is sufficient to apply Theorem 1 in order to get the desired claim. □

Though Lemma 7 summarizes some known results on $\mathcal{H}(\mathbf{B})$ the proofs based on the separation technique of Lemma 6 seem to be new. It is worth mentioning that the proofs have another common feature namely, each of them reduces the problem of $\mathcal{H}(\mathbf{B})$ to a problem on spherical tilings. Though the separation technique works in general the spherical reduction fails to lead to the required upper bound on $\mathcal{H}(\mathbf{B})$ in case of general convex bodies $\mathbf{B} \subset \mathbf{E}^d, d \geq 3$ (see Remark 4). Thus in a sense we are lucky in these cases and as it is shown in the next sections we need deeper information about separation by hyperplanes to attack Hadwiger's conjecture.

Lemma 7. (a) (Levi (1955), Gohberg and Markus (1960) and Boltjansky (1960)). *If $\mathbf{B}$ is a planar convex domain other than a parallelogram, then $\mathcal{H}(\mathbf{B}) = 3$; if $\mathbf{B}$ is a parallelogram, then $\mathcal{H}(\mathbf{B}) = 4$.*
(b) (Martini (1985)). *If $\mathbf{B}$ is a zonotope of $\mathbf{E}^d, d \geq 3$ other than a parallelotope, then $\mathcal{H}(\mathbf{B}) \leq 3 \cdot 2^{d-2}$; if $\mathbf{B}$ is a parallelotope of $\mathbf{E}^d$, then $\mathcal{H}(\mathbf{B}) = 2^d$.*
(c) (Boltjansky (1960)). *For any convex body $\mathbf{B}$ of $\mathbf{E}^d, d \geq 3$ with no more than d corners (i.e. boundary points with non-uniquely determined supporting hyperplanes) $\mathcal{H}(\mathbf{B}) = d + 1$.*

Proof. (a) First we suppose that $\mathbf{B} \subset \mathbf{E}^2$ is a convex domain other than a parallelogram. Let $\mathcal{C}$ be a circle centered at $O(\in \text{int } \mathbf{B})$ with positive radius. Let $\hat{F}_i^*$ be the central projection of the face F_i^* of the polar set $\mathbf{B}^*$ from $O \in \text{int } \mathbf{B}^*$ onto $\mathcal{C}$. (Here $\{F_i^* | i \in I\}$ denotes the set of all faces of $\mathbf{B}^*$.) Every $\hat{F}_i^*$ is a closed arc of $\mathcal{C}$ the central angle of which is smaller than π (Fig. 6). Because of Lemma 6 it is sufficient to find lines h_1, h_2, h_3 which strictly separate O from the arcs $\{\hat{F}_i^* | i \in I\}$ of $\mathcal{C}$. Without loss of generality we may suppose that the endpoints of the 1-dimensional arcs belong to $\{\hat{F}_i^* | i \in I\}$ as 0-dimensional arcs. Either there exist a point $\hat{F}_k^*$ and a 1-dimensional arc $\hat{F}_l^*$ with $O \in \text{int } [\text{conv } (\hat{F}_k^* \cup \hat{F}_l^*)]$ (Fig. 6) or the set of 0-dimensional arcs of $\{\hat{F}_i^* | i \in I\}$ is symmetrical about O. In the first case there are lines h_1 and h_2 which strictly separate O from any arc $\hat{F}_i^* (i \in I)$ of $\mathcal{C}$ except $\hat{F}_l^*$ (see Fig. 6). Since $\hat{F}_l^*$ possesses a central angle smaller that π it can be strictly separated by a line h_3 from O. In the second case we proceed similarly, i.e. we consider the points $\hat{F}_k^*$ and $\hat{F}_l^*$ which are symmetrical about O. Since at both points at most two 1-dimensional arcs of $\{\hat{F}_i^* | i \in I\}$ meet and since $\mathbf{B}^*$ cannot be a quadrangle whose diagonals meet at O therefore we may suppose that the 1-dimensional arcs of $\{\hat{F}_i^* | i \in I\}$ meeting at $\hat{F}_l^*$ can be strictly separated by a line h_3 from O. The rest of the proof is the same as before. Hence we have proved that for a convex domain $\mathbf{B} \subset \mathbf{E}^d$ other than a parallelogram $\mathcal{H}(\mathbf{B}) = 3$. Finally, suppose that $\mathbf{B} \subset \mathbf{E}^2 (O \in \text{int } \mathbf{B})$ is a parallelogram. Then $\mathbf{B}^* \subset \mathbf{E}^2$ is a quadrangle whose diagonals meet at O. Consequently, it is obvious that the smallest number of lines which strictly separate the faces of $\mathbf{B}^*$ from O is four i.e. $\mathcal{H}(\mathbf{B}) = 4$.

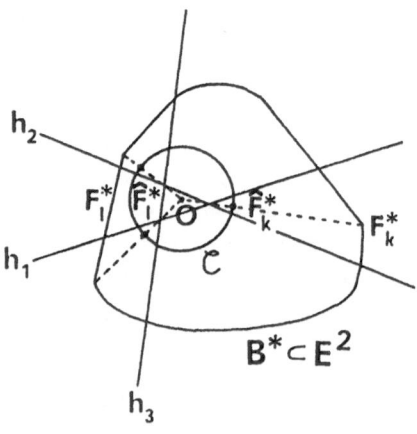

Fig. 6

(b) Let $\mathbf{B}$ be a zonotope of $\mathbf{E}^d, d \geq 3$ and let $\mathbf{S}^{d-1}$ be a $(d-1)$-dimensional sphere of $\mathbf{E}^d$ centered at $O \in \text{int } \mathbf{B}$ with positive radius. Since $\mathbf{B}$ is Minkovski sum of finitely many line segments therefore the central projection of the faces of the polar polytope $\mathbf{B}^*$ from O onto $\mathbf{S}^{d-1}$ is a tiling of $\mathbf{S}^{d-1}$ generated by finitely many $(d-2)$-dimensional great spheres. Lemma 6 implies that it is sufficient to prove the following. Let $\mathcal{T}$ be a tiling of $\mathbf{S}^{d-1}, d \geq 2$ generated by finitely many $(d-2)$-dimensional great spheres such that each face of $\mathcal{T}$ lies on an open hemisphere (i.e. each face of $\mathcal{T}$ has (spherical) diameter less than π). Then either there are at most $3 \cdot 2^{d-2}$ open hemispheres of $\mathbf{S}^{d-1}$ such that each face of $\mathcal{T}$ lies on at least one of the open hemispheres or $\mathcal{T}$ is a dissection of $\mathbf{S}^{d-1}$ by d linearly independent $(d-2)$-dimensional great spheres which we call a crosstiling of $\mathbf{S}^{d-1}$. We prove this statement by induction on d. Case $d = 2$ follows from (a). Then suppose that the claim is true for any tiling of dimension $< d - 1$ and take the tiling $\mathcal{T}$ which is the dissection of $\mathbf{S}^{d-1}$ by the $(d-2)$-dimensional great spheres $GS_1, GS_2, \ldots, GS_n$ $(n \geq d)$. Obviously, $\mathcal{T} \cap GS_i$ $(1 \leq i \leq n)$ is a tiling of $\mathbf{S}^{d-2}$ generated by the $(d-3)$-dimensional great spheres $GS_i \cap GS_j$, $j \neq i$, $1 \leq j \leq n$ such that each face of $\mathcal{T} \cap GS_i$ lies on an open hemisphere of $\mathbf{S}^{d-2}$. If $\mathcal{T} \cap GS_i$ is a crosstiling of $\mathbf{S}^{d-2}$ for any $i \in \{1, 2, \ldots, n\}$, then $\mathcal{T}$ is a crosstiling of $\mathbf{S}^{d-1}$. Finally, if $\mathcal{T} \cap GS_i$ is other than a crosstiling of $\mathbf{S}^{d-2}$ for some $i \in \{1, 2, \ldots, n\}$, then by induction there are at most $3 \cdot 2^{d-3}$ open hemispheres of $\mathbf{S}^{d-2}$ such that each face of $\mathcal{T} \cap GS_i$ lies on at least one of the open hemispheres of $\mathbf{S}^{d-2}$. As any face of $\mathcal{T}$ is not bisected by GS_i the existence of at most $2 \cdot (3 \cdot 2^{d-3}) = 3 \cdot 2^{d-2}$ open hemispheres of $\mathbf{S}^{d-1}$ with the required property follows very easily.

(c) Consider a convex body $\mathbf{B}$ of $\mathbf{E}^d$, $d \geq 3$ with at most d corners and $O \in$ int $\mathbf{B}$. Since $\mathbf{B}$ has at most d boundary points where the supporting hyperplane is not uniquely determined therefore the polar set $\mathbf{B}^*$ ($O \in$ int $\mathbf{B}^*$) possesses at most d faces of dimension ≥ 1. Let us denote these faces by $F_1^*, F_2^*, \ldots, F_n^*$ ($0 \leq n \leq d$) and their central projections from O onto a $(d-1)$-dimensional sphere $\mathbf{S}^{d-1}$ of $\mathbf{E}^d$ centered at O with positive radius by $\hat{F}_1^*, \hat{F}_2^*, \ldots, \hat{F}_n^*$. Because of Lemma 6 it is sufficient to show that $\hat{F}_1^*, \hat{F}_2^*, \ldots, \hat{F}_n^*$ and the points of $\mathbf{S}^{d-1}$ which are not covered by any of $\hat{F}_1^*, \hat{F}_2^*, \ldots, \hat{F}_n^*$ can be strictly separated from O by $(d+1)$ hyperplanes of $\mathbf{E}^d$. Let H_i, $1 \leq i \leq n$ be a hyperplane of $\mathbf{E}^d$ which strictly separates O from $\hat{F}_i^*$. Then shifting and slight turning of the hyperplanes $H_1, H_2, \ldots, H_n$ yield the hyperplanes $\mathbf{H}_1, \mathbf{H}_2, \ldots, \mathbf{H}_n$ which together with the hyperplanes $\mathbf{H}_{n+1}, \ldots, \mathbf{H}_{d+1}$ bound a simplex S of $\mathbf{E}^d$ whose interior contains O such that each $\hat{F}_i^*$, $1 \leq i \leq n$ is strictly separated from O by $\mathbf{H}_i$. Finally, a small homothetic copy of S with homothety-center O has the property that the hyperplanes of its facets strictly separate O from $\hat{F}_1^*, \hat{F}_2^*, \ldots, \hat{F}_n^*$ and the points of $\mathbf{S}^{d-1}$ which are not covered by any of $\hat{F}_1^*, \hat{F}_2^*, \ldots, \hat{F}_n^*$. $\square$

Remark 4. The proof of (a) gives the following Gallai-type statement: Let the circle $\mathcal{C}$ of $\mathbf{E}^2$ with a system $\{\hat{F}_i | i \in I\}$ of arcs of central angles $\{\varphi_i | 0 < \varphi_i \leq \pi, i \in I\}$ be given such that for any $i \neq j$, $\{i, j\} \subset I$ there exists a pair $\{P_{ij}^1, P_{ij}^2\}$ of antipodel points of $\mathcal{C}$ with $P_{ij}^1 \in \hat{F}_i$ and $P_{ij}^2 \in \hat{F}_j$. Then there are points N_1, N_2, N_3, N_4 of $\mathcal{C}$ having the property that for every $i \in I$ $\{N_1, N_2, N_3, N_4\} \cap$ relint $_c\hat{F}_i \neq \emptyset$. (Here relint $_c\hat{F}_i \subset \hat{F}_i$ means the open arc of $\mathcal{C}$ determined by the endpoints of $\hat{F}_i$.) It is not hard to see that the analogous statement on the d-sphere, $d \geq 2$ is false.

4. On Convex Bodies with Finitely Many Corner Points

We prove

Theorem 9 (Charazishvili (1973)). *If $\mathbf{B}$ is a convex body of $\mathbf{E}^3$ with at most 4 corner points, then $\mathcal{H}(\mathbf{B}) = 4$. Moreover, there exists a convex body $\mathbf{B}$ of $\mathbf{E}^d$, $d \geq 4$ ($\tilde{\mathbf{B}}$ of $\mathbf{E}^d$, $d \geq 3$, resp.) with $d+1$ ($d+2$, resp.) corner points such that $\mathcal{H}(\mathbf{B}) = d+2$ ($\mathcal{H}(\tilde{\mathbf{B}}) = d+2$ resp.).*

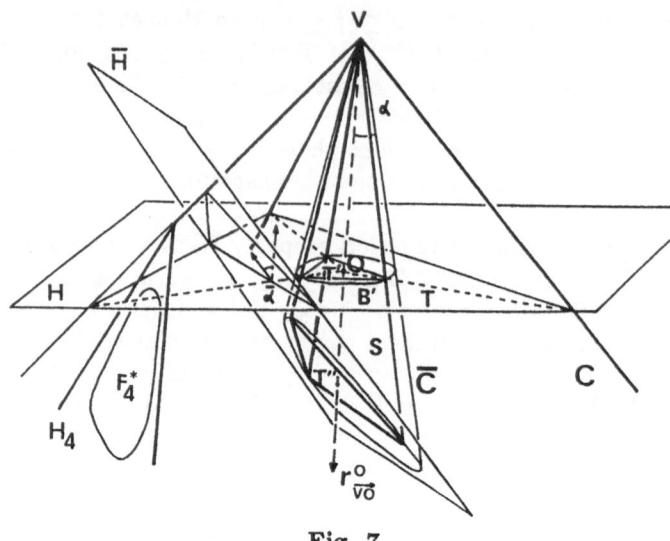

Fig. 7

Proof. Let the origin O be an interior point of the convex body $\mathbf{B}$ of $\mathbf{E}^3$ with at most four corner points. Take the polar body $\mathbf{B}^*$ of $\mathbf{B}$. Since $\mathbf{B}$ has at most 4 corner points therefore the convex body $\mathbf{B}^*$ possesses at most 4 faces of dimension ≥ 1. Let us denote these faces by $F_1^*, F_2^*, \ldots, F_n^*$ $(0 \leq n \leq 4)$ and the (not necessarily unique) supporting planes of them by $H_1, H_2, \ldots, H_n$. (Of course, $F_i^* = H_i \cap \mathbf{B}^*$ for any $1 \leq i \leq n$.) Let H_i^+ denote the (closed) supporting half-space determined by H_i $(1 \leq i \leq n)$. ($\mathbf{B}^* \subset H_i^+$ for any $1 \leq i \leq n$.) Since $\mathbf{B}$ has finitely many corner points therefore $0 \leq \mathrm{card}\ (F_i^* \cap F_j^*) \leq 1$ for any $i \neq j$, $1 \leq i \leq n$, $1 \leq j \leq n$. Because of Lemma 6 in order to prove $\mathcal{H}(\mathbf{B}) = 4$ it is sufficient to show the following.

Lemma 8. *If $\mathbf{B}^* \subset \mathbf{E}^3$ is a convex body with $O \in \mathrm{int}\ \mathbf{B}^*$ and with at most 4 faces say, $F_1^*, F_2^*, \ldots, F_n^*$ $(0 \leq n \leq 4)$ of dimension ≥ 1 which in addition possess the property that $0 \leq \mathrm{card}\ (F_i^* \cap F_j^*) \leq 1$ $(i \neq j, \{i, j\} \subset \{1, 2, \ldots, n\})$, then O can be strictly separated by 4 planes from the faces of $\mathbf{B}^*$.*

Proof. $\bigcap_{i=1}^{n} H_i^+$ is a closed convex set of $\mathbf{E}^3$ which contains $\mathbf{B}^*$. We distinguish the following three cases:

(a) $\mathrm{card}[\ \mathrm{vert}\ (\bigcap_{i=1}^{n} H_i^+)] \geq 2$,
(b) $\mathrm{card}[\ \mathrm{vert}\ (\bigcap_{i=1}^{n} H_i^+)] = 1$,
(c) $\mathrm{card}[\ \mathrm{vert}\ (\bigcap_{i=1}^{n} H_i^+)] = 0$.

Case (a). Obviously, $n = 4$ and if V is a vertex of $\bigcap_{i=1}^{4} H_i^+$, then V belongs to 3 facets of $\bigcap_{i=1}^{4} H_i^+$ say, $V \in H_1 \cap H_2 \cap H_3$. Thus $V \notin H_4$. Let us consider $\mathbf{C} = H_1^+ \cap H_2^+ \cap H_3^+$ which is a closed convex cone with apex V. Then there exists a plane H passing through O which intersects $\mathbf{C}$ in a triangle $T(= H \cap \mathbf{C})$. Without loss of generality we may suppose that $\overrightarrow{OV}$ is orthogonal to H. Take the plane $\overline{H}$ which strictly separates F_4^* from O and intersects the open ray $r_{\overrightarrow{VO}}^O$ emanating from O and having direction vector $\overrightarrow{VO}$. Let $\overline{\alpha}$ denote the angle of $r_{\overrightarrow{VO}}^O$ and $\overline{H}$ (Fig. 7). Finally, let us consider the smaller homothetic copy T' of T with homothety-center O which has the property that the angle α of the cone $\overline{C}$ with apex V and base B' where B' is the circumscribed circle of T' is smaller than $\overline{\alpha}$. Thus $\overline{C} \cap \overline{H}$ is a planar compact convex set possessing the inscribed triangle T'' which is the central projection of T' from V onto $\overline{H}$. Let $S = \mathrm{conv}(\{V\} \cup T'')$. S is a 3-dimensional simplex and it has the smaller homothetic copy $\overline{S}$ with homothety center O the planes of the facets of which strictly separate the faces of $\mathbf{B}^*$ from O.

Case (b). Let V be the only vertex of $\bigcap_{i=1}^n H_i^+$. If $n = 3$ or one of the faces $F_1^*, F_2^*, F_3^*, F_4^*$ say, F_4^* does not contain V, then one can use the method of case (a) to prove the statement. Thus we suppose that $n = 4$ and $V \in F_1^* \cap F_2^* \cap F_3^* \cap F_4^*$. Then $\mathbf{C} = \bigcap_{i=1}^4 H_i^+$ is a closed convex cone with apex V having 4 unbounded facets. Take the plane H passing through O which intersects $\mathbf{C}$ in a quadrangle $Q(= H \cap \mathbf{C})$. We may suppose that $\overrightarrow{OV}$ is orthogonal to H. Since $0 \le \mathrm{card}(F_i^* \cap F_j^*) \le 1$ $(i \ne j, 1 \le i \le 4, 1 \le j \le 4)$ therefore $F_j^* \cap (H_i \cap H_j) = \{V\}$ for any $i \ne j$, $1 \le i \le 4$, $1 \le j \le 4$. Choose a plane $\overline{H}$ parallel to H very close to V such that it strictly separates V from H. Let $\overline{H}^+$ denote the closed half-space bounded by $\overline{H}$ which is disjoint from V. Obviously, $F_i^* \cap \overline{H}^+$ is a planar compact convex set with $(F_i^* \cap \overline{H}^+) \cap (H_i \cap H_j) = \emptyset$ for any $1 \le i \le 4$ and $j \ne i$, $1 \le j \le 4$. Finally, let the closed (line) segment s_i be the central projection of $F_i^* \cap \overline{H}^+$ from V onto $H_i \cap H$ $(1 \le i \le 4)$. The segments s_1, s_2, s_3 and s_4 are pairwise disjoint. (In fact $s_i \cap (H_i \cap H_j) = \emptyset$, $i \ne j$, $1 \le i \le 4$, $1 \le j \le 4$.) We claim that there are planes $\mathbf{H}_1, \mathbf{H}_2, \mathbf{H}_3$ which strictly separate O from the faces F_1^*, F_2^*, F_3^* and F_4^* such that they bound the closed half-spaces $\mathbf{H}_1^+, \mathbf{H}_2^+$ and $\mathbf{H}_3^+$ whose intersection is a closed convex cone containing O as an interior point. Having this the rest of the proof is obvious. So let us see the proof of the above claim.

If O is not the intersection point of the diagonals of Q, then the statement is a simple corollary of Lemma 7 (a). Thus we suppose that O lies on both diagonals of Q (Fig. 8). Consider the closed segments s_i and s_j on two consecutive sides of Q. Then there exists a line $l_1 \subset H$ very close to O which strictly separates O from $s_i \cup s_j$ in H.

Consequently, if H is considered to be a horizontal plane, then the almost vertical plane $\mathbf{H}_1$ which passes through l_1 strictly separates $F_i^* \cup F_j^*$ from O. Finally, the existence of $\mathbf{H}_2$ and $\mathbf{H}_3$ with the required property is obvious.

Case (c). Since $\mathrm{card}[\mathrm{vert}(\bigcap_{i=1}^n H_i^+)] = 0$ therefore there exists a line l of $\mathbf{E}^3$ which is parallel to any of the planes $H_1, H_2, \ldots, H_n$. Let H be a plane orthogonal to l and let $\overline{B}$ be the orthogonal projection of $\mathbf{B}^*$ onto H. Since $O \in \mathrm{int}\, \mathbf{B}^*$ therefore the orthogonal projection $\overline{O}$ of O onto H is a relative

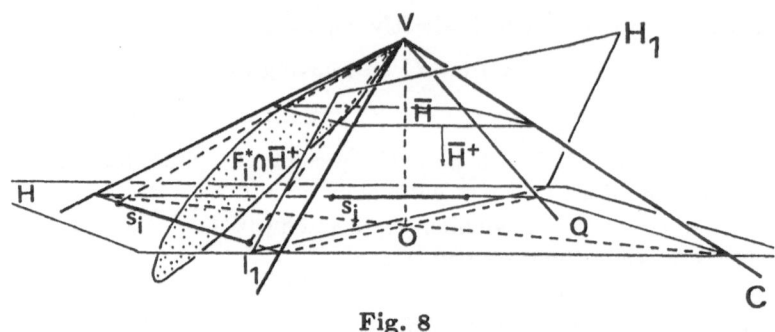

Fig. 8

interior point of $\overline{B}$. Thus Lemma 7 (a) implies the existence of the lines l_1, l_2, l_3 and l_4 of H which strictly separate $\overline{O}$ from the faces of $\overline{B}$. Consider H a horizontal plane. Then the vertical planes $\mathbf{H}_1, \mathbf{H}_2, \mathbf{H}_3$ and $\mathbf{H}_4$ which contain the lines l_1, l_2, l_3 and l_4 strictly separate the faces $F_1^*, F_2^*, \ldots, F_n^*$ of $\mathbf{B}^*$ from O. Shifting and slight turning of the planes $\mathbf{H}_1, \mathbf{H}_2, \mathbf{H}_3$ and $\mathbf{H}_4$ yield a narrow simplex around O the planes of the facets of which strictly separate O from the faces of $\mathbf{B}^*$.

This completes the proof of Lemma 8. $\square$

Constructions. We construct a special $\tilde{\mathbf{B}}$ which then can be used in the construction of $\mathbf{B}$.

Let $\mathbf{S}$ be a $(d-1)$-dimensional simplex of the hyperplane H of $\mathbf{E}^d, d \geq 3$ with vertices $V_1, V_2, \ldots, V_d$ and centroid O. Let F_i be the facet of $\mathbf{S}$ opposite to V_i with centroid O_i, $1 \leq i \leq d$. Take the $(d-2)$-dimensional simplices $\tilde{F}_i = \{X \in \mathbf{E}^d |$ there exists $Y \in F_i$ with $\overline{O_iX} = \frac{1}{3} \cdot \overline{O_iY}\}$, $1 \leq i \leq d$ of H (Fig. 9). It is easy to check that any two of the sets $\tilde{F}_i$, $1 \leq i \leq d$ have a line passing through O which intersects both sets. We choose $\frac{1}{2} < \lambda < 1$ such that the points $\tilde{V}_i$, $1 \leq i \leq d$ satisfy $\overrightarrow{O\tilde{V}_i} = \lambda \cdot \overrightarrow{OV_i}$, $1 \leq i \leq d$ and lie outside conv$[\cup\{\tilde{F}_i | 1 \leq i \leq d\}]$. Finally, let $A_iB_i(C_iD_i$, resp. $)$ be the segment (of $\mathbf{E}^d$) of length 2 $(2\lambda(d-1)$, resp. $)$ with midpoint $O_i(\tilde{V}_i$, resp. $)$ being orthogonal to H, $1 \leq i \leq d$. We may choose the notation such that the segments A_iD_i, B_iC_i contain O as a relative interior point, $1 \leq i \leq d$. As $\lambda \cdot (d-1) > 1$ the points C_i, D_i lie further from H than the points A_i, B_i, $1 \leq i \leq d$. Thus there exists a smooth convex body $\tilde{\mathbf{B}}^*$ of $\mathbf{E}^d$ whose faces of dimension ≥ 1 are conv$[\tilde{F}_i \cup \{A_i, B_i\}]$, $1 \leq i \leq d$ and conv$[\{C_i | 1 \leq i \leq d\}]$, conv$[\{D_i | 1 \leq i \leq d\}]$. As any two of these $(d+2)$ facets of $\tilde{\mathbf{B}}^*$ have a line passing through O which intersects both facets the smallest number of hyperplanes of $\mathbf{E}^d$ which strictly separate O from the faces of $\tilde{\mathbf{B}}^*$ is $d+2$. Thus by Lemma 6 we get that $\mathcal{H}(\tilde{\mathbf{B}}) = d+2$ for the convex body $\tilde{\mathbf{B}}$ of $\mathbf{E}^d$ the polar body of which is $\tilde{\mathbf{B}}^*$ and $\tilde{\mathbf{B}}$ has $(d+2)$ corner points.

Finally, we construct a convex body $\mathbf{B}$ of $\mathbf{E}^d, d \geq 4$ with $d+1$ corner points such that $\mathcal{H}(\mathbf{B}) = d+2$. We can get the polar body $\mathbf{B}^*$ of $\mathbf{B}$ in the following way. We take the convex body $\tilde{\mathbf{B}}^*$ of $\mathbf{E}^{d-1}$ constructed above lying

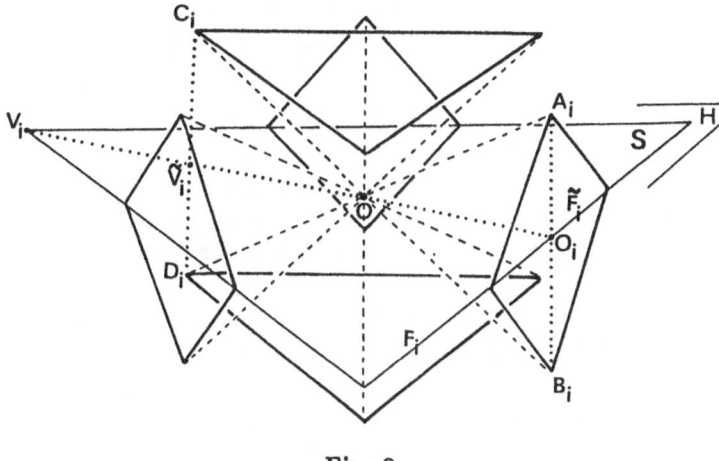

Fig. 9

in a hyperplane $\tilde{H}$ of $\mathbf{E}^d$ and then we choose a point $V \in \mathbf{E}^d \backslash \tilde{H}$. It is not hard to see that there exists a convex body $\mathbf{B}^*$ of $\mathbf{E}^d$ the number of the faces of dimension ≥ 1 of which is $d+1$ in addition whose mentioned faces are facets (on bd $\mathbf{B}^*$) and each of them contains V and one of the $(d-2)$-dimensional faces of $\tilde{\mathbf{B}}^* \subset \tilde{H}$ and whose boundary is smooth everywhere except at V. Obviously, we need $(d+2)$ hyperplanes of $\mathbf{E}^d$ in order to strictly separate the faces of $\mathbf{B}^*$ from $O \in \mathrm{relint}\ \tilde{\mathbf{B}}^* \cap \mathrm{int}\ \mathbf{B}^*$. Thus for the polar body $\mathbf{B}$ of $\mathbf{B}^*$ we have $\mathcal{H}(\mathbf{B}) = d+2$ (see Lemma 6) and $\mathbf{B}$ has $d+1$ corner points.   □

Remark 5. In fact, one can prove that there exists a constant $2 > c > 1$ such that for any sufficiently large d there is a convex body $\mathbf{B}$ of $\mathbf{E}^d$ which finitely many corner points and $\mathcal{H}(\mathbf{B}) \geq c^d$ (see also Erdős and Füredi (1983)).

5. Solution of Hadwiger-Levi's Covering Problem for Convex Polyhedra with Affine Symmetry

Let $\mathbf{P} \subset \mathbf{E}^3$ be a convex polyhedron with symmetry i.e. let us suppose that the symmetry group of $\mathbf{P}$ consists of the identity and at least one other isometry of $\mathbf{E}^3$. Then it is easy to see that the symmetry group of $\mathbf{P}$ contains either

(1) a reflection in a plane of $\mathbf{E}^3$ or
(2) a reflection in a point of $\mathbf{E}^3$ or
(3) a rotation about a line of $\mathbf{E}^3$ through angle $\frac{2\pi}{n}$, $n \in \{2, 3, \ldots\}$.

In this section we prove

Theorem 10. *If $\mathbf{P}$ is a convex polyhedron of $\mathbf{E}^3$ with symmetry, then* $\mathcal{H}(\mathbf{P}) \leq 8$.

Proof. We distinguish the above mentioned three cases and use the terminology of vertices, edges and faces which refer to 0-, 1- and 2-dimensional faces of convex polyhedra.

(1) Let $\mathbf{P} \subset \mathbf{E}^3$ be a convex polyhedron which is symmetrical in the plane H of $\mathbf{E}^3$. Without loss of generality we may suppose that the origin O of $\mathbf{E}^3$ belongs to $H \cap \text{int } \mathbf{P}$. Let us consider the polar body $\mathbf{P}^*$ of $\mathbf{P}$. Obviously, the convex polyhedron $\mathbf{P}^*$ ($O \in \text{int } \mathbf{P}^*$) is symmetrical in H. We want to prove that $\mathcal{H}(\mathbf{P}) \leq 8$. Lemma 6 implies that we have to find 8 planes of $\mathbf{E}^3$ which strictly separate O from the faces of $\mathbf{P}^*$. Let $\{F_i^* | i \in I\}$ be the set of the faces of $\mathbf{P}^*$ which have at least one point in common with H. The collection of the remaining faces of $\mathbf{P}^*$ is denoted by $\{F_j^* | j \in J\}$. Also, let $\overline{H}_i$ ($\overline{H}_i^+$, resp.) be the supporting plane (the supporting half-space, resp.) of $\mathbf{P}^*$ with $\overline{H}_i \supset F_i^*$ (boundary $\overline{H}_i$, resp.), $i \in I$. Finally, let us consider the polyhedral set $\overline{\mathbf{P}} = \bigcap_{i \in I} \overline{H}_i^+$. Obviously, $\overline{\mathbf{P}}$ is symmetrical in H. The set of the faces of $\overline{\mathbf{P}}$ can be labelled by $\{\overline{F}_i | i \in I\}$ with $\overline{F}_i \supset F_i^*$ for any $i \in I$. As we know (see Lemma 6) $\mathcal{H}(\mathbf{P})$ is the minimum number of planes which strictly separate O from the faces of $\mathbf{P}^*$.

Obviously, it is sufficient to strictly separate O from the faces $\{\overline{F}_i | i \in I\}$ of $\overline{\mathbf{P}}$ by 8 planes provided that these planes strictly separate O from the faces $\{F_j^* | j \in J\}$ of $\mathbf{P}^*$ as well.

If $\overline{\mathbf{P}}$ is unbounded, then $\overline{\mathbf{P}}$ is a cylinder with base $\mathbf{P}^* \cap H$. Hence by Lemma 7 (a) we get the inequality $\mathcal{H}(\mathbf{P}) \leq 6$. Thus we are left with the case when $\overline{\mathbf{P}}$ is bounded, i.e. when $\overline{\mathbf{P}}$ is a convex polyhedron which is symmetrical in H with $O \in H \cap \text{int } \overline{\mathbf{P}}$.

Let T be the convex n-gon $H \cap \mathbf{P}^* = H \cap \overline{\mathbf{P}}$ with vertices $V_1, V_2, \ldots, V_n$. Let H_+ and H_- be the closed half-spaces bounded by H. Then $H_+ \cap \overline{\mathbf{P}}$ and $H_- \cap \overline{\mathbf{P}}$ are convex polyhedra which we denote by $\overline{\mathbf{P}}^+$ and $\overline{\mathbf{P}}^-$. (Obviously, if we reflect $\overline{\mathbf{P}}^+$ in H, then we get $\overline{\mathbf{P}}^-$ and vica versa.) Let the sides $V_1V_2, V_2V_3, \ldots, V_{n-1}V_n$ and V_nV_1 of T be denoted by $s_1, s_2, \ldots, s_{n-1}$ and s_n. If s_k, $1 \leq k \leq n$ is an edge of $\overline{\mathbf{P}}$, then there are faces $\overline{F}_k^+ \subset H_+$ and $\overline{F}_k^- \subset H_-$ of $\overline{\mathbf{P}}$ meeting along s_k. (In fact, in this case $\overline{F}_k^+$ is a face of $\overline{\mathbf{P}}^+$ and $\overline{F}_k^-$ is a face of $\overline{\mathbf{P}}^-$.) If s_k, $1 \leq k \leq n$ is not an edge of $\overline{\mathbf{P}}$, then for the face of $\overline{\mathbf{P}}$ which contains s_k we use both of the notations $\overline{F}_k^+$ and $\overline{F}_k^-$. However, in this case the plane of $\overline{F}_k^+(\overline{F}_k^-)$ must be orthogonal to H and the orthogonal projection of $\overline{F}_k^+(\overline{F}_k^-)$ onto H is s_k. Finally, let l be the line passing through O and being orthogonal to H.

Proposition 1. *If the line passing through V_{k_1} and O bisects the side s_{k_2} of T ($\{k_1, k_2\} \subset \{1, 2, \ldots, n\}$), then $\mathcal{H}(\mathbf{P}) \leq 8$.*

Proof. Without loss of generality we may suppose that $k_2 = 1$. Then let the line $l_1 \subset H$ ($l_2 \subset H$, resp.) strictly separate the vertices $V_{k_1}, \ldots, V_n, V_1$ ($V_2, \ldots, V_{k_1}$, resp.) from O in H (Fig. 10). Imagine that H is a horizontal plane.

Then consider the almost horizontal plane $H_1(H_2$, resp.) for which $H_1 \cap H = l_1$ ($H_2 \cap H = l_2$, resp.) and reflect the plane $H_1(H_2$, resp.) in H getting the almost horizontal plane H_3 (H_4, resp.) with $H_3 \cap H = l_1(H_4 \cap H = l_2$, resp.). Finally, let $H_5(H_6$, resp.) be the vertical plane with $H_5 \cap H = l_1$ ($H_6 \cap H = l_2$, resp.). It is very easy to see that the faces $\{\overline{F}_i | i \in I\}$ of $\mathbf{P}$ and the faces $\{F_j^* | j \in J\}$ of $\mathbf{P}^*$ are strictly separated from O by the planes $H_1, H_2 \ldots H_6$ except the faces $\overline{F}_1^+, \overline{F}_1^-$. But $\overline{F}_1^+$ and $\overline{F}_1^-$ can be strictly separated by two planes from O. This completes the proof of the inequality $\mathcal{H}(\mathbf{P}) \leq 8$.   □

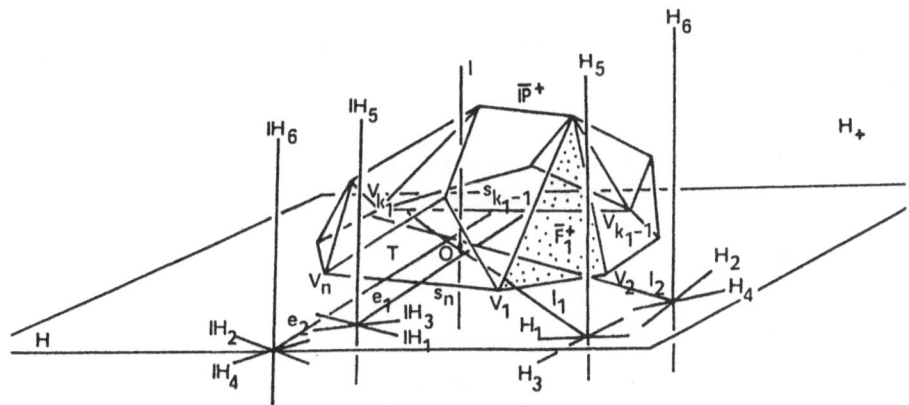

Fig. 10

Proposition 2. *If s_{k_1} and s_{k_2}, $\{k_1, k_2\} \subset \{1, 2, \ldots, n\}$ are two sides of T with $O \in \mathrm{relint}\,[\mathrm{conv}(s_{k_1} \cup s_{k_2})]$, $\overline{F}_{k_1}^+ \cap l = \emptyset$ and $\overline{F}_{k_2}^+ \cap l = \emptyset$, then $\mathcal{H}(\mathbf{P}) \leq 8$.*

Proof. Without loss of generality we may suppose that the two sides in question are $V_{k_1-1}V_{k_1}$ and V_1V_n (Fig. 10). Let e_1 and e_2 be two parallel lines of H between which O lies such that both of them bisect $V_{k_1-1}V_{k_1}$ and V_1V_n. Let $\mathbf{H}_1(\mathbf{H}_2$, resp.) be the almost horizontal plane for which $\mathbf{H}_1 \cap H = e_1(\mathbf{H}_2 \cap H = e_2$, resp.) and finally, reflect the plane $\mathbf{H}_1(\mathbf{H}_2$, resp.) in H getting the almost horizontal plane $\mathbf{H}_3(\mathbf{H}_4$, resp.) with $\mathbf{H}_3 \cap H = e_1$ ($\mathbf{H}_4 \cap H = e_2$, resp.). (Again H is considered to be a horizontal plane.) If $\mathbf{H}_5(\mathbf{H}_6$, resp.) is the vertical plane with $\mathbf{H}_5 \cap H = e_1(\mathbf{H}_6 \cap H = e_2$, resp.), then it is obvious that the faces $\{\overline{F}_i | i \in I\}$ of $\overline{\mathbf{P}}$ and the faces $\{F_j^* | j \in J\}$ of $\mathbf{P}^*$ are strictly separated from O by the planes $\mathbf{H}_1, \mathbf{H}_2, \ldots, \mathbf{H}_6$ except the faces $\overline{F}_{k_1-1}^+, \overline{F}_{k_1-1}^-, \overline{F}_n^+, \overline{F}_n^-$. As $\overline{F}_{k_1-1}^+ \cap l = \emptyset$ (therefore $\overline{F}_{k_1-1}^- \cap l = \emptyset$) and $\overline{F}_n^+ \cap l = \emptyset$ (therefore $\overline{F}_n^- \cap l = \emptyset$) there are vertical planes $\mathbf{H}_7$ and $\mathbf{H}_8$ which strictly separate the faces $\overline{F}_{k_1-1}^+, \overline{F}_{k_1-1}^-$ and $\overline{F}_n^+, \overline{F}_n^-$ of $\overline{\mathbf{P}}$ from O.   □

Corollary 4. *Further on we may suppose that $n = 2m$ ($m \geq 2$) and the diagonals $V_k V_{k+m}$ of T ($1 \leq k \leq m$) intersect each other at O and $l \cap (\overline{F}_k^+ \cup \overline{F}_{k+m}^+) \neq \emptyset$ for any $1 \leq k \leq m$.*

If $m = 2$, then it is easy to see that $\mathcal{H}(\mathbf{P}) \leq 8$ consequently, further on we suppose that $m \geq 3$. With relabelling we can manage to have $l \cap \overline{F}_k^+ \neq \emptyset$ for any $k \in \{1, 2, \ldots, m\}$. (Among others this means that the sides $s_1, s_2, \ldots, s_{2m-1}, s_{2m}$ of T are not necessarily consecutive sides but s_k and s_{k+m}, $1 \leq k \leq m$ are "opposite" sides i.e. $O \in \mathrm{relint}\,[\mathrm{conv}(s_k \cup s_{k+m})]$.) Hence there exists a vertex V of $\overline{\mathbf{P}}^+$ with $l \cap (\bigcap_{1 \leq k \leq m} \overline{F}_k^+) = \{V\}$.

Proposition 3. *If*

(i) *there are faces $\overline{F}_k^+$ and $\overline{F}_{k+m}^+$ of $\overline{\mathbf{P}}$ ($1 \leq k \leq m$) for which $V \in \overline{F}_k^+ \cap \overline{F}_{k+m}^+$*

or

(ii) *the faces $\overline{F}_1^+, \overline{F}_2^+, \ldots, \overline{F}_m^+$ of $\overline{\mathbf{P}}$ have the property that $O \notin \mathrm{relint}\,[\mathrm{conv}(s_1 \cup s_2 \cup \ldots \cup s_m)]$,*

then $\mathcal{H}(\mathbf{P}) \leq 8$.

Proof. Both (i) and (ii) can be proved very easily with the help of Fig. 11. We leave the easy details to the reader.   □

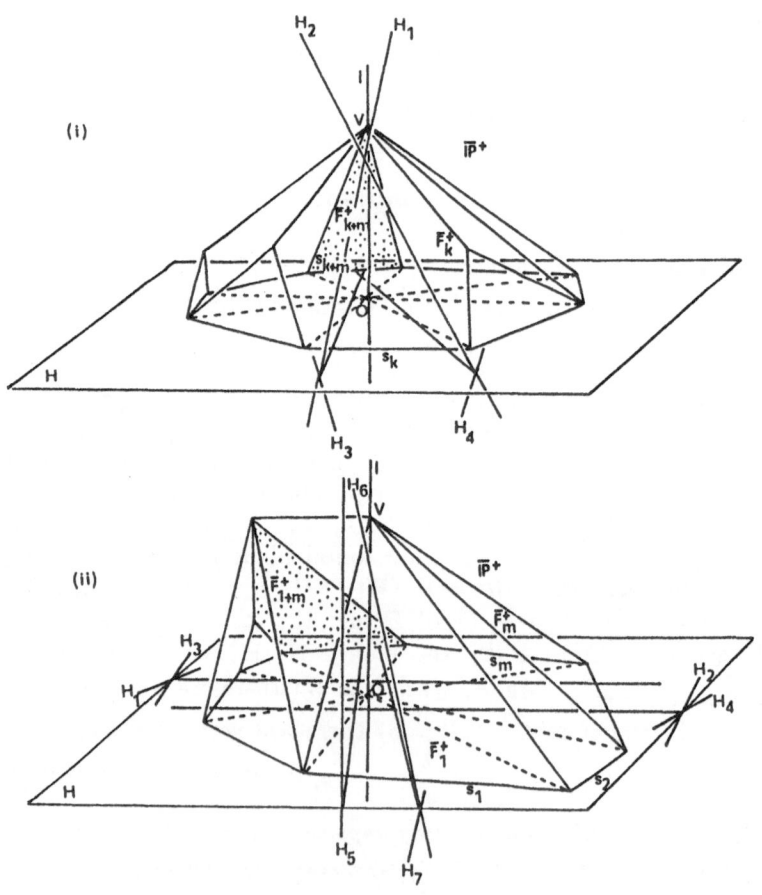

Fig. 11

The following statement completes the proof of Theorem 10 in case (1).

Proposition 4. *Suppose that neither (i) nor (ii) occur. Then* $\mathcal{H}(\mathbf{P}) \leq 8$.

Proof. We distinguish two cases: (iii) There are faces $\overline{F}_k^+$ and $\overline{F}_{k+m}^+$ of $\overline{\mathbf{P}}$ $(1 \leq k \leq m)$ such that neither of them is orthogonal to H_j. (iv) $\overline{F}_k^+ \cap l \neq \emptyset$ and $\overline{F}_{k+m}^+$ is orthogonal to H for any $k \in \{1, 2, \ldots, m\}$.

(iii) Without loss of generality we may suppose that the faces $\overline{F}_1^+$ and $\overline{F}_{1+m}^+$ of $\overline{\mathbf{P}}$ are not orthogonal to H. Since neither (i) nor (ii) occur therefore there are faces of $\overline{\mathbf{P}}$ say, $\overline{F}_2^+$ and $\overline{F}_3^+$ for which $V \in \overline{F}_1^+ \cap \overline{F}_2^+ \cap \overline{F}_3^+$ and $O \in$ relint $[\text{conv}(s_1 \cup s_2 \cup s_3)]$ but $O \notin$ relint $[\text{conv }(s_{k_1} \cup s_{k_2})]$ for any $\{k_1, k_2\} \subset \{1, 2, 3\}$. Let $l_1 \subset H$; $l_2 \subset H$; $l_3 \subset H$ be the lines which strictly separate O from s_3 and some points of s_1; s_1 and some points of s_2; s_2 and some points of s_3 (Fig. 12). Then consider the almost horizontal planes H_1, H_2 and H_3 with $H_1 \cap H = l_1$, $H_2 \cap H = l_2$ and $H_3 \cap H = l_3$ and reflect them in H getting the planes H_4, H_5 and H_6 with $H_4 \cap H = l_1$, $H_5 \cap H = l_2$ and $H_6 \cap H = l_3$. (Again H is considered to be a horizontal plane.) Finally, let l_7 and l_8 be two parallel lines of H between which O lies such that both of them bisect s_1 and s_{1+m}. It is easy to see that the vertical planes H_7 and H_8 for which $H_7 \cap H = l_7$ and $H_8 \cap H = l_8$ together with the almost horizontal planes $H_1, H_2, \ldots, H_6$ strictly separate O from the faces $\{\overline{F}_i | i \in I\}$ of $\overline{\mathbf{P}}$ as well as from the faces $\{F_j^* | j \in J\}$ of $\mathbf{P}^*$. Thus $\mathcal{H}(\mathbf{P}) \leq 8$.

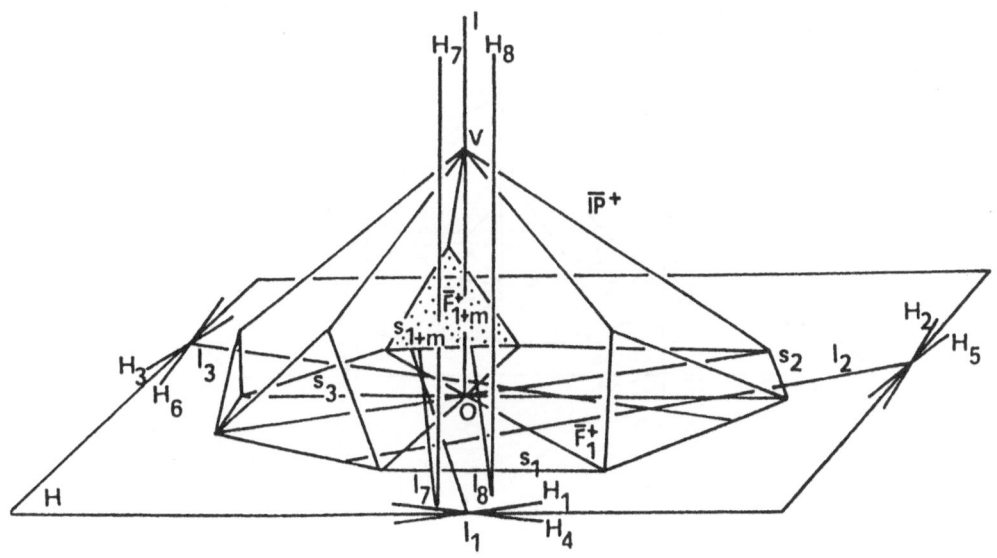

Fig. 12

(iv) In order to make the proof easier it is useful to distinguish the following two subcases:

(a) No two of the sides $s_1, s_2, \ldots, s_m$ are consecutive sides of T. (So $n = 2(2m'+1)$.)

(b) There are integers $k_1 < k_2$ with $1 \leq k_1 \leq k_2 \leq m$ such that s_{k_1} and s_{k_2} are consecutive sides of T. Since both (a) and (b) can be proved easily (see Fig. 13) we leave the details to the reader.   □

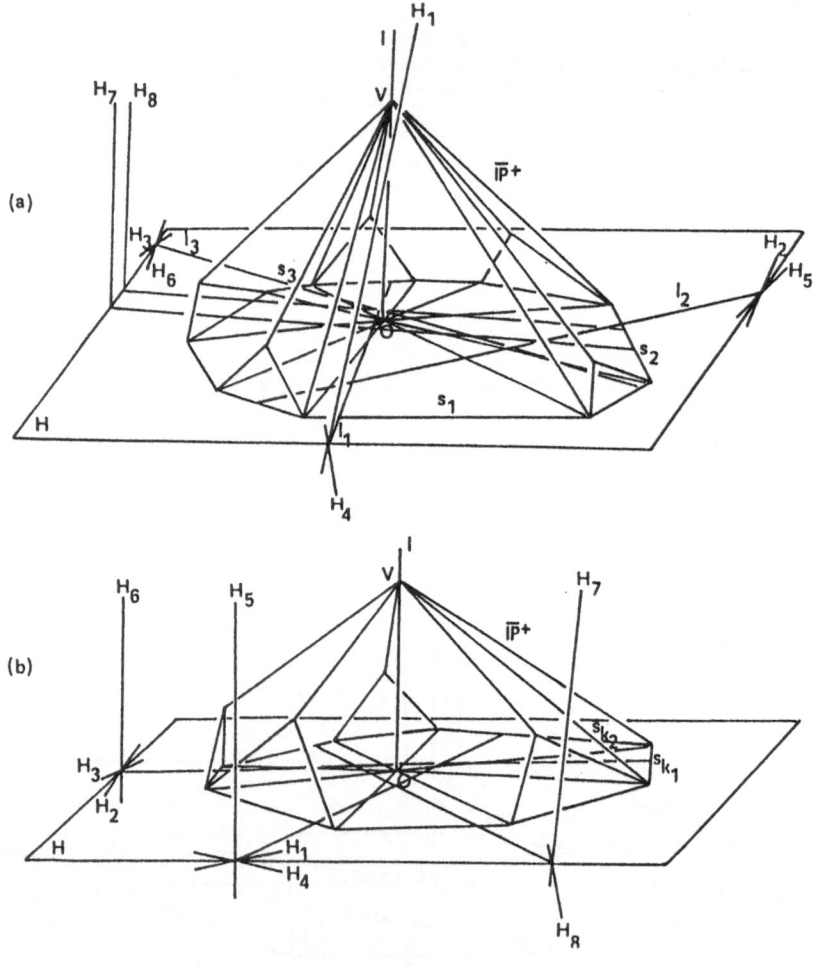

Fig. 13

(2) Let $\mathbf{P}$ be a centrally symmetric convex polyhedron of $\mathbf{E}^3$. Without loss of generality we may suppose that the origin O of $\mathbf{E}^3$ is the center of $\mathbf{P}$. Let us consider the polar body $\mathbf{P}^*$ of $\mathbf{P}$. Obviously, $O \in \text{int } \mathbf{P}^*$ and $\mathbf{P}^*$

is a centrally symmetric convex polyhedron with center O. We want to show that $\mathcal{H}(\mathbf{P}) \leq 8$. Lemma 6 implies that we have to find 8 planes of $\mathbf{E}^3$ which strictly separate O from the faces of $\mathbf{P}^*$. We prove the existence of four pairs of pairwise parallel planes which strictly separate the faces of $\mathbf{P}^*$ from O.

Let H be a plane of $\mathbf{E}^3$ which passes through O and does not contain any vertex of $\mathbf{P}^*$. Let $\{F_i^*|i \in I\}$ be the set of the faces of $\mathbf{P}^*$ which are intersected by H and let $\{F_j^*|j \in J\}$ be the set of the remaining faces of $\mathbf{P}^*$. The faces $\{F_j^*|j \in J\}$ can be strictly separated by two parallel planes from O. Thus we have to consider the faces $\{F_i^*|i \in I\}$ of $\mathbf{P}^*$ only. However, we may not use more than 6 planes in order to strictly separate them from O.

Let us denote the plane of F_i^* by $\overline{H}_i$, $i \in I$. Since $\overline{H}_i$ is a supporting plane of $\mathbf{P}^*$ it bounds a supporting half-space $\overline{H}_i^+$ of $\mathbf{P}^*$. Let us consider the polyhedral set $\overline{\mathbf{P}} = \bigcap_{i \in I} \overline{H}_i^+$. Obviously, $\overline{\mathbf{P}}$ is symmetrical about O and the faces of $\overline{\mathbf{P}}$ can be labelled by $\{\overline{F}_i|i \in I\}$ with $\overline{F}_i \supset F_i^*$ for any $i \in I$. Thus it is sufficient to strictly separate the faces $\{\overline{F}_i|i \in I\}$ from O by 3 pairs of pairwise parallel planes. If $\overline{\mathbf{P}}$ is unbounded, then it can be done very easily namely, as $\overline{\mathbf{P}}$ is a cylinder Lemma 7 (a) implies the existence of 3 pairs of pairwise parallel planes which strictly separate the faces $\{\overline{F}_i|i \in I\}$ from O. The following statement completes the proof of Theorem 10 in case (2). (The convex set $S \subset \mathbf{E}^3$ is bisected by the plane H if and only if $S \cap H_+ \neq \emptyset$ and $S \cap H_- \neq \emptyset$ for the open half-spaces H_+ and H_- bounded by H.)

Lemma 9. *If $\overline{\mathbf{P}} \subset \mathbf{E}^3$ is a centrally symmetric convex polyhedron all faces of which are bisected by a plane of $\mathbf{E}^3$, then there are 3 pairs of pairwise parallel planes which strictly separate the faces of $\overline{\mathbf{P}}$ from the center point of $\overline{\mathbf{P}}$.*

Proof. Let the center point of $\overline{\mathbf{P}}$ be the origin O of $\mathbf{E}^3$ and let $\{\overline{F}_1, \overline{F}_2, \ldots, \overline{F}_{2n-1}, \overline{F}_{2n}\}, n \geq 3$ be the set of the faces of $\overline{\mathbf{P}}$. All these faces are bisected by the plane H of $\mathbf{E}^3$. Without loss of generality we may suppose that H passes through O and does not contain any vertex of $\overline{\mathbf{P}}$. Let s_i be the side $\overline{F}_i \cap H$ of the centrally symmetric convex (2n)-gon $\overline{\mathbf{P}} \cap H$ with center O, $1 \leq i \leq 2n$. Finally, let H_+ be one of the two open half-spaces bounded by H.

If $V_1 \in \overline{\mathbf{P}} \cap H_+$ is the vertex of $\overline{\mathbf{P}}$ which lies furthest from H, then the faces $\{\overline{F}_l|l \in L\}$ of $\overline{\mathbf{P}}$ containing V_1 have the property that $O \in$ relint $[\text{conv}(\bigcup_{l \in L} s_l)]$. As $\overline{\mathbf{P}}$ is centrally symmetric about O there are 3 faces of $\overline{\mathbf{P}}$ say, $\overline{F}_1, \overline{F}_3$ and $\overline{F}_5$ meeting at V_1 such that $O \in$ relint $[\text{conv}(s_1 \cup s_3 \cup s_5)]$ and $O \notin$ relint $[\text{conv}(s_i \cup s_j)]$ for any $i \neq j$, $i \in \{1, 3, 5\}$, $j \in \{1, 3, 5\}$. Let $\overline{H}_i$ be the plane of $\overline{F}_i$, $i \in \{1, 3, 5\}$ and let $\overline{H}_i^+$ be the supporting half-space of $\overline{\mathbf{P}}$ bounded by $\overline{H}_i$. Finally, reflect $\overline{F}_1, \overline{F}_3, \overline{F}_5$ and $\overline{H}_1^+, \overline{H}_3^+, \overline{H}_5^+$ about O. Let the images be denoted by $\overline{F}_4, \overline{F}_6, \overline{F}_2$ and $\overline{H}_4^+, \overline{H}_6^+, \overline{H}_2^+$. Obviously, the convex polyhedron $\mathbf{Q} = \bigcap_{1 \leq i \leq 6} \overline{H}_i^+$ is a parallelopiped. Without loss of generality we may suppose that $\mathbf{Q}$ is a cube the faces of which are the squares $Q_1, Q_2, \ldots, Q_6$ with $Q_i \supset \overline{F}_i$ for any $1 \leq i \leq 6$. (Use affinity if it is necessary.) As we know the faces $\overline{F}_1, \overline{F}_3, \overline{F}_5$ of $\overline{\mathbf{P}}$ meet at the vertex V_1 of $\overline{\mathbf{P}}$. Let

V_2 be the vertex in common of the faces $\overline{F}_2, \overline{F}_4, \overline{F}_6$ of $\overline{P}$. Obviously, O is the midpoint of the segment $V_1 V_2$ and H bisects every face of Q. So $H \cap Q$ is a hexagon with center O and with sides $W_1 W_2, W_2 W_3, W_3 W_4, W_4 W_5, W_5 W_6,$ and $W_6 W_1$ where $Q_1 \cap H = W_1 W_2 \supset s_1 = \overline{F}_1 \cap H; Q_2 \cap H = W_2 W_3 \supset s_2 = \overline{F}_2 \cap H; Q_3 \cap H = W_3 W_4 \supset s_3 = \overline{F}_3 \cap H; Q_4 \cap H = W_4 W_5 \supset s_4 = \overline{F}_4 \cap H; Q_5 \cap H = W_5 W_6 \supset s_5 = \overline{F}_5 \cap H; Q_6 \cap H = W_6 W_1 \supset s_6 = \overline{F}_6 \cap H$ (Fig. 14).

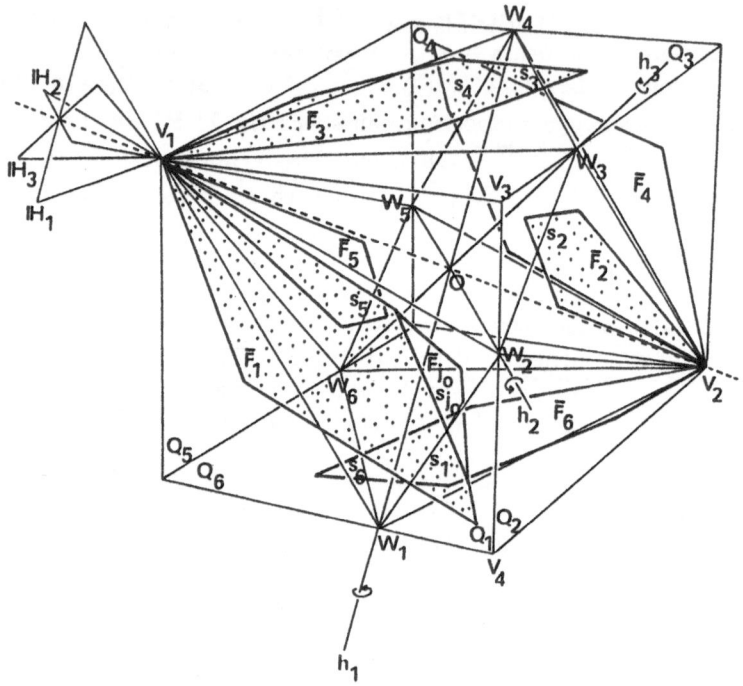

Fig. 14

Take the planes H_1, H_2 and H_3 determined by the non- colinear point-triplets $\{V_1, W_1, W_4\}, \{V_1, W_2, W_5\}$ and $\{V_1, W_3, W_6\}$. Let $h_i = H_i \cap H$, $1 \leq i \leq 3$. (Obviously, $\{O, V_2\} \subset H_1 \cap H_2 \cap H_3$.) In order to strictly separate the faces $\{\overline{F}_j | 1 \leq j \leq 2n\}$ of $\overline{P}$ from O by 3 pairs of pairwise parallel planes it is sufficient to find 3 planes $\overline{H}_1, \overline{H}_2$ and $\overline{H}_3$ passing through O with the property that each face $\overline{F}_j$ ($1 \leq j \leq 2n$) of $\overline{P}$ is disjoint from at least one of the planes $\overline{H}_1, \overline{H}_2$ and $\overline{H}_3$. We claim that if $\overline{H}_i$ is a slightly rotated copy of H_i about h_i ($1 \leq i \leq 3$) with the property $\overline{H}_1 \cap (Q_5 \cup Q_2) = \emptyset$, $\overline{H}_2 \cap (Q_3 \cup Q_6) = \emptyset$ and $\overline{H}_3 \cap (Q_1 \cup Q_4) = \emptyset$, then any face $\overline{F}_j$ ($1 \leq j \leq 2n$) of $\overline{P}$ is disjoint from at least one of the planes $\overline{H}_1, \overline{H}_2$ and $\overline{H}_3$. In order to prove this claim let us choose an arbitrary face $\overline{F}_{j_o}$ of $\overline{P}$ and find a described rotated copy $\overline{H}_{i_o}$ of H_{i_o} with $\overline{F}_{j_o} \cap \overline{H}_{i_o} = \emptyset$ for a certain $i_o \in \{1, 2, 3\}$. Without loss of generality we may suppose that s_{j_o} lies between the sides s_1 and s_2 with respect to an orientation of the boundary of $\overline{P} \cap H$. (See also Fig. 14.) The cases $\overline{F}_{j_o} = \overline{F}_1$ and $\overline{F}_{j_o} = \overline{F}_2$ are considered as well.

If $\overline{F}_{j_o} = \overline{F}_1 (\overline{F}_{j_o} = \overline{F}_2$, resp.), then a slightly rotated copy $\overline{\mathbf{H}}_3$ of $\mathbf{H}_3$ about h_3 ($\overline{\mathbf{H}}_1$ of $\mathbf{H}_1$ about h_1, resp.) possesses the property that $\overline{\mathbf{H}}_3 \cap \overline{F}_1 = \emptyset$ ($\overline{\mathbf{H}}_1 \cap \overline{F}_2 = \emptyset$, resp.). In the remaining cases $\overline{F}_{j_o} \cap (h_1 \cup h_3) = \emptyset$ and $\overline{F}_{j_o} \subset \text{conv}(\{V_1, V_2, W_1, W_3, V_3, V_4\})$ where V_3 and V_4 are the vertices of $\mathbf{Q}$ with $V_3 = Q_1 \cap Q_2 \cap Q_3$ and $V_4 = Q_1 \cap Q_2 \cap Q_6$. Consequently, either (a) $\overline{F}_{j_o} \cap (\mathbf{H}_1 \cup \mathbf{H}_3) = \emptyset$ or (b) $\overline{F}_{j_o} \cap (\mathbf{H}_1 \cup \mathbf{H}_3) = \{V_1\}$ or (c) $\overline{F}_{j_o} \cap (\mathbf{H}_1 \cup \mathbf{H}_3) = \{V_2\}$. In case (a) we are done. In case (b) a slightly rotated copy $\overline{\mathbf{H}}_3$ of $\mathbf{H}_3$ about h_3 has the property that $\overline{\mathbf{H}}_3 \cap \overline{F}_{j_o} = \emptyset$. Finally, in case (c) a slightly rotated copy $\overline{\mathbf{H}}_1$ of $\mathbf{H}_1$ about h_1 possesses the property that $\overline{\mathbf{H}}_1 \cap \overline{F}_{j_o} = \emptyset$. This completes the proof of Lemma 9.   □

(3) Let the rotation $R(l, \frac{2\pi}{n})$ about the line l through angle $\frac{2\pi}{n}$, $n = 2, 3, \ldots$ be a symmetry of the convex polyhedron $\mathbf{P}$ of $\mathbf{E}^3$. Illuminate $\mathbf{P}$ by two opposite unit vectors parallel to l. If the vertices of $\mathbf{P}$ which are not illuminated by these two directions lie in a plane, then Lemma 6 and Lemma 7 (a) imply that $\mathcal{H}(\mathbf{P}) \leq 6$. Hence we suppose that the mentioned vertices of $\mathbf{P}$ are not planar and choose the origin $O \in l$ in the interior of their convex hull. Then consider the polar body $\mathbf{P}^*$ of $\mathbf{P}$. Obviously, $O \in \text{int } \mathbf{P}^*$ and $\mathbf{P}^*$ is a convex polyhedron of $\mathbf{E}^3$ whose symmetry is $R(l, \frac{2\pi}{n})$. We want to show that $\mathcal{H}(\mathbf{P}) \leq 8$. Lemma 6 implies that we have to find 8 planes of $\mathbf{E}^3$ which strictly separate O from the faces of $\mathbf{P}^*$.

Let H be the plane of $\mathbf{E}^3$ which passes through O and is orthogonal to l. Let $\{F_i^* | i \in I\}$ be the set of the faces of $\mathbf{P}^*$ which have at least one point in common with H. The collection of the remaining faces of $\mathbf{P}^*$ is denoted by $\{F_j^* | j \in J\}$. Also, let $\overline{H}_i (\overline{H}_i^+$, resp.) be the supporting plane (the supporting half-space, resp.) of $\mathbf{P}^*$ with $\overline{H}_i \supset F_i^*$ (boundary $\overline{H}_i$, resp.), $i \in I$. Finally, let $\overline{P} = \bigcap_{i \in I} \overline{H}_i^+$. Obviously, $\overline{P}$ is a convex polyhedron of $\mathbf{E}^3$ (see Corollary 2) whose symmetry is $R(l, \frac{2\pi}{n})$. The set of the faces of $\overline{P}$ can be labelled by $\{\overline{F}_i | i \in I\}$ with $\overline{F}_i \supset F_i^*$ for any $i \in I$. Obviously, it is sufficient to strictly separate O from the faces $\{\overline{F}_i | i \in I\}$ of $\overline{P}$ by 8 planes provided that these planes strictly separate O from the faces $\{F_j^* | j \in J\}$ of $\mathbf{P}^*$ as well. It is clear that we need to prove this only for (a) $n = 2k + 1$, $k = 1, 2, 3, \ldots$ and (b) $n = 2$. The proof of Case (b) uses a technique similar to the following method of Case (a). However, the details are rather complicated so we omit them here. For the complete proof the reader should consult Bezdek (1991/a).

Case (a). Obviously, there are two vertices of $\overline{P}$ say, V and $\mathbf{V}$ which belong to l. The faces of $\overline{P}$ meeting at V ($\mathbf{V}$, resp.) have at least one point in common with H. It is natural to distinguish the following two cases:

(i) there are faces $V \in \overline{F}_1$ and $\mathbf{V} \in \overline{\mathbf{F}}_1$ of $\overline{P}$ which are bisected by H; (ii) either no face of $\overline{P}$ at V or no face of $\overline{P}$ at $\mathbf{V}$ is bisected by H.

(i) Take the transformation $R(l, \frac{2\pi}{2k+1})$ $(m-1)$ times , $1 \leq m \leq 2k+1$ and let $\overline{F}_m (\overline{\mathbf{F}}_m$ resp.) be the image of $\overline{F}_1 (\overline{\mathbf{F}}_1$, resp.). Let $\overline{H}_m^+ (\overline{\mathbf{H}}_m^+$, resp.) denote the supporting half-space of the face $\overline{F}_m (\overline{\mathbf{F}}_m$, resp.) of $\overline{\mathbf{P}}$. Let the consecutive vertices of the convex regular $(2k+1)$-gon $H \cap (\bigcap_{1 \leq m \leq 2k+1} \overline{H}_m^+)$ $(H \cap (\bigcap_{1 \leq m \leq 2k+1} \overline{\mathbf{H}}_m^+)$, resp.) with center O be denoted by $A_1, A_2, \ldots, A_{2k+1}$ $(B_1, B_2, \ldots, B_{2k+1}$, resp.) such that $H \cap \overline{F}_1 \subset A_1 A_2, H \cap \overline{F}_2 \subset A_2 A_3, \ldots, H \cap \overline{F}_{2k+1} \subset A_{2k+1} A_1$ $(H \cap \overline{\mathbf{F}}_1 \subset B_1 B_2, H \cap \overline{\mathbf{F}}_2 \subset B_2 B_3, \ldots, H \cap \overline{\mathbf{F}}_{2k+1} \subset B_{2k+1} B_1$, resp.). Each side of any of the two regular $(2k+1)$-gons bisects two (consecutive) sides of the other regular polygon (see Fig. 15 for $k = 1$). Without loss of generality we may suppose the existence of the following intersection points $C_1 = A_1 A_2 \cap B_{2k+1} B_1$, $D_1 = A_1 A_2 \cap B_1 B_2$; $C_2 = A_2 A_3 \cap B_1 B_2$, $D_2 = A_2 A_3 \cap B_2 B_3$; $\ldots$; $C_{2k+1} = A_{2k+1} A_1 \cap B_{2k} B_{2k+1}$, $D_{2k+1} = A_{2k+1} A_1 \cap B_{2k+1} B_1$. With relabelling we can manage to have a line l_1 in H passing through O such that the points $C_1, D_1, C_2, D_2, \ldots, C_{k+1}, D_{k+1}$ lie on the same side of l_1 (i.e. in a closed half-plane of H bounded by l_1). So the segments $\overline{F}_1 \cap H$ and $\overline{F}_{k+1} \cap H$ lie on the same side of l_1. We say that the face $\overline{F}$ of $\overline{\mathbf{P}}$ lies between $\overline{F}_1$ and $\overline{F}_{k+1}$ if $V \notin \overline{F}$ and the orientation of the boundary of the convex polygon $\overline{\mathbf{P}} \cap H$ generated by the ordering $\overline{F}_1 \cap H$, $\overline{F} \cap H$, $\overline{F}_{k+1} \cap H$ corresponds to the direction of rotation $R(l, \frac{2\pi}{2k+1})$. Now let H_1 be the plane spanned by l_1 and l. It is easy to prove that for any face $\overline{F}$ of $\overline{\mathbf{P}}$ lying between $\overline{F}_1$ and $\overline{F}_{k+1}$ either $\overline{F} \cap H_1 = \emptyset$ or $\overline{F} \cap H_1 = \{\mathbf{V}\}$. Thus there exists a slightly rotated copy H_1' of H_1 about l_1 such that any face of $\overline{\mathbf{P}}$ which lies between $\overline{F}_1$ and $\overline{F}_{k+1}$ is disjoint from H_1'. In other words there exists a plane $\mathbf{H}_1$ of $\mathbf{E}^3$ which strictly separates O from the faces of $\overline{\mathbf{P}}$ lying between $\overline{F}_1$ and $\overline{F}_{k+1}$. As $k \geq 1$ we can find further planes $\mathbf{H}_2$ and $\mathbf{H}_3$ (by simple rotations of $\mathbf{H}_1$ about l) such that each face of $\overline{\mathbf{P}}$ which does not contain V is strictly separated from O by at least one of the planes $\mathbf{H}_1, \mathbf{H}_2$ and $\mathbf{H}_3$. Obviously, the faces of $\overline{\mathbf{P}}$ meeting at V can be strictly separated from O by further planes $\mathbf{H}_4, \mathbf{H}_5$ and $\mathbf{H}_6$ (see also Lemma 7 (a)). Finally, the faces $\{F_j^* | j \in J\}$ of $\mathbf{P}^*$ can be strictly separated from O by the planes $\mathbf{H}_7$ and $\mathbf{H}_8$ (which are parallel to H.) This completes the proof of case (i).

(ii) Here either we can use the method of case (i) without any modification or we need a slight modification of it which means among others that the planes $\mathbf{H}_7$ and $\mathbf{H}_8$ strictly separate O not only from the faces $\{F_j^* | j \in J\}$ of $\mathbf{P}^*$ but from some faces of $\overline{\mathbf{P}}$ and so one of the planes $\mathbf{H}_4, \mathbf{H}_5$ and $\mathbf{H}_6$ can be used to strictly separate O from some further faces of $\overline{\mathbf{P}}$ which do not contain V. $\qquad\square$

We are now in a position to prove the following corollary of Theorem 10.

Theorem 11. *If* $\mathbf{P}$ *is a convex polyhedron of* $\mathbf{E}^3$ *with affine symmetry (i.e. the affine symmetry group of* $\mathbf{P}$ *consists of the identity and at least one other affinity of* $\mathbf{E}^3$*), then* $\mathcal{H}(\mathbf{P}) \leq 8$.

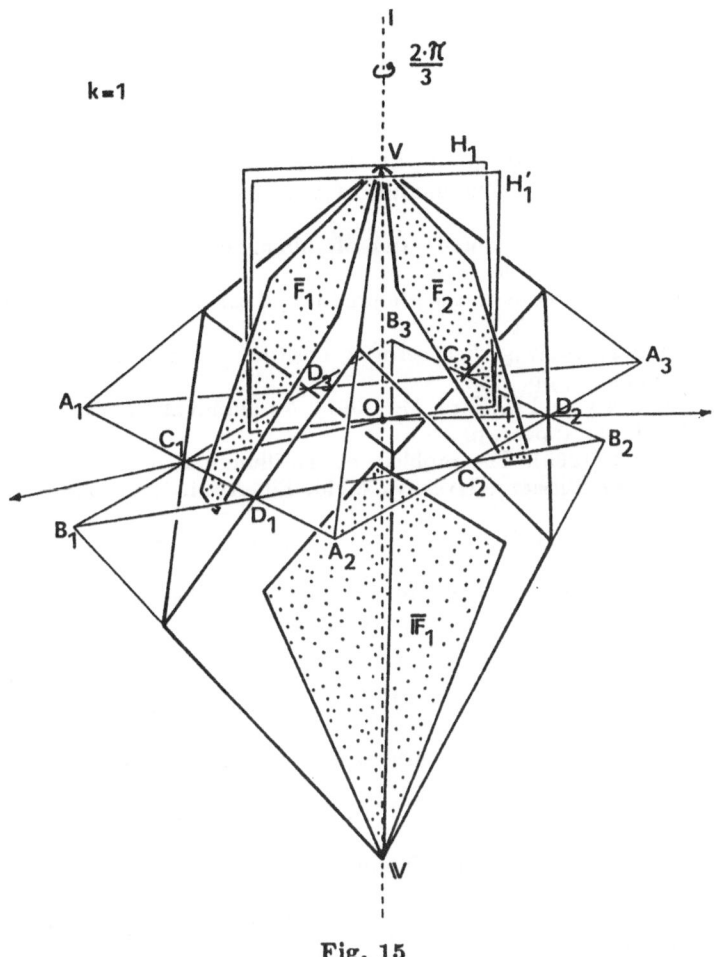

Fig. 15

Proof. Let A_1 be an affine symmetry (different from the identity) of **P**. It is known (see Grötschel, Lovász and Schrijver (1988)) that there exists a unique ellipsoid **E** called Löwner-John ellipsoid of **P** with minimum volume containing **P**. Let A_2 be an affinity of $\mathbf{E}^3$ (i.e. an affine isomorphism of $\mathbf{E}^3$) such that $A_2(\mathbf{E})$ is a ball of $\mathbf{E}^3$. Obviously, $A_2(\mathbf{P})$ is a convex polyhedron with affine symmetry $A_2 A_1 A_2^{-1}$. As the Löwner-John ellipsoid of $A_2(\mathbf{P})$ is the ball $A_2(\mathbf{E})$ it is obvious that $A_2 A_1 A_2^{-1}$ is an affine symmetry of $A_2(\mathbf{E})$. Thus $A_2 A_1 A_2^{-1} = Q$ is an isometry of $\mathbf{E}^3$ and we have obtained that $A_2(\mathbf{P})$ is a convex polyhedron with symmetry Q. As $\mathcal{H}[A_2(\mathbf{P})] = \mathcal{H}(\mathbf{P})$ Theorem 10 implies that $\mathcal{H}(\mathbf{P}) \leq 8$.   □

Acknowledgement. The work was supported by Hung. Nat. Found. for Sci. Research No. 326-0413.

References

Bezdek, K. (1983): Über einige Kresüberdeckungen. Beiträge zur Alg. und Geom. 14, 7–13

Bezdek, K. (1991a): The problem of illumination of the boundary of a convex body by affine subspaces. Mathematika 38, 362–375

Bezdek, K. (1991b): On the illumination of smooth convex bodies. Arch. Math. 58, 611–614

Bezdek, K. (1991c): On the illumination of unbounded closed convex sets. Israel J. Math., to appear, pp. 1–8

Bezdek, K. (1991d): A note on the illumination of convex bodies. Geometriae Dedicata, to appear, pp. 1–3

Bezdek, K. (1991e): Hadwiger's covering conjecture and its relatives. Amer. Math. Monthly, to appear, pp. 1–4

Bezdek, K., Kiss, Gy. and Mollard, M. (1991): An illumination problem for zonoids. Israel J. Math., to appear, pp. 1–6

Boltjansky, V. G. (1960): The problem of the illumination of the boundary of a convex body (in Russian). Izvestiya Mold. Fil. Akad. Nauk SSSR, 76, no. 10, 77–84

Boltjansky, V. G. and Gohberg, I. (1985): Results and problems in combinatorial geometry. Cambridge Univ. Press

Boltjansky, V. G. and Soltan, P. S. (1978): Combinatorial geometry of miscellaneous classes of convex sets (in Russian). Stinca, Kishinev

Boltjansky, V. G. and Soltan, P. S. (1990): A solution of Hadwiger's problem for zonoids. Combinatorica, to appear.

Borsuk, K. (1933): Drei Sätze über die n-dimensionale euklidische Sphäre. Fund. Math. 20, 177–190

Buchman, E. and Valentine, F. A. (1970): A characterization of the parallelopiped in E^n. Pacific J. Math. 35, 53–57

Buchman, E. and Valentine, F. A. (1976): External visibility. Pacific J. Math. 64, 333–340

Charazishvili, A. B. (1973): On the illumination problem (in Russian). Soobsc. Akad. Nauk Gruzin. SSR 71, 289–291

Erdős, P. and Füredi, Z. (1983): The greatest angle among n points in the d-dimensional Euclidean space. Ann. Discr. Math. 17, 275–283

Fejes Tóth L. (1977): Illumination of convex discs. Acta Math. Acad. Sci. Hungar. 29, 355–360

Gohberg, I. and Markus, A. S. (1960): A certain problem about the covering of convex sets with homothetic ones (in Russian). Izvestiya Mold. Fil. Akad. Nauk SSSR 76, no 10, 87–90

Grötschel, M., Lovász, L. and Schrijver, A. (1988): Geometric algorithms and combinatorial optimization. Springer

Grünbaum, B. (1963): Borsuk's problem and related questions. Convexity, Proc. of Symp. in Pure Math. (AMS, Providence), Vol. 7, 271–284

Grünbaum, B. (1967): Convex polytopes. Wiley & Sons, New York

Hadwiger, H. (1945/46): Überdechung einer Menge durch Mengen kleineren Durchmessers. Comm. Math. Helv. 18, 73–75

Hadwiger, H. (1957): Ungelöste Probleme, Nr. 20. Elem. Math. 12, 121

Hadwiger, H. (1960): Ungelöste Probleme, Nr. 38. Elem. Math. 15, 130–131

Kahn, J. and Kalai, G. (1992): A counterexample of Borsuk's conjecture. Preprint, pp.1–4

Lassak, M. (1984): Solution of Hadwiger's covering problem for centrally symmetric convex bodies in E^3. J. London Math. Soc. 30, no 2, 501–511

Lassak, M. (1985): Covering plane convex bodies with smaller homothetical copies. Coll. Math. Soc. J. Bolyai, Vol. 48, Intuitive Geometry, 331–337

Lassak, M. (1986): Covering a plane convex body by four homothetical copies with the smallest positive ratio. Geom. Dedicata **21**, 151–167

Lassak, M. (1988): Covering the boundary of a convex set by tiles. Proc. Amer. Math. Soc. **104**, no. 1, 269–272

Lay, S. R. (1982): Convex sets and their applications. Pure and Applied Math., Wiley & Sons, New York

Levi, F. W. (1955): Überdeckung eines Eibereiches durch Parallelverschiebungen seines offenen Kerns. Arch. Math. **6** 369–370

McMullen, P. and Shephard, G. C. (1971): Convex polytopes and the upper bound conjecture. Cambridge Univ. Press

Martini, H. (1985): Some results and problems around zonotopes. Coll. Math. Soc. J. Bolyai, Vol. 48, Intuitive Geometry, 383–418

Schramm, O. (1988): Illuminating sets of constant width. Mathematika **35**, no. 2, 180–189

Soltan, P. S. (1963): Towards the problem of covering and illumination of convex sets (in Russian). Izvestiya Akad. Nauk Mold. SSR, No.1, 49–57

Soltan, P. S. and Soltan, V. P. (1986): On the X-raying of convex bodies (in Russian). Sov. Math. Dokl. **33** 42–44; translation from Dokl. Akad. Nauk SSSR **286**, 50–53

Weissbach, B. (1981): Eine Bemerkung zur Überdeckung beschränkter Mengen durch Mengen kleineren Durchmessers. Beiträge zur Alg. und Geom. **11**, 119–122

Chapter IX. Geometric and Combinatorial Applications of Borsuk's Theorem

Imre Bárány

1. Introduction

In this chapter geometric and combinatorial applications and extensions of the Borsuk theorem [Bo] are presented. This theorem has several equivalent forms. Let S^d denote the d-dimensional sphere, i.e., the boundary of the Euclidean unit ball in R^{d+1}.

Theorem 1.1. *If $f : S^d \rightarrow R^d$ is continuous, then there exists an $x \in S^d$ with $f(x) = f(-x)$.*

Theorem 1.2. *If S^d is covered by $d+1$ sets that are all open (or all closed) in S^d, then one of the sets contains antipodal points.*

Theorem 1.3. *If $f : S^d \rightarrow S^n$ is continuous and antipodal (i.e., $f(x) = -f(-x)$ for every $x \in S^d$), then $d \leqq n$.*

Theorem 1.4. *If $f : S^d \rightarrow S^d$ is continuous and antipodal, then the degree of f is odd.*

Theorem 1.1 which was conjectured by Ulam and proved by Borsuk is often called the Borsuk-Ulam theorem. Theorem 1.2 was also found by Lyusternik and Shnirel'man [LS]. Theorems 1.2 and 1.3 are due to Borsuk.

The choice of material in this chapter reflects the author's taste and so it is in no way a complete survey around Borsuk's theorem. The interested reader is advised to consult some other survey papers: Björner [Bj] writes on topological methods in combinatorics, Alon [Al] gives some nice recent applications, Steinlein [Ste] presents an extensive list of applications and extensions in topology, functional analysis, differential equations, approximation theory, geometry, etc., and Bogatyi [Bg] is concerned with sphere coverings and conjectures of Hadwiger, Kneser and Knaster.

2. Van Kampen-Flores Type Results

We start with a geometric application. Radon's theorem [DGK] is a very simple fact: Any set $A \subset R^d$ of $d+2$ points contains two disjoint subsets A_1 and A_2 with conv $A_1 \cap$ conv $A_2 \neq \phi$. Equivalently, given the $(d+1)$-dimensional simplex $\Delta^{d+1} \subset R^{d+1}$ and a linear map $f : \Delta^{d+1} \to R^d$, there are two disjoint faces, A_1 and A_2, of Δ^{d+1} with $f(A_1) \cap f(A_2) \neq \phi$. The interesting fact here is that one can replace the word "linear" by "continuous".

Theorem 2.1 [BB]. *Given a continuous map $f : \Delta^{d+1} \to R^d$, there are two disjoint faces, A_1 and A_2, of Δ^{d+1} with $f(A_1) \cap f(A_2) \neq \phi$.*

This theorem is an easy consequence of Borsuk's theorem (1.1) and the following fact. (The support of $x \in \Delta^{d+1}$ is the minimal face of Δ^{d+1} containing x.)

Theorem 2.2 [BB]. *There is a continous map $g : S^d \to \Delta^{d+1}$ such that the supports of $g(x)$ and $g(-x)$ are disjoint for each $x \in S^d$.*

Theorem 2.1 shows that it is not only the linear but the topological structure of R^d as well that implies Radon's theorem. Actually, an even stronger statement is true. Given a convex compact body $C \subset R^{d+1}$ with nonempty interior, two points of C are said to be opposite if they lie on two different parallel supporting hyperplanes of C.

Theorem 2.3 [BB]. *Let $C \subset R^{d+1}$ be a convex compact set with nonempty interior. Given a continuous map $f : \partial C \to R^d$, there are opposite points a_1 and a_2 with $f(a_1) = f(a_2)$.*

This theorem implies 2.1 because, when $C = \Delta^{d+1}$, the supports of opposite points are disjoint faces of Δ^{d+1}. It contains Borsuk's theorem as well: take $C = B^{d+1}$, the unit ball of R^{d+1}. On the other hand, Theorem 2.3 is proved by using Borsuk's theorem.

There is a generalization of Radon's theorem, due to Tverberg [Tv] which is far from being trivial. It says that given a set A of $(d+1)(p-1)+1$ points in R^d, there is a partition $\{A_1, \ldots, A_p\}$ of A with $\bigcap_{i=1}^{p}$ conv $A_i \neq \phi$. In the spirit of the above reformulation of Radon's theorem, Tverberg's theorem has an equivalent form: Given a linear map $f : \Delta^N \to R^d$ where $N = (d+1)(p-1)$, there are pairwise disjoint faces $A_1, \ldots, A_p$ of Δ^N with $\bigcap_{i=1}^{p} f(A_i) \neq \phi$. The topological generalization of this theorem is established only when p is prime:

Theorem 2.4 [BSS]. *Given a continuous map $f : \Delta^N \to R^d$ where $N = (d+1)(p-1)$ and p is prime, there are pairwise disjoint faces $A_1, \ldots, A_p$ of Δ^N with $\bigcap_{i=1}^{p} f(A_i) \neq \phi$.*

The proof is based on appropriate generalizations of Theorems 1.1 and 2.2. In order to state them we need some preparations.

First we define a CW-complex $X_{d,p}$. Take p disjoint copies of the $d(p-1)$-dimensional ball and identify their boundaries. This is the CW-complex $X = X_{d,p}$. The identified boundary is $S^{d(p-1)-1}$ which we denote by S for brevity. Suppose the cyclic group Z_p acts freely on S and let ω denote the action of its generator. (Z_p acts freely on S if the orbit of every $x \in S$, $\{x, \omega x, \ldots, \omega^{p-1}x\}$ consists of p distinct points.) Any such ω can be extended from S to X as follows: If (y, r, q) denotes the point of X from the q-th ball with radius r and S-coordinate y, then set $\omega(y, r, q) = (\omega y, r, q+1)$ where $q+1$ is reduced mod p. The extended ω is, obviously, a free Z_p action on X.

We remark here that $X = X_{d,p}$ is defined for every d, $p \geq 1$. It is clear, moreover, that $\dim X_{d,p} = d(p-1)$ and that $X_{d,p}$ is $[d(p-1)-1]$-connected.

Now we specify $\omega : S \to S$. $S = S^{d(p-1)-1}$ is the unit sphere of the substance of $\prod_{i=1}^{p} R^d$ orthogonal to the "diagonal" $D = \{(x, \ldots, x) \in \prod_1^p R^d\}$. So S is the set of d by p matrices (x_{ij}) with

$$\sum_{j=1}^{p} x_{ij} = 0, \quad i = 1, \ldots, d, \quad \sum_{i,j} x_{ij}^2 = 1.$$

Now ω acts on S by $\omega((x_{i,j})) = (x_{i,j+1})$ where $j+1$ is again reduced mod p. Thus ω cyclically shifts the columns of these matrices. Clearly, it is a free Z_p action on S.

Now we are in a position to give the two ingredients of the proof of Theorem 2.4.

Theorem 2.5 [BSS]. *There is a continuous map $g : X \to \Delta^N$ such that for every $x \in X$ the supports of the points $g(x), g(\omega x), \ldots, g(\omega^{p-1}x)$ are pairwise disjoint.*

Theorem 2.6 [BSS]. *For any continuous map $h : X \to R^d$ there exists an $x \in X$ with $h(x) = h(\omega w) = \ldots = h(\omega^{p-1}x)$.*

Theorem 2.5 follows from the fact that the CW-complex Y defined as

$$\{(y_1, \ldots, y_p) : y_i \in \Delta^N, \text{ the supports of } y_1, \ldots, y_p \text{ are pairwise disjoint}\}$$

is $[d(p-1)-1]$-connected. Theorem 2.6 is a generalization of 1.1 to the case when, instead of the antipodal map, we have a free Z_p action on the appropriate topological space. Its proof is derived from the following Lemma which appears in [KZ] and [Do] as well.

Lemma 2.7. *Assume $d \geq 1$, $p \geq 1$, and θ is a free Z_p action on S^d. If $\alpha : S^d \to S^d$ is continuous and commutes with θ, then it has degree 1 modulo p.*

Sarkaria notices [Sa1] that both 2.1 and 2.4 are van Kampen-Flores type results. Van Kampen [Ka] and Flores [Fl] proved more than fifty years ago that there is no continuous one-to-one map from Δ_{k-1}^{2k}, the $(k-1)$-dimensional skeleton of Δ^{2k}, onto the Euclidean space $R^{2(k-1)}$. In fact, both of them proved a little more: for any continuous map $f : \Delta_{k-1}^{2k} \to R^{2(k-1)}$ there are disjoint simplices A_1 and A_2 of Δ_{k-1}^{2k} such that $f(A_1) \cap f(A_2) \neq \phi$. Sarkaria [Sa1] proves a generalization of this last statement and 2.1 and 2.4. To state this generalization let k, j, p, d and n be integers with $1 \leq k \leq n+1$, $2 \leq j \leq p$, $0 \leq d$ with p a prime. A p-tuple of sets is called j-wise disjoint if the intersection of any j of the sets is empty.

Theorem 2.8. *Assume that either (a) $n = k - 1$ and $(n + 1)(j - 1) > (d + 1)(p - 1)$ or (b) $n > k - 1$ and $(m + 1)(p - 1) + pk \geq (n + 1)(j - 1) > (d + m + 2)(p - 1)$ for some non-negative integer m. Then, for any continuous map $f : \Delta_{k-1}^n \to R^d$, there exists a j-wise disjoint p-tuple $(A_1, \ldots, A_p)$ of simplices of Δ_{k-1}^n such that $\bigcap_{i=1}^p f(A_i) \neq \phi$.*

This gives 2.4 when $n = k - 1 = (d + 1)(p - 1)$, p a prime and $j = 2$. The proof of 2.8 uses the powerful technique of deleted joins and deleted products from combinatorial topology, developed by Sarkaria [Sa1,2,3]. This technique may give the solution of one of the main problems in this field, Theorem 2.3 for non-prime p:

Conjecture 2.9. *Let $d \geq 2$, $p \geq 2$ and $N = (d + 1)(p - 1)$. Then for any continuous map $f : \Delta^N \to R^d$ there exist disjoint faces $A_1, \ldots, A_p$ of Δ^N with $\bigcap_{i=1}^p f(A_i) \neq \phi$.*

M. Özaydin [Öz] proved Conjecture 2.9 in the case when p is a prime power.

The following colored version of Tverberg's theorem has recently been proven by Živaljević and Vrećica [ŽV2]. Let $C_1, \ldots, C_{d+1} \subset R^d$ be finite sets considered "colors". A $(d + 1)$-tuple S is multicolored if S contains one point of each color exactly. Bárány, Füredi, Lovász [BFL] raise the question whether there exists a constant $T = T(r, d)$ such that if each color class contains T points, then there exist r multicolored and pairwise disjoint $(d + 1)$-tuples $S_1, \ldots, S_r$ whose convex hulls intersect, i.e., $\bigcap_{j=1}^r \operatorname{conv} S_j \neq \phi$. Živaljević and Vrećica prove that such $T(r, d)$ exists and in fact is rather small:

Theorem 2.10. $T(r, d) \leq 2p(r) - 1$ *where $p(r)$ is the smallest prime greater than or equal to r.*

In the proof of this nice result the complex K of multicolored $(d + 1)$-tuples is considered. It is shown tha the r-fold deleted join of K is highly connected (while the r-fold deleted product is not). Then the scheme of the proof of Theorem 2.4 is used. Without going into details I mention that the proof gives a topological variant (in the sense of 2.4) of Theorem 2.10 as well.

It is shown in [BFL] that $T(3,2) \leq 7$ and in [BL] that $T(r,2) = r$. The methods of [BFL], together with 2.10 prove that the number of ways in which an n-set in R^d can be bisected by a hyperplane into two sets of cardinality $n/2$ is $O(n^{d-\varepsilon_d})$ with $\varepsilon_d = (4d)^{-(d+1)}$. This is the first $o(n^d)$ bound for the problem in the case $d > 3$. The interested reader should read [BFL].

3. The Ham-Sandwich Theorem

One of the classical consequences of the Borsuk theorem is the following:

Theorem 3.1. *Given d bounded and measurable sets in R^d there exists a hyperplane that simultaneously bisects them all.*

A very interesting generalization of this result has been proved by Živaljević and Vrećica [ŽV1].

Theorem 3.2. *Let $M_1, \ldots, M_k$ be bounded and measurable subsets of R^d $(k = 1, 2, \ldots d)$. Then there exists a $(k-1)$-dimensional affine subspace $L \subset R^d$ with the following property. For every closed halfspace H containing L and for every $i = 1, 2, \ldots k$, the measure of $M_i \cap H$ is at least $(d - k + 2)^{-1}$.*

When $k = d$, this is the ham-sandwich Theorem 3.1. When $k = 1$ it gives a theorem of Rado [Ra] (cf. [DGK] as well) saying that for any measurable set $E \subset R^d$ there exists a point $x \in R^d$ such that any halfspace H containing x contains at least a $\frac{1}{d+1}$-fraction of E.

Akiyama and Alon use 3.1 to establish the following theorem.

Theorem 3.3 [AA]. *Let $E_1, E_2, \ldots, E_d$ be d pairwise disjoint subsets of R^d, each containing exactly n points, and suppose that the points in $E = \bigcup_1^d E_i$ are in general position. Then there is a partition of E into pairwise disjoint sets $F_1, \ldots, F_n$, each containing exactly one point from each E_i such that the simplices conv $F_1, \ldots,$ conv F_n are pairwise disjoint.*

Using 3.1, one can show the existence of a hyperplane $H \subset R^d$ such that for the two open halfspaces H^+ and H^- determined by H

$$|H^- \cap E_i| = \lfloor \frac{n}{2} \rfloor \text{ and } |H^+ \cap E_i| = \lfloor \frac{n}{2} \rfloor \text{ for all } i = 1, \ldots, d.$$

Notice that when n is odd this implies that H contains precisely one point from each E_i. The proof proceeds by induction on n.

The ham-sandwich theorem is used by Lovász and Simonovits [LS] for the proof of a "convex" isoperimetric inequality.

Theorem 3.4. *Let $K \subset R^d$ be a convex set of diameter D. Let $K = A \cup B$ be a decomposition of K into two closed sets with no common interior point. Assume, further, that $A \cap B$ has $(d-1)$-dimensional measure. Then*

$$\min(\operatorname{vol} A, \operatorname{vol} B) < D \operatorname{vol}_{d-1} A \cap B.$$

In the proof one cuts A and B by a ham-sandwich cut H, and defines H^+ as the halfspace bounded by H that contains more of $A \cap B$. Then one iterates this for the convex set $K \cap H^+$ with decomposition $A \cap H^+$ and $B \cap H^+$. For the details see [LS].

A very nice application of Borsuk's theorem in geometric transversal theory has been found by Pollack and Wenger [PW]: for details see the paper [GPW] in this volume.

4. Centrally Symmetric Polytopes

Given a polytope $P \subset R^d$, let $f_i(P)$ be the number of i-dimensional faces of P. The f-vector of P is $(f_0(P), \dots, f_d(P))$. There has been a significant progress in understanding the structure of the f-vector [McM], [St1], [BiL], for instance a complete characterization of the f-vector of simplicial polytopes was found by Stanley and Billera and Lee. Much less is known about the f-vector of centrally symmetric polytopes. An interesting result is due to Figiel, Lindenstrauss and Milman [FLM]. It says that if P is a centrally symmetric, d-dimentsional polytope, then

$$f_0(P)f_{d-1}(P) \geq c^d$$

where $c > 1$ is a universal constant. The proof is based on the concentration of the measure phenomena (see [FLM]).

Another result the proof of which uses Borsuk's theorem will be given in more detail here.

Theorem 4.1 [BL]. *Any centrally symmetric, d-dimensional, simplicial polytope has at least 2^d facets. (A polytope is simplicial if all of its facets are simplices.)*

This theorem follows from either Theorem 4.2 or 4.4 below. In order to state them we need some preparation.

Given a convex polytope $K \subset R^n$ and $x \in \partial K$, we define $T(x, K)$ to be the set of outer unit normals to K at x. Clearly $T(x, K) \subset S^{n-1}$ is nonempty. The d-dimensional outer angle of K at x is defined as

$$\alpha_d(x, K) = \mu_{d-1}(T(x, K))$$

where μ_{d-1} is the $(d-1)$-dimensional Lebesgue measure in R^n (or on S^{n-1}) normalized so that $\mu_{d-1}(S^{d-1}) = 1$ when S^{d-1} is considered to be isometrically embedded into S^{n-1}. Obviously, $\alpha_d(x, K) = 0$ if the support of x is more than $(n-d)$-dimensional and $\alpha_d(x, K) = \infty$ if it is less than $(n-d)$-dimensional. Finally, let $\mathcal{A}^{(d)}$ denote the set of all d-dimensional subspaces of R^n.

Theorem 4.2 [BL]. *Let K be a centrally symmetric d-dimensional polytope and $A \in \mathcal{A}^{(d)}$. Then*

$$\sum \alpha_d(x, K) \geq 1$$

where the summation is taken over all vertices x of the polytope $K \cap A$.

The proof is based on the following extension of Borsuk's theorem.

Lemma 4.3 [BL]. *If $\varphi : S^d \to S^k$ is antipodal, then*

$$\mu_d(\varphi(S^d)) \geq \mu_d(S^d) = 1 .$$

Now let $\mathcal{L} = \mathcal{L}^{n-d}(K)$ be the set of all $(n-d)$-dimensional faces of the polytope $K \subset R^n$. We define a map $\varphi : S^{n-d} \to \mathrm{skel}_{n-d}K$ to be *special* if

(i) φ is antipodal
(ii) for each $L \in \mathcal{L}$, either $L \in \varphi(S^{n-d})$ or $(\mathrm{rel\,int}\, L) \cap \varphi(S^{n-d}) = \phi$.

Let us denote by M the minimum size of a centrally symmetric set of $(n-d)$-faces of K meeting all special images of S^{n-d}. In other words, M is the value of the following discrete linear program:

minimize

$$\sum_{L \in \mathcal{L}} x(L)$$

subject to

$$x(L) = 0 \text{ or } 1 \ (\forall L),$$
$$x(L) = x(-L) \ (\forall L),$$

$$\sum_{L \in \mathcal{L}} x(L) \geq 2 \quad (\forall \varphi \text{ special}).$$
$$L \subset \varphi(S^{n-d})$$

Theorem 4.4. *Assume K is a centrally symmetric, n-dimensional polytope and $A \in \mathcal{A}^{(d)}$ has no point in common with the $(n-d-1)$-dimensional faces of K. Then the polytope $A \cap K$ has at least M vertices.*

Here is a *sketch of the proof.* Set $x_A(L) = 1$ if $A \cap L \neq \phi$ and $x(L) = 0$ otherwise (for $L \in \mathcal{L}$). Clearly $\sum x_A(L)$ equals the number of vertices of $A \cap K$. We show that $x_A(L)$ satisfies the conditions of the above discrete linear program. The first two conditions are easy. For the third one we have to show that $A \cap \varphi(S^{n-d})$ contains at least two points (for every special map φ). Let π denote the orthogonal projection from R^n to the orthogonal complement of A, $A^\perp$. Since φ is antipodal, $A \cap \varphi(S^{n-d})$ contains two points iff $0 \in \pi \circ \varphi(S^{n-d})$. But $\pi \circ \varphi : S^{n-d} \to A^\perp \cong R^{n-d}$ so by Borsuk's theorem $\pi \circ \varphi(z) = 0$ for some $z \in S^{n-d}$.

Theorem 4.4 yields a stronger version of 4.1: under the same assumption $f_{d-1}(P) \geq 2^d$ with equality iff P is (a linear image of) the d-dimensional cross polytope.

The above results were significantly generalized by Stanley [Sta2] using the theory of Cohen-Macaulay rings and toric varieties. His results are best stated in terms of the h-vector which is defined from the f-vector by

$$h_i(P) = \sum_{j=0}^{i} (-1)^{i-j} \binom{d-j}{d-i} f_{j-1}(P).$$

(Here $f_{-1} = 1$.) $f(P)$ determines $h(P)$ uniquely and vice versa. Experience shows that it is easier to work with the h-vector than with the f-vector. The Dehn-Sommerville equations take the form $h_i(P) = h_{d-i}(P)$. Theorem 4.1 is equivalent to

$$h_0(P) + h_1(P) + \ldots + h_d(P) \geq 2^d$$

for any centrally symmetric, simplicial polytope. Answering a question of A. Björner, Stanley proved the following:

Theorem 4.5 [Sta2]. *For a centrally symmetric simplicial d-polytope P*

$$h_i \geq \binom{d}{i}, \quad 0 \leq i \leq d, \quad \text{and}$$

$$h_i - h_{i-1} \geq \binom{d}{i} - \binom{d}{i-1}, \quad 1 \leq i \leq \left\lfloor \frac{d}{2} \right\rfloor.$$

We finish this section with an open problem. Given a centrally symmetric simplicial d-polytope P with $2n$ vertices, what is the largest possible value of $f_i(P)$? Even a plausible conjecture is not known. $f_j(P)$ is certainly maximized for $j \leq \lfloor \frac{d}{2} \rfloor - 1$ if every set of $j+1$ vertices, no two of them antipodal, forms a j-face. Then $f_j(P) = 2^{j+1}\binom{n}{j+1}$. But the obvious conjecture that $f_j(P)$ is maximal for $j \geq \lfloor \frac{d}{2} \rfloor$ when $f_j(P) = 2^{j+1}\binom{n}{j+1}$ is false. [MS].

5. Kneser's Conjecture

The chromatic number of a graph $G(V, E)$ is the smallest integer c such that the vertices of the graph can be colored by c colors in such a way that the endpoints of every edge have different colors.

It turns out that there is a connection between Borsuk's theorem and the chromatic number. Namely, Lovász [L1] proved in 1976 a result which relates the chromatic number to the neighborhood complex $N(G)$ of the graph G. The vertex set of $N(G)$ is just V, and vertices $v_1, \ldots, v_m$ form a simplex of $N(G)$ if they have a common neighbor.

Theorem 5.1 [L1]. *If $N(G)$ is $(k-1)$-connected, then G is not $(k+1)$-colorable.*

The basic idea of the proof is the following. Let $N'(G)$ denote the baricentric subdivision of $N(G)$. So a vertex of $N'(G)$ is identified with the center of a simplex $\{v_1, \ldots, v_m\}$ of $N(G)$. Define the ν-image of this vertex as the center of the simplex formed by all common neighbors of $v_1, v_2, \ldots, v_m$. Then ν maps the vertices of $N'(G)$ to the vertices of $N'(G)$. It is not difficult to show that ν can be extended to a simplicial map $\nu : N'(G) \to N'(G)$. Moreover, ν behaves like an antipodal map, for instance $\nu^3 = \nu$. Assume now that there is a coloring of G with $k+1$ colors. This is equivalent to the existence of a graph homomorphism $G \to K_{k+1}$ (the definition will be given after Theorem 5.4). This homomorphism induces a simplicial map $N(G) \to N(K_{k+1})$ which is equivariant with respect to ν. It is easy to see that $|N(K_{k+1})| \cong S^{k-1}$ with $\nu = -id$ essentially. Using the $(k-1)$-connectedness of $N(G)$ one can construct a map $\psi : S^k \to |N(G)|$ which is equivariant, i.e., the following diagram:

$$
\begin{array}{ccccccc}
S^k & \xrightarrow{\psi} & |N(G)| & \longrightarrow & |N(K_{k+1})| & \longrightarrow & S^{k-1} \\
{\scriptstyle -id}\downarrow & & {\scriptstyle \nu}\downarrow & & {\scriptstyle \nu}\downarrow & & {\scriptstyle -id}\downarrow \\
S^k & \xrightarrow{\psi} & |N(G)| & \longrightarrow & |N(K_{k+1})| & \longrightarrow & S^{k-1}
\end{array}
$$

is commutative. This contradicts Borsuk's theorem 1.3.

The other proofs of 5.1 (due to Walker [Wa] and Sarkaria [Sa3]) make use of Borsuk's theorem, too. Lovász used this result to determine the chromatic number of the Kneser graph G_n^m. The vertices of this graph are the n-tuples of the set $\{1, 2, \ldots, m\}$, and two vertices form an edge if the corresponding n-tuples are disjoint. Assume $m \geq 2n$ and set $k = m - 2n$. A good $(k+2)$-coloring of G_n^m is this. An n-tuple $\{i_1, \ldots, i_n\}$ with $1 \leq i_1 < i_2 \ldots < i_n \leq m$ gets color i_1 if $i_1 \leq k+1$, otherwise it gets color $k+2$. It is easy to see that disjoint n-tuples have got different colors. The conjecture raised by Kneser [Kn] in 1955 was that the chromatic number of G_n^m is $k+2$. Lovász proved this by showing that $N(G_n^m)$ is $(k-1)$-connected. One can give a simple form to this result.

Theorem 5.2. *If the n-tuples of a $(2n + k)$-set are split into $k + 1$ classes, then one of the classes contains two disjoint n-tuples.*

Shortly after Lovász proved Kneser's conjecture I found the following simple proof [Bá]. A theorem of Gale [Ga] says that for n, $k \geq 1$ there is a set $V \subset S^k$ with $2n + k$ points such that each open hemisphere contains at least n points of V. Denote by $H(a)$ the hemisphere of S_k with center $a \in S^k$, i.e. $H(a) = \{x \in S^k : a \cdot x > 0\}$. Identify the $2n + k$ points with the set V from Gale's theorem. Suppose its n-tuples are split into $k+1$ classes, i.e., they have a $(k + 1)$-coloring. This defines a $(k + 1)$-coloring of S^k in the following way. A point $a \in S^k$ gets the color of each n-tuple in $H(a) \cap V$. So some points may have several colors and, by Gale's theorem, every point has a color. It is easy to check that the points of the same color form an open set in S^k.

On applying Borsuk's theorem 1.2 we get a point $a \in S^k$ such that a and $-a$ have the same color. The point a got this color from an n-tuple in $H(a) \cap V$ and the point $-a$ got it from an n-tuple in $H(-a) \cap V$. But these two n-tuples are disjoint and of the same color.

I mention here that Lovász' topological theorem 5.1 has other applications as well [L2] and is of greater general interest. His method can be extended to another situation, namely, to the following generalized "Kneser" conjecture of Erdős.

Theorem 5.3 [AFL]. *Let n, $t \geq 1$, $k \geq 0$. If the n-tuples of a $[tn + (t-1)k]$-element set are split into $k + 1$ classes then one of the classes will contain t pairwise disjoint n-tuples. The number $[tn + (t - 1)k]$ is best possible.*

The proof of this is analogous to Lovász' proof of the Kneser conjecture. First, for a general t-uniform hypergraph H a suitable neighborhood complex $N(H)$ is defined. Then it is shown that if t is a prime and $N(H)$ is $[k(t-1)-1]$-connected then H is not $(k+1)$-colorable. This is done by using Theorem 2.6 (instead of Borsuk's theorem) when t is an odd prime. Then a neat argument shows that if 5.3 holds for two values of t then it holds for their product as well. Finally, one has to show that the underlying neighborhood complex is $[k(t-1)-1]$-connected.

Recently, K.S. Sarkaria showed [Sa3] that results of the above type can be obtained, in an almost canonical way, using the powerful technique of deleted joins and deleted products. He proved, among other related things, the following generalization of 5.3.

Theorem 5.4 [Sa3]. *Let k, n, $t \geq 1$, $j \geq 2$ and $N \geq \frac{1}{j-1}((t-1)k+tn+j-2)$. If the n-tuples of an N-set are colored by $k + 1$ colors then there will be t n-tuples that are j-wise disjoint and of the same color. The lower bound on N is best possible.*

In connection with Theorem 5.2 the following is of interest. A map $\varphi :$ $G \to H$ from the graph $G(V, E)$ to the graph $H(U, F)$ is a *homomorphism* if

$\varphi : V \to U$ and $(u, v) \in E$ implies $(\varphi(u), \varphi(v)) \in F$. Call a homomorphism $\varphi :$ $G \to G_p^r$ *sharp* if there is no homomorphism $G \to G_p^{r-1}$. Then the fact that G is k-chromatic can be expressed by saying that there is a sharp homomorphism $G \to G_1^k$. In this terminology the Kneser-Lovász theorem says that there is a sharp homomorphism

$$G_n^m \to G_1^{m-2n+2}$$

if $m \geq 2n$. Stahl [St] proves that there is a homomorphism

$$G_n^m \to G_{n-1}^{m-2},$$

and that there is a sharp homomorphism

$$G_n^m \to G_{qn}^{qm}$$

provided $m \geq 2n$ and $q \geq 1$. So we have a chain of homomorphisms

$$G_n^m \to G_{qn}^{qm} \to G_{qn-1}^{qm-2} \to \cdots \to G_{qn-(n-1)}^{qm-2(n-1)}.$$

Stahl conjectures that every induced homomorphism here is sharp. For this it is enough to prove

Conjecture 5.5 [St]. *The homomorphism*

$$G_n^m \to G_{(q-1)n+1}^{qm-2(n-1)}$$

is sharp provided $m \geq 2n$, $q \geq 1$.

When $q = 1$, this is Theorem 5.2.
Our last example comes from combinatorics.

Theorem 5.6 [A2]. *Assume an open necklace has ka_i beads of color i, $1 \leqq i \leqq t$, $k \geq 2$. Then it is possible to cut it in $(k-1)t$ places and partition the resulting intervals into k collections, each containing exactly a_i beads of color i.*

It is easy to give an example where $(k-1)t-1$ cuts do not suffice. The case $k = 2$ is proved in [AW] using Borsuk's theorem. The proof of the general case is based on Theorems 2.5 and 2.6 and on the fact that the CW-complex Y (defined after Theorem 2.6) is highly connected. A continuous generalization of this result, due to N. Alon again, is as follows.

Theorem 5.7 [A2]. *Let $\lambda_1, \ldots \lambda_t$ be t continuous probability measures on the unit interval. Then it is possible to cut the interval at $(k-1)t$ places and partition the $(k-1)t + 1$ intervals obtained into k collections $C_1, \ldots, C_k$ so that $\lambda_i(UC_j) = 1/k$ for all i and j.*

The number of cuts $(k-1)t$ is best possible again. When $k = 2$ the last theorem is known as the Hobby-Rice theorem [HR] on L_1 approximation.

6. Sphere Coverings

Extending Theorem 1.2 Hopf [Ho] showed that if S^d is convered by $d+1$ closed sets $F_1, \ldots, F_{d+1}$, then for any angle $0 \le \varphi \le \pi$ some F_i will contain a pair of points that are seen from the center of S^d at angle φ. Hadwiger [Ha1] conjectured that all angles between 0 and π are realized in a single F_i. Let $h = h(d)$ denote the largest integer r such that if $F_1, \ldots, F_r$ are closed sets covering S^d, then there is an F_i that contains pairs at all angles between 0 and π. In this formulation Hadwiger's conjecture is that $h(d) = d+1$. The inequality $h(d) \le d+1$ is trivial. Hadwiger [Ha2] proved that $h(1) = 2$ and $h(2) = 3$. In 1967 Larman [La] showed that $h(d) \ge d/2$, his proof uses a method of Lyusternik and Shnirel'man [LS]. A recent result of Bogatyi [Bg] shows $h(d) \ge d$:

Theorem 6.1. *If S^d is covered by d closed sets, then one of the sets contains pairs of points at all angles between 0 and π.*

The proof uses Borsuk's theorem and an analogue of the Lyusternik-Shnirel'man theorem on the category of a projective space. Bogatyi's result settles Hadwiger's conjecture almost completely: $d \le h(d) \le d+1$.

Another old conjecture, related to Borsuk's theorem is due to Knaster [Kna]. It says that if $A \subset S^d$ is a set of $d - m + 2$ points and $f : S^d \to R^m$ is continuous, then there is a rotation ρ of S^d such that $f(\rho(A))$ is a single point. Borsuk's theorem is a special case of this conjecture with $m = d$ and A a pair of antipodal points. Also, Hopf's theorem cited above is another special case of the conjecture.

Knaster's conjecture would imply the following. If $A \subset S^d$ is a set of $d - m + 2$ points and $F_1, \ldots, F_{m+1}$ is a closed cover of S^d; then some F_i contains a congruent copy of A, i.e., $\rho(A) \subset F_i$ for some rotation ρ of S^d. To prove this implication consider the map $f = (f_1, \ldots, f_m) : S^d \to R^m$ defined by $f_i(x) = \text{dist}(x, F_i)$. Then, by Knaster's conjecture, $f(\rho(A))$ is a single point for some rotation ρ. If $f_i(\rho(A)) = 0$ for some $i = 1, \ldots, m$, then $\rho(A) \subset F_i$. If $f_i(\rho(A)) > 0$ for all $i = 1, \ldots, m$, then $\rho(A) \subset F_{m+1}$.

The following special case is of interest. Let V^{m+1} stand for the $(m+1)$ vertices of the regular simplex with circumscribed ball having radius 1. We think of this simplex as inscribed in S^k, $k \ge m$.

Conjecture 6.3 [Bg]. *If $F_1, \ldots, F_m$ is a closed cover of S^d, then some F_i contains a congruent copy of V^{d-m+3}.*

This conjecture has been proved for $m = d+1$ (Borsuk's theorem), $m = d$ (Yang's theorem [Ya]), and $m = 2$ by Bogatyi and Khimshiashvili [BK].

It is also possible that, just like in Hadwiger's conjecture, one F_i will contain every $(d - m + 3)$-point configuration.

Conjecture 6.4 [Bg]. *If $F_1, \ldots, F_m$ is a closed cover of S^d, then here is an F_i containing a congruent copy of every set $A \subset S^d$ of size $d - m + 3$.*

Acknowledgement. This work has been supported by the Program in Discrete Mathematics and its Application at Yale and by Hungarian National Foundation for Scientific Research Grant No. 1812

References

[A1]   Alon, N.: Some recent combinatorial applications of Borsuk-type theorems. In: Algebraic, Extremal and Metric Combinatorics. M.M. Deza, P. Frankl, D.G. Rosenberg (eds.). Cambridge University Press, Cambridge (1988), pp. 1–12

[A2]   Alon, N.: Splitting necklaces. Adv. Math. **63** (1987) 247–253

[AA]   Akiyama, J. and N. Alon: Disjoint simplices and geometric hypergraphs. In: Combinatorial Mathematics – Proceedings of the Third International Conference 1985 (ed. G. Bloom et al.) Ann. New York Acad. Sci. **555** (1989) 1–3

[AFL] Alon, N., P. Frankl and L. Lovász: The chromatic number of Kneser hypergraphs. Trans. Amer. Math. Soc, **298** (1986) 359–370

[AW]   Alon, N. and D.B. West: The Borsuk-Ulam theorem and bisection of necklaces. Proc. Amer. Math. Soc. **98** (1986) 623–628

[BB]   Bajmóczy, E.G. and I. Bárány: On a common generalization of Borsuk's and Radon's theorem. Acta Math. Hung. **34** (1979) 347–350

[Bá]   Bárány, I.: A short proof of Kneser's conjecture. J. Comb. Theory A **25** (1978) 325–326

[BáL] Bárány, I. and D.G. Larman: A colored version of Tverberg's theorem, to appear in J. London Math. Soc. (2) **45** (1992) 314–320

[BFL] Bárány, I., Z. Füredi and L. Lovász: On the number of halving planes. Combinatorica **10** (1990) 175–183

[BL]   Bárány, I. and L. Lovász: Borsuk's theorem and the number of facets of centrally symmetric polytopes. Acta Math. Acad. Sci. Hungar. **40** (1982) 323–329

[BSS] Bárány, I., S.B. Shlosman and A. Szücs: On a topological generalization of a theorem of Tverberg. J. London Math. Soc. (2) **23** (1981) 158–164

[Bg]   Bogatyi, S.A.: Topological method in combinatorial problems. Usp. Mat. Nauk. **41** (1986) 37–48 (in Russian)

[BK]   Bogatyi, S.A. and G.N. Khimshiashvili: On a problem of Knaster about mappings from spheres into the line. In: Proc. 5th Tiraspol' Symposium on General Topology and its Applications. Shtiintsa, Kishinev (1985) 28–30 (in Russian)

[BiL] Billera, L.J. and C.W. Lee: Sufficiency of McMullen's conditions for f-vectors of simplicial polytopes. Bull. Amer. Math. Soc. **2** (1980) 181–185

[Bj]   Björner, A.: Topological methods. Chapter in: Handbook of Combinatorics (R. Graham, M. Grötschel, L. Lovász eds.) to appear 1992

[Bo]   Borsuk, K.: Drei Sätze über die n-dimensionale euklidische Sphäre. Fundamenta Math. **20** (1933) 177–190

[DGK] Danzer, L., B. Grünbaum and V. Klee: Helly's theorem and its relatives. Convexity, Proc. Symp. Pure Math. **7** AMS, Providence (1963)

[Do]   Dold, A.: Simple proofs of some Borsuk-Ulam results. Contemp. Math. **19** (1983) 65–69

[Dv]   Dvoretzky, A.: Some results on convex bodies and Banach spaces. Proc. Int.
         Symp. on Linear Spaces, Jerusalem (1961) 123–160
[FLM]   Figiel, T., F. Lindenstrauss and V.D. Milman: The dimension of almost
         spherical sections of convex bodies. Acta Math. **139** (1977) 53–94
[Fl]   Flores, A.: Über n-dimensionale Komplexe die in R_{2m+1} absolut selbstver-
         schlungen sind. Ergeb. Math. Kolloq. **6** (1933/34) 4–7
[Ga]   Gale, D.: Neighboring vertices on a convex polyhedron. In: Linear Inequal-
         ities and Related Systems (H.W. Kuhn and A.W. Tucker eds.). Princeton
         University Press, Princeton (1956)
[GPW]   Goodman, J.E., R. Pollack, R. Wenger: Geometric transversal theory. This
         volume 1992
[Ha1]   Hadwiger, H.: Ein Überdeckungssatz für den euklidschen Raum. Portugal
         Math. **4** (1944) 140–144
[Ha2]   Hadwiger, H.: Eine Bemerkung zum Borsukschen Antipodensatz. Viertel-
         jschr. Naturforsch. Ges., Zürich **89** 211–214
[HB]   Hobby, C.R., and J.R. Rice: A moment problem in L_1 approximation. Proc.
         Amer. Math. Soc. **16** (1932) 665–670
[Ho]   Hopf, H.: Eine Verallgemeinerung bekannter Abbildungs- und Überdeckungs-
         sätze. Portugal. Math. **4** (1944) 129–139
[Ka]   van Kampen, E.R.: Komplexe in euklidischen Räumen. Abh. Math. Sem. **9**
         (1932) 72–78 and 152–153
[Kn]   Kneser, M.: Aufgabe 300. Jber. Deutsch. Math. Verein. **58** (1958)
[Kna]   Knaster, B.: Problem 4. Colloq. Math. **1** 30
[KZ]   Krasnoselsky, M.A. and P.P. Zabrejko: Geometritcheskie zadatchi neline-
         jnogo analiza. Nauka, Moscow 1975 (in Russian)
[La]   Larman, D.G.: On the realization of distances with coverings of an n-sphere.
         Mathematika **14** (1967) 203–206
[L1]   Lovász, L.: Kneser's conjecture, chromatic number and homotopy. J. Comb.
         Theory A **25** (1978) 319–324
[L2]   Lovász, L.: Self-dual polytopes and the chromatic number of distance graphs
         on the sphere. Acta Sci. Math. (Szeged) **45** (1983) 317–323
[LoS]   Lovász, L. and M. Simonovits: Mixing rate of Markov chains, an isoperimet-
         ric inequality, and computing the volume. Proc. 31st FOCS, (1990) 346–355
[LS]   Lyusternik, L.A. and L.G. Shnirel'man: Topological methods in variational
         problems and their applications to the differential geometry of surfaces.
         Uspekhi Mat. Nauk **2** (1947) 166–217 (in Russian)
[McM]   McMullen, P.: The number of vertices of simplicial polytopes. Israel J. Math.
         9 (1971) 559–570
[MS]   McMullen, P. and G.C. Shepard: Diagrams for centrally symmetric poly-
         topes. Mathematika **15** (1968) 123–128
[Öz]   Özaydin, M.: Equivariant maps for the symmetric group. Preprint (1990)
[PW]   Pollack, R. and R. Wenger: Necessary and sufficient conditions for hyper-
         plane transversal. Combinatorica **10** (1990)
[Ra]   Rado, R.: A theorem on general measure. J. London Math. Soc. **26** (1946)
         291–300
[Sa1]   Sarkaria, K.S.: A generalized van Kampen-Flores theorem. Proc. Amer.
         Math. Soc. **III** (1991) 559–565
[Sa2]   Sarkaria, K.S.: Kneser colorings of polyhedra. Ill. Jour. Math. **33** (1989)
         592–620
[Sa3]   Sarkaria, K.S.: A generalized Kneser conjecture. J. Comb. Theory B **49**
         (1990) 236–240
[St]   Stahl, S.: n-tuple colorings and associated graphs. J. Comb. Theory B **20**
         (1976) 185–203

[Sta1] Stanley, R.P.: The upper bound conjecture and Cohen-Macaulay rings. Studies in Appl. Math. **54** (1975) 135–142

[Sta2] Stanley, R.P.: On the number of faces of centrally-symmetric simplicial polytopes. Graphs and Combinatorics **3** (1987) 55–66

[Ste]   Steinlein, H.: Borsuk's antipodal theorem and its generalizations and applications: A survey. In: Méthodes Topologiques en Analyse Non Linéaire. Coll. Sém. de Math. Sup., Vol. 95. A. Granas (ed.). Univ. de Montréal Press, Montréal (1985), pp. 166–235

[Tv]   Tverberg, H.: A generalization of Radon's theorem. J. London Math. Soc. **41** (1966) 123–128

[Wa]   Walker, J.W.: From graphs to ortholattices and equivariant maps. J. Combinat. Th. B **35** (1983) 171–192

[Ya]   Yang, C.T.: On maps from spheres to euclidean spaces. Amer. J. Math. **79** (1957) 725–732

[ŽV1]   Živaljević, R.T. and S.T. Vrećica: An extension of the ham sandwich theorem. Bull. London Math. Soc. **22** (1990)

[ŽV2]   Živaljević, R.T. and S.T. Vrećica: The colored Tverberg problem and complexes of injective functions, to appear in J. Comb. Theory B (1992)

Chapter X. A Survey of Recent Results in the Theory of Packing and Covering

Gábor Fejes Tóth and Wlodzimierz Kuperberg

1. Introduction

The theory of packing and covering, originated as an offspring of number theory and crystallography early in this century, has quickly gained interest of its own and is now an essential part of discrete geometry. The theory owes its early development to its aesthetic appeal and its classical flavor, but more recently, some of its topics have been found related to the rapidly developing areas of mathematics connected with computer science, and the theory of packing and covering has been boosted by a renewed interest.

The influence from combinatorics and computer science was two-fold: new problems were introduced, mostly of a computational nature, and new, combinatorial techniques were adopted. In this survey we shall try to expose the computational and combinatorial aspects of some of the recent advancements in the theory.

Continuing an earlier trend, most of the research concentrated on packing and covering of the plane, although some progress was achieved in the difficult investigations of the analogous problems in three and, to some extent, higher dimensions. The selection of topics in this survey is not intended to form an account of all research accomplishments in the area, but rather to illustrate some of the recent trends and new techniques. The contents of this survey have also been affected by the authors' points of view and personal tastes, thus certain results, even some very interesting ones, have not been included. Also, certain topics, such as tiling, packing with spheres in all dimensions, finite packing and covering with spheres, and packing and covering with sequences of convex sets, although suitable for the purpose of this survey, are not included because they have been reviewed extensively elsewhere quite recently (see Grünbaum and Shephard [1986], Conway and Sloane [1987], Gritzmann and Wills [1986], and Groemer [1987], respectively). Readers interested in a more thorough treatment of the theory or in the topics which are only mentioned here are referred to the "books and survey papers" part of the listed literature.

2. Preliminaries and Basic Concepts

Let $\mathcal{F}$ be a family of sets and U a region in space. We say that $\mathcal{F}$ is a *packing in* U if every member of $\mathcal{F}$ is a subset of U and all members of $\mathcal{F}$ have mutually disjoint interiors, and we say that $\mathcal{F}$ is a *covering of* U if U is contained in the union $\bigcup \mathcal{F}$ of all members of $\mathcal{F}$. Our space will usually be the Euclidean d-dimensional space E^d, with $d = 2$ or $d = 3$ most of the time, and each member of the family $\mathcal{F}$ will usually be a *convex body*, i.e. a convex compact set with an interior point. A plane convex body will be referred to as a *convex disk*. If U is the whole space, then a packing in U [a covering of U] will be simply called a *packing* [a *covering*]. The volume of a (measurable) set S will be denoted by $V(S)$. In the two-dimensional case, "volume" of S will mean the area of S, and, accordingly, it will be denoted by $A(S)$. Since we shall deal with sets and families of sets of a rather simple nature, such as convex bodies, polyhedra, etc., we need not be concerned with the problem of measurability.

If U is a bounded region, then the *density of $\mathcal{F}$ relative to* U is defined as

$$d(\mathcal{F}, U) = \frac{1}{V(U)} \sum_{F \in \mathcal{F}} V(F \cap U).$$

If $\mathcal{F}$ is a packing (in the whole space), then the (*upper*) *density* of $\mathcal{F}$ is defined as

$$d(\mathcal{F}) = \limsup_{r \to \infty} d(\mathcal{F}, B(r)),$$

where $B(r)$ denotes the ball of radius r centered at the origin.

If $\mathcal{F}$ is a covering, then the (*lower*) *density* of $\mathcal{F}$ is defined as

$$d(\mathcal{F}) = \liminf_{r \to \infty} d(\mathcal{F}, B(r)).$$

We would like to point out that some modifications in the above definitions are possible, leading to various, slightly different notions of density (see for instance Rogers [1964]), but in this context, all of these notions would turn out to be equivalent.

One of the classical problems of the theory of packings and coverings is to find a densest packing and a thinnest covering with congruent copies of a given convex body K. Specifically, it is to find the *packing density* of K and the *covering density* of K, denoted by $\delta(K)$ and $\vartheta(K)$, respectively, which are defined as follows: $\delta(K)$ is the supremum of the densities $\delta(\mathcal{P})$ among all packings $\mathcal{P}$ with congruent copies of K and $\vartheta(K)$ is the infimum of the densities $d(\mathcal{C})$ among all coverings $\mathcal{C}$ with congruent copies of K. If we restrict the arrangements of copies of K to those which consist of parallel translates of K, then we obtain a similarly defined notion of the *translation packing density* of K and that of the *translation covering density* of K, denoted by $\delta_T(K)$ and $\vartheta_T(K)$, respectively. Further, if only lattice arrangements are considered, we obtain the *lattice packing density* of K and the *lattice covering density* of K, denoted by $\delta_L(K)$ and $\vartheta_L(K)$, respectively. By a *lattice* we mean a set of points (vectors) in E^d obtained as all integer-linear combinations of some basis for E^d, and a *lattice arrangement* of K is a set of translates of K by the vectors of a lattice. It is easy to see that the densities $\delta_L(K)$ and $\vartheta_L(K)$ are attained in the corresponding extreme lattice arrangements, i.e. they can be defined not just as the supremum and the infimum, but as the maximum and the minimum, respectively. A similar statement is true for each of the other densities assigned to K, as it follows from general results of Groemer [1963] and [1968].

It is evident that $\delta_L(K) \leq \delta_T(K) \leq \delta(K) \leq 1 \leq \vartheta(K) \leq \vartheta_T(K) \leq \vartheta_L(K)$ for every convex body K. However, finding the values of these densities of K is, in general, a very difficult problem, especially in dimension $d \geq 3$. To quote Rogers [1964]: "Despite considerable advances in the Geometry of Numbers since Minkowski's time, the problem of determining the value of $\delta_L(K)$ for a given convex 3-dimensional body K remains a formidable task". This statement can be applied to the other densities assigned to K as well. The words, written in 1963, fairly accurately describe the state of the theory today.

3. A Review of Some Classical Results in the Plane

One of the earliest theorems in the theory of packings states that the maximum density of any packing in the plane with congruent circular disks equals $\pi/\sqrt{12} = 0.90689968\ldots$, attained in the natural, "honeycomb" arrangement of the circles (see Thue [1910]). The analogous result for coverings is the following theorem of Kershner [1939], now considered classical as well, which states that the minimum density of any covering of the plane with congruent circular disks equals $2\pi/\sqrt{27} = 1.20919957\ldots$, also attained in the honeycomb arrangement of the circles. The honeycomb pattern is obtained by tiling the plane with congruent regular hexagons. For the packing, the circles are inscribed in the hexagons, and for the covering, they are circumscribed about them (see Figs. 3.1 and 3.2).

It should be noted that although the problem of l a t t i c e packings with circular disks was considered and solved long before Thue (see Lagrange [1773]), he solved the problem for a r b i t r a r y packings with congruent circular disks, and that was a considerably more difficult task. Thus, for the circular disk B^2, $\delta(B^2) = \delta_L(B^2)$ and $\vartheta(B^2) = \vartheta_L(B^2)$.

For a given convex disk K, let $H(K)$ denote the minimum area of a hexagon containing K and let $h(K)$ denote the maximum area of a hexagon contained in K. In [1950], L. Fejes Tóth proved that

$$\delta(K) \leq A(K)/H(K)$$

for each convex disk K. Concerning coverings, he conjectured that $\vartheta(K) \geq A(K)/h(K)$. A partial result supports this conjecture. Say that two convex disks, K_1 and K_2, *cross*, if each of the sets $K_1 \backslash K_2$ and $K_2 \backslash K_1$ is disconnected. L. Fejes Tóth [1950] proved that if $\mathcal{C}$ is a crossing-free covering with congruent copies of a convex disk K, then $d(\mathcal{C}) \geq A(K)/h(K)$. Since two translates of a convex disk cannot cross, it follows that

$$\vartheta_T(K) \geq A(K)/h(K)$$

for every convex disk K.

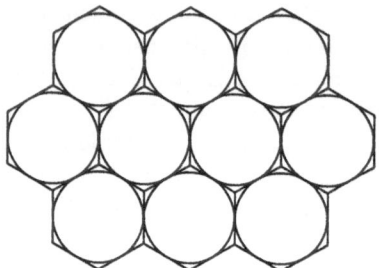
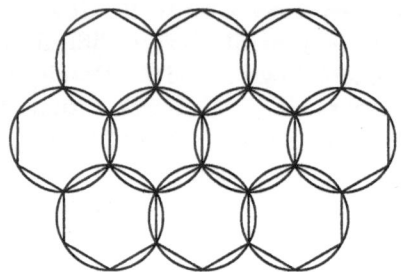

Fig. 3.1 A densest circle packing (Thue).

Fig. 3.2 A thinnest circle covering (Kershner).

By a theorem of Dowker [1944], if K is a centrally symmetric convex disk, then $h(K)$ and $H(K)$ can be obtained by inscribing in, resp. circumscribing about K a centrally symmetric hexagon. Since translates of any centrally symmetric hexagon can be arranged to form a plane lattice tiling, translation copies of K generated by the corresponding lattice produce a lattice covering, resp. a lattice packing, whose density equals to the bound given above. Thus, for each centrally symmetric convex disk K, we get:

$$\delta(K) = \delta_L(K) = A(K)/H(K)$$

and

$$\vartheta_T(K) = \vartheta_L(K) = A(K)/h(K).$$

In the following discussion of the translation packing density, we employ the "difference-body trick" invented by Minkowski in [1904]. For a pair of compact sets K and L in the Euclidean space E^d and a number λ, let $\lambda K = \{\lambda x : x \in K\}$ and $K + L = \{x + y : x \in K, y \in L\}$. We write $K - L$ instead of $K + (-1)L$. In this notation, Minkowski's trick can be described as follows: Let $\mathcal{V}$ be a set of vectors and let $\mathcal{F}$ be a family of translates of a convex body by the vectors from $\mathcal{V}$. Then $\mathcal{F}$ is a packing if and only if the family $\mathcal{F}^*$ of translates of the convex body $\frac{1}{2}(K - K)$ by the vectors from $\mathcal{V}$ is a packing as well. Observe that, for a given K, the densities of the packings $\mathcal{F}$ and $\mathcal{F}^*$ depend on $\mathcal{V}$ only, and $d(\mathcal{F})/d(\mathcal{F}^*) = V(K)/V(\frac{1}{2}(K - K))$. Thus, $d(\mathcal{F}) = \delta_T(K)$ if and only if $d(\mathcal{F}^*) = \delta_T(\frac{1}{2}(K - K))$. Observe now that the body $\frac{1}{2}(K - K)$

is centrally symmetric. This, and the previously quoted statements yield the following result of Rogers [1951]:

$$\delta_T(K) = \delta_L(K) \text{ for each convex disk } K.$$

The conjecture that $\vartheta_T(K) = \vartheta_L(K)$ in general (i.e. for non-centrally-symmetric disks as well) remains open.

Certain other classical results of the theory of packings and coverings will be quoted in the following sections as a background for the more recent developments. The reader interested in a more complete account of the classical part of the theory is referred to L. Fejes Tóth [1972] and Rogers [1964]. The first reference presents results in two and three dimensions, and the second one represents essentially the current status of the theory of packings and coverings concerning general results without restrictions on dimension.

4. Economical Packing in and Covering of the Plane

In this section we describe results concerning the lower bounds for the various kinds of packing densities and the upper bounds for the covering densities. As we mentioned it before, historically, the theory of packing and covering developed in connection with number theory (geometry of numbers, problems on diophantine approximations). In that context, packings with translates, especially lattice packings, of a centrally symmetric convex body occur naturally. Thus, the early results are about the centrally symmetric case, translations and lattice arrangements. As one of the first general results of this kind, Mahler [1947] proved the inequality

$$\delta_L(K) \geq \sqrt{3}/2 = 0.86602540\ldots$$

for all centrally symmetric convex disks K. The conjectured greatest lower bound for $\delta_L(K)$ over all centrally symmetric convex disks K is

$$\delta_L(K_0) = (8 - \sqrt{32} - \ln 2)/(\sqrt{8} - 1) = 0.90241418\ldots,$$

attained by the so-called "smoothed octagon" K_0 constructed by Reinhardt in [1934] (see also L. Fejes Tóth [1972], p. 104) and shown here on Fig. 4.1. In [1961], Ennola improved Mahler's inequality to

$$\delta_L(K) \geq (3\sqrt{2} + \sqrt{3} - \sqrt{6})/4 = 0.88130043\ldots$$

and he suggested that a refinement of his method would increase the bound to about 0.8925, which was in fact accomplished by Tammela in [1970]. Reinhardt's conjecture on the smoothed octagon remains open.

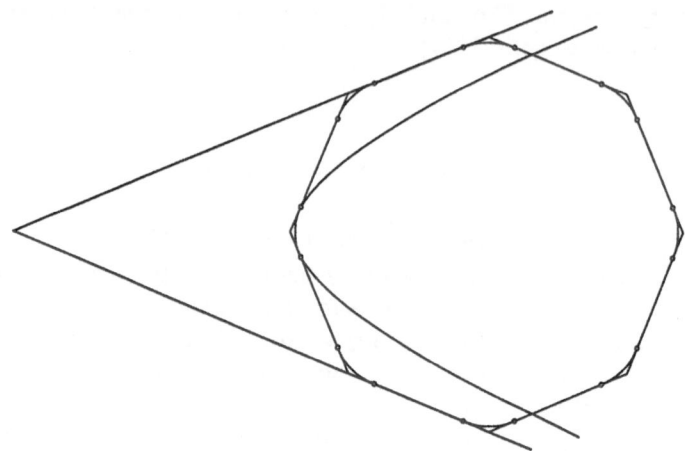

Fig. 4.1 Reinhardt's smoothed octagon.

Surprisingly, the analogous problem for coverings in the case of central symmetry of K has been settled as a corollary of a seemingly unrelated result. In [1939], Sas proved that for every integer $n \geq 3$ and for every convex disk K the maximum area n-gon P_n inscribed in K satisfies $A(P_n) \geq \frac{n}{2\pi} \sin \frac{2\pi}{n} A(K)$ and the equality occurs only if K is an ellipse. Moreover, in case K is centrally symmetric and n is even, the proof of Sas yields the existence of a centrally symmetric n-gon inscribed in K with area not less than this bound. In particular, every centrally symmetric convex disk K contains a centrally symmetric hexagon h with $A(h) \geq \frac{2\pi}{\sqrt{27}} A(K)$. Since h tiles the plane in a lattice-like manner, it follows that

$$\vartheta_L(K) \leq 2\pi/\sqrt{27}$$

for each centrally symmetric disk K and the equality holds if (and only if) K is an ellipse, which shows that the above inequality is sharp.

As we noted before, $\delta_L(K) = \delta(K)$ for every centrally symmetric convex disk K, therefore the problems of finding lower bounds for $\delta_L(K)$, $\delta_T(K)$ and $\delta(K)$ are mutually equivalent in the case of central symmetry of K. In the same case, the equality $\vartheta_L(K) = \vartheta(K)$ is just a conjecture, but since the upper bound of $2\pi/\sqrt{27}$ for $\vartheta_L(K)$ is attained by a circular disk B^2, and since $\vartheta(B^2) = \vartheta_L(B^2)$, the sharp inequality

$$\vartheta(K) \leq 2\pi/\sqrt{27}$$

follows for each centrally symmetric convex disk K.

For non-centrally symmetric convex disks K, the densities $\delta_L(K)$ and $\delta(K)$ differ in general, and so do the densities $\vartheta_L(K)$ and $\vartheta(K)$, while $\delta_L(K) = \delta_T(K)$ is known and $\vartheta_L(K) = \vartheta_T(K)$ is conjectured.

In [1950], Fáry gave the following, sharp bounds for $\delta_L(K)$ and $\vartheta_L(K)$ in the plane:

$$\delta_L(K) \geq 2/3 \text{ and } \vartheta_L(K) \leq 3/2$$

with equality attained in each case if and only if K is a triangle. Courant [1965] gave an elegant, elementary proof of the first inequality of Fáry.

Chakerian and Lange proved in [1971] that every convex disk K is contained in a quadrilateral Q with $A(Q) \leq \sqrt{2}A(K)$. Since every quadrilateral tiles the plane, they concluded that

$$\delta(K) \geq \sqrt{2}/2 = 0.70710678\ldots.$$

Using inscribed quadrilaterals and the theorem of Sas, one obtains, in a similar way, an upper bound for covering densities:

$$\vartheta(K) \leq \pi/2 = 1.57079632\ldots.$$

In order to improve upon these bounds, it is natural to consider more general types of tiles inscribed in and circumscribed about K. Obviously, a small area tile containing K yields a dense packing and a large area tile contained in K yields a thin covering. The type of tile that proved useful in this method is a *p-hexagon*, which is a convex hexagon with a pair of parallel opposite sides of equal length. In this definition, "opposite" means separated by two other sides, and the extreme (degenerate) cases are allowed (see Fig. 4.2).

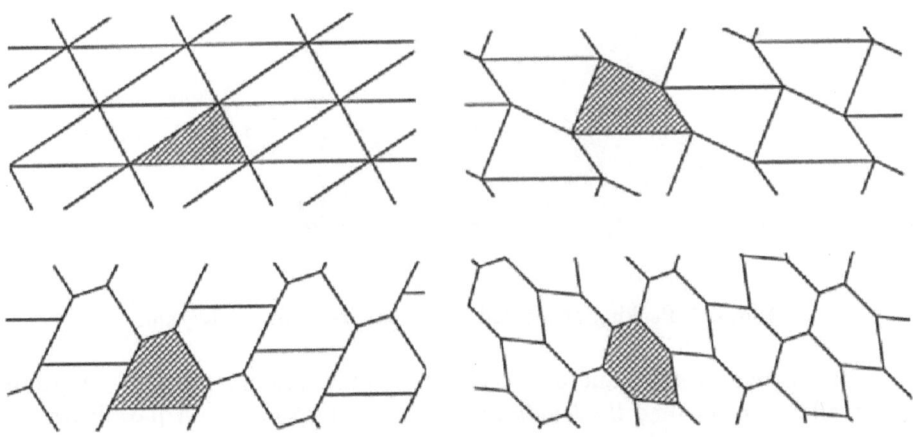

Fig. 4.2 *p*-hexagons and the corresponding tilings.

The above idea and its refinements have produced gradual improvements of the bounds for the discussed densities (see W. Kuperberg [1987a], [1987b], and [1989]), resulting in

$$\delta(K) \geq 25/32 = 0.78125$$

and

$$\vartheta(K) \leq 8(2\sqrt{3} - 3)/3 = 1.23760430\ldots.$$

In addition, the p-hexagon technique has been used by W. Kuperberg in [1987b] to obtain the sharp inequality

$$\delta(K)/\vartheta(K) \geq 3/4\,,$$

previously known only under the assumption of central symmetry of K (see L. Fejes Tóth [1972] p. 106).

While the upper bound for $\vartheta(K)$ quoted above remains the best known, the one for $\delta(K)$ has been improved by G. Kuperberg and W. Kuperberg in [1990] to

$$\delta(K) \geq \sqrt{3}/2 = 0.86602540\ldots$$

by a different approach. Again, this inequality has been previously known under the assumption of central symmetry of K (see Mahler [1946]). Because of the relevance of the technique used in the proof of the above inequality to the computational aspects of the problem, we present some details here.

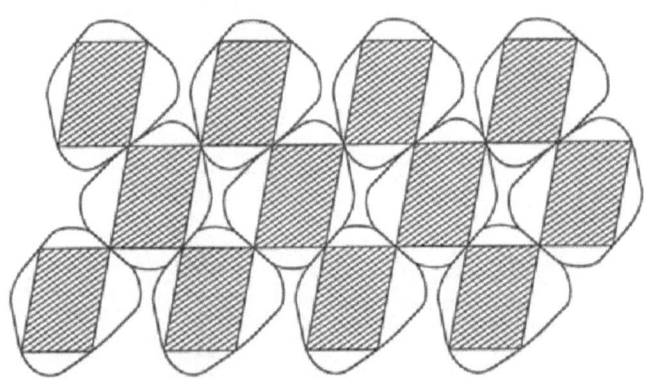

Fig. 4.3 Packing generated by an extensive parallelogram.

For a convex disk K and a direction v, the maximum length of a chord of K parallel to v is called *the length of K in the direction of v*. A parallelogram inscribed in K is said to be *extensive* if each of its sides' length is at least one-half of the length of K in the direction of the side. If P is an extensive parallelogram inscribed in K then P generates a packing with congruent copies of K in the following way: place four copies of K around K by reflecting K in (rotating K by 180° about) each vertex of P. Reflect each new copy of K in each free vertex of the corresponding copy of P and continue this procedure indefinitely. Since P is extensive, no two copies of K overlap (see Fig. 4.3).

The density of this packing equals $A(K)/2A(P)$, thus an extensive parallelogram of a small area yields a dense packing. The crucial part of the proof is to show that for every convex disk K and every direction v there exists an extensive parallelogram $P(v)$ inscribed in K, with a pair of sides parallel to v and such that

$$A(P(v)) \leq A(K)/\sqrt{3},$$

from which the inequality $\delta(K) \geq \sqrt{3}/2$ follows. Observe that v is arbitrary, thus it is perhaps possible to raise the bound of $\sqrt{3}/2$ through a suitable choice of v. Also, this method makes an algorithm possible for an explicit construction of $P(v)$ for a given polygonal K and for selecting v so as to minimize the area of $P(v)$, resulting in a construction of a fairly dense (in some cases the densest) packing with copies of K.

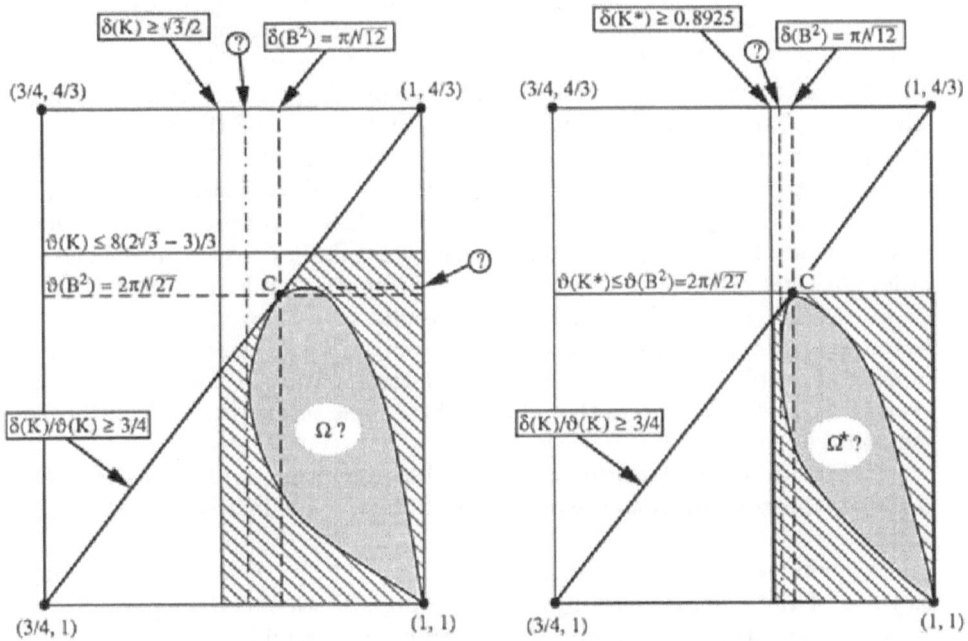

Fig. 4.4 Polygons enclosing Ω and Ω^*, respectively.

The density bounds reviewed in this section can be summarized by a graphic illustration as follows. Let Ω [resp. Ω^*] denote the subset of the Cartesian plane R^2 consisting of all points (x, y) such that $x = \delta(K)$ and $y = \vartheta(K)$ for some [centrally symmetric] convex disk K. The inequalities discussed above produce the regions containing Ω and Ω^*, respectively illustrated on Fig. 4.4 as the lightly shaded polygons. The (hypothetical) silhouettes of the sets Ω and Ω^* are shaded darker.

The problem of a complete and explicit description of the sets Ω and Ω^* seems to be too general, considering the diversity among, and the often encountered difficulty in, the problems of computing the packing and covering densities of specific convex disks. Some simpler questions about the nature of the sets Ω and Ω^* may have a better chance of being answered first, such as: Is the set Ω [resp. Ω^*] closed? Is it convex? Is the point $C = (\pi/\sqrt{12}, 2\pi/\sqrt{27})$, corresponding to the circular disk, the only point of Ω [resp. Ω^*] which lies on the line $y = (4/3)x$?

5. Multiple Packing and Covering

Let k be a positive integer. A family $\mathcal{F}$ of sets is said to be a *k-fold packing* if each point belongs to the interior of at most k members of $\mathcal{F}$, and $\mathcal{F}$ is called a *k-fold covering* if each point belongs to at least k members of $\mathcal{F}$. Thus we have natural generalizations of the notions of (simple) packing and covering, which are included here by setting $k = 1$. Many of the questions about simple packings and coverings generalize easily to multiple packings and coverings. The classical problems concerning the extreme densities and the arrangements that produce them play as important role here as in the context of simple packings and coverings. In the survey paper of G. Fejes Tóth [1983], nearly all results of this type obtained before 1983 are listed. We present here a few results of the classical type and we emphasize a technique based on the higher order Dirichlet cells, since this tool plays an important role in computational geometry. We also review some recent results of a strongly combinatorial flavor, concerning the problem of decomposing a k-fold arrangement into a number of simple (1-fold) ones. The combinatorial nature of this topic prevails even in the methods used in the proofs, where some results from graph theory are utilized.

Some additional, algorithmic type results on multiple arrangements, dealing with the computational aspects of the theory, are included in the next section.

We define the densities of multiple packings and coverings in the same way as those of the (simple) packings and coverings. Thus, to every convex body K, we assign the corresponding k-fold packing and k-fold covering densities: general, restricted to translations, and restricted to lattice type arrangements, denoted by: $\delta^k(K)$, $\delta_T^k(K)$, $\delta_L^k(K)$ and $\vartheta^k(K)$, $\vartheta_T^k(K)$, $\vartheta_L^k(K)$, respectively. The natural problems concerning lower bounds for multiple packing densities and upper bounds for multiple covering densities for arbitrary convex bodies form a vast, almost untouched research area. Among the few obtained results, we mention here the following. Erdős and Rogers proved in [1962] that E^d can be covered by translates of any convex body K with density at most $d \log d + d \log(\log d) + 4d$ so that no point is covered more than $e(d \log d + d \log(\log d) + 4d)$ times. This implies that for any dimension d there exists a constant c_d such that

$$\delta_T^k(K) \geq c_d k$$

for every convex body K in E^d.

In [1976], Cohn proved that

$$\vartheta_L^k(K) \leq [(k+1)^{1/d} + 8d]^d$$

for each d-dimensional convex body K. This inequality should be regarded as an asymptotic estimate as k or d tends to infinity. For instance, for $d = 2$, Cohn's bound is greater than the trivial bound of $(3/2)k$ for all $k \leq 1914$. The trivial bound of $(3/2)k$ is derived from the obvious inequality $\vartheta_L^k(K) \leq k\vartheta_L(K)$ and Fáry's inequality $\vartheta_L(K) \leq 3/2$ for convex disks K.

Bolle [1989] gives asymptotic bounds for $\delta_L^k(K)$ and $\vartheta_L^k(K)$ as $k \to \infty$ and K remains a fixed convex disk: For every convex disk K there exist positive constants c_1 and c_2 depending on K only, such that

$$\delta_L^k(K) \geq k - c_1 k^{2/5}$$

and

$$\vartheta_L^k(K) \leq k + c_2 k^{2/5} .$$

For convex disks K of a special kind which Bolle calls "acute" disks, and which include all convex polygons, each of the above two inequalities is strenghtened by replacing the exponent of $2/5$ with $1/3$.

Concerning k-fold packings with congruent balls in E^d, Few [1964] observed that if each ball in a (simple) packing with unit balls is replaced with a concentric ball of radius $2k/(k+1)$, then the resulting arrangement of balls is a k-fold packing. This observation yields

$$\delta^k(B^d) \geq [2k/(k+1)]^{d/2} \delta(B^d)$$

and

$$\delta_L^k(K) \geq [2k/(k+1)]^{d/2} \delta_L(B^d) .$$

Few's observation, by its very nature, cannot yield a good lower bound for $\delta^k(B^d)$ when k is large, but it supplements the bound of Erdős and Rogers (at least for k-fold packings of balls) which requires k to be large.

We now turn to the density bounds problems from the opposite side: for a given convex body K or a given class of convex bodies K, find a reasonable upper bound for the k-fold packing density and a reasonable lower bound for the k-fold covering density. Florian [1978], generalizing a result of Schmidt [1961], proved that if the boundary of K is smooth, then $\delta^k(K) < k$ and $\vartheta^k(K) > k$. Obviously, for each specific smooth body K, one expects specific bounds $\alpha < k$ and $\beta > k$ such that $\delta^k(K) \leq \alpha$ and $\vartheta^k(K) \geq \beta$, but Florian's method does not yield such specific bounds, even when K is a ball. Such specific bounds for $\delta^k(B^d)$, at least for certain pairs of k and d, have been obtained first by Few [1964]. He proved that

$$\delta^k(B^d) \leq (1 + d^{-1})[(d+1)^k - 1][k/(k+1)]^{d/2} .$$

Observe that this inequality is significant (i.e. gives an upper bound better than the trivial upper bound of k) only if d is much greater than k. Few obtained his result by estimating the multiplicity of an arrangement obtained from a k-fold packing with unit balls by enlarging the radius of each ball by the factor of $1 + 1/k$. This technique can be traced back to Blichfeldt's idea introduced in [1929] in order to estimate the packing density $\delta(B^d)$.

By a further elaboration of this idea for $k = 2$, in [1968] Few obtained a stronger inequality

$$\delta^2(B^d) \leq \frac{4}{3}(d+2)\left(\frac{2}{3}\right)^{d/2}$$

which is better than the trivial one for all $d \geq 11$.

Next we will discuss a technique which generalizes the concept of Dirichlet cells, and which yields non-trivial upper bounds for $\delta^k(B^d)$ for all k and d, and non-trivial lower bounds for $\vartheta^k(B^d)$ for all k and d as well. For a discrete set of points S in E^d, a partition of E^d into cells is defined, assigning to each point of S the set of points $D(p)$ of E^d which are closer to that point of S than to any other point of S. The cell $D(p)$ assigned to $p \in S$ is an open convex (possibly non-bounded) polytopal set called the (open) *Dirichlet* cell of p. Obviously, $p \in D(p)$. Moreover, if S is the set of centers of congruent non-overlapping balls (as in a packing), then $D(p)$ contains the ball centered at p. This property of Dirichlet cells makes them useful in estimating the packing density of balls. In fact, Thue's theorem on congruent circle packings in E^2 can be proved by observing that the Dirichlet cell for a circle in such a packing is of area at least as large as that of a regular hexagon circumscribed about the circle (see L. Fejes Tóth [1972] p. 62).

We now define the generalized concept of the k-th order Dirichlet cells. Again, let S be a discrete subset of E^d and let k be a positive integer. To each point $p \in S$ assign the set $D^k(p)$ consisting of points x such that at most $k-1$ points of S are closer to x than p is. The set $D^k(p)$ is called the (open) *k-th order Dirichlet cell* of p. The family $\mathcal{D}^k = \{D^k(p) : p \in S\}$ has the following properties:

(1) Each point of E^d belongs to at most k members of $\mathcal{D}^k$ and the set of points of E^d which do not belong to k members of $\mathcal{D}^k$ is contained in the union of a countable family of $(d-1)$-dimensional hyperplanes in E^d. For short, we say that $\mathcal{D}^k$ is a k-fold tiling of E^d.

(2) Each $D^k(p)$ is star-shaped with respect to p.

(3) If S is the set of centers of congruent balls in a k-fold packing, then $D^k(p)$ contains the ball centered at p.

(4) If S is the set of centers of congruent balls in a k-fold covering, then $D^k(p)$ is contained in the ball centered at p.

At this point we show a connection between the k-th order Dirichlet cells and the similarly defined k-th order Voronoi diagram, which may be of interest to computational geometers. The k-th order Voronoi diagram, associated with the discrete set S is defined as follows: For any k-element subset $T = \{p_1, p_2, \ldots, p_k\}$ of S let $V^k(T)$ be the set of points x in E^d such that, for every $p \in S \backslash T$,

$$\text{dist}(p, x) \geq \max_{1 \leq i \leq k} \text{dist}(p_i, x).$$

It is easy to check that each cell $V^k(T)$ is a closed convex polyhedral set and the collection of the cells is a tiling. The *k-th order Voronoi diagram* is obtained as the cell complex of all intersections of the cells $V^k(T)$. For a more detailed description and computational aspects of the higher order Voronoi diagrams see sections 13.3–13.5 of Edelsbrunner [1987]. The connection between the cells V^k and the cells D^k is expressed by:

$$V^k(T) = \bigcap_{p \in T} D^k(p)$$

and

$$D^k(p) = \bigcup_{T \ni p} V^k(T)$$

We now outline a method of estimating the k-fold packing and covering densities of the ball B^d which makes use of the k-th order Dirichlet cells. Assume that a k-fold packing [covering] with unit balls is saturated [reduced] to a sufficiently high degree, i.e. that for some sufficiently large number N no N balls of the arrangement can be replaced by a greater [smaller] number of balls without destroying its property of being a k-fold packing [covering]. This assumption implies the existence of an upper bound (independent from the arrangement) for the number of cells of the corresponding k-th order Voronoi diagram which make up each $D^k(p)$ as well as an upper bound for the number of faces in each $V^k(T)$. These upper bounds yield an upper [lower] bound for the volume of each $D^k(p)$ and a density bound follows. For example, this technique (with some elaboration) results in the inequality

$$\delta^2(B^2) \leq 1.826 \ldots.$$

It is interesting to observe that while Dirichlet cells constituted a basic technique from the very beginning of the theory of packings and coverings, the analogous notions of the k-th order Dirichlet cells and k-th order Voronoi diagrams are fairly recent. It seems that k-th order Dirichlet cells were first studied by G. Fejes Tóth [1976] (see also G. Fejes Tóth [1979]). The concept of k-th order Voronoi diagrams was introduced independently by Shamos and Hoey [1975] and G. Fejes Tóth [1976].

In contrast to the problems discussed above, the results concerning decompositions of multiple arrangements into simple ones give insight into the combinatorial structure of such arrangements leaving the density measure aside. Following is a basic problem of this type: Decompose a k-fold packing [covering] into as few [as many] as possible simple packings [coverings]. Typically, this leads to finding common bounds for the number of simple arrangements into which every k-fold arrangement from a given class can be decomposed. Research in this direction was initiated by Pach in [1980]. He proved that any 2-fold packing with positively-homothetic copies of a convex disk (in E^2) can be decomposed into 4 (simple) packings. He observed that the nerve of such a 2-fold packing is a planar graph, therefore the decomposition into 4 packings follows directly from the 4-color theorem. (The *nerve* of an arrangement is defined by associating a vertex with each member of the arrangement and joining two vertices by an edge if and only if the two corresponding members intersect). For a convex disk K, let $r(K)$ denote the circum-radius of K. In the same paper [1980], Pach proved that if $\mathcal{F}$ is a k-fold packing with (not necessarily congruent) convex disks such that $\pi r^2(K)/A(K) \leq L$ for each $K \in \mathcal{F}$, then $\mathcal{F}$ can be decomposed into $n \leq 9Lk$ packings.

Concerning coverings, Mani-Levitska and Pach [1990] showed that every 33-fold covering of the plane with congruent circles can be decomposed into 2 coverings. For any polygonal centrally symmetric convex disk P, Pach [1986] proved that if r is a positive integer, then there exists an integer $k = k(P, r)$ such that every k-fold covering with translates of P can be decomposed into r coverings. Unfortunately, the number $k(P, r)$ in Pach's theorem increases with the number of sides of P, thus the "natural" attempt to extend this result to all centrally symmetric convex disks through polygonal approximation fails. Also, as examples of Mani-Levitska and Pach [1990] show, in 3 dimensions, statements analogous to the above two theorems are not true.

6. Some Computational Aspects of Packing and Covering

We present here a few results concerning algorithms for determining certain packing and covering densities of convex discs and the corresponding arrangements. In some cases, the classical results (see Sect. 3) produce such an algorithm or at least a theoretical foundation for one, leaving the computational complexity to be determined. In other cases, the problem of finding such an algorithm is addressed.

Mount and Silverman [1991] consider the computational complexity of the following two problems: For a centrally symmetric convex polygon P with n vertices, construct a thinnest covering of the plane with translates of P; and, for a convex polygon P with n vertices, construct a densest packing of the plane with translates of P. They show that each of the problems can be solved by a linear time algorithm, which means that there exists a constant k such that the number of elementary operations dictated by the algorithm is at most kn. The covering problem is solved by inscribing a maximum area centrally symmetric hexagon in P, obtained by inscribing a maximal triangle T in P and then reflecting T in the center of symmetry of P. Since there exists a linear time algorithm finding T (see Dobkin and Snyder [1979]), the covering problem can be solved in linear time as well. For the packing problem, the authors use the "difference body trick" of Minkowski (see Sect. 3 above) to reduce the problem to centrally symmetric convex polygons whereby the number of vertices increases by a factor of 2 at most. Further, in view of the equality $\delta(K) = \delta_L(K)$, (see again Sect. 3 above) it suffices to consider lattice arrangements only. Next, they use another observation of Minkowski: the problem of finding the densest lattice packing with translates of a centrally symmetric convex disk K is equivalent to the problem of finding the minimum area central parallelogram in K, where a *central parallelogram* in K is defined as a parallelogram whose one vertex is the symmetry center of K and the remaining vertices lie on the boundary of K. One can easily observe the connection between the minimum area central parallelogram in K and the minimum area centrally symmetric hexagon containing K, hence the connection with the packing problem. The key part of Mount and Silverman's solution of the packing problem is a linear time algorithm producing the minimum area central parallelogram for P.

In some cases, Mount and Silverman's result can be applied to lattice packings and coverings of greater multiplicity by means of results linking multiple packing and covering densities with simple ones. For instance, a theorem of Dumir and Hans-Gill [1972a] stating that $\vartheta_L^2(K) = 2\vartheta_L(K)$ for every centrally symmetric convex disk K combined with the above algorithm yields a linear time construction of the densest lattice 2-fold covering for any centrally symmetric convex polygon. Analogously, for packings, the equality $\delta_L^k(K) = k\delta_L(K)$ holds for $k = 2$ (Dumir and Hans-Gill [1972a]) and for $k = 3$ and 4 (G. Fejes Tóth [1983]) for every centrally symmetric convex disk, which, again, enables one to apply Mount and Silverman's algorithm in the context of k-fold packings for $k = 2, 3,$ and 4.

Linhart [1983] described algorithms which compute $\delta_L^k(B^2)$ and $\vartheta_L^k(B^2)$ with any prescribed accuracy. Using these algorithms he determined the values of $\delta_L^k(B^2)$ and $\vartheta_L^k(B^2)$ with an accuracy of at least two decimals for all $k \leq 20$.

Another approach for the determination of the densities $\delta_L^k(B^2)$ was proposed by Temesvári, Horváth and Yakovlev [1987] and for the determination of the densities $\vartheta_L^k(B^2)$ by Temesvári [1988]. In contrast to Linhart, these authors do not give explicit algorithms, but lay the theoretical foundations for algorithms determining $\delta_L^k(B^2)$ and $\vartheta_L^k(B^2)$ in finite time for all given k. For each k, they reduce the problems of finding $\delta_L^k(B^2)$ and $\vartheta_L^k(B^2)$ to the determination of the extreme values of finitely many explicitly given functions of one variable. Unfortunately, the number of functions which have to be considered increases rapidly with k.

7. Restrictions on the Number of Neighbors in a Packing

What is the maximum number of unit balls in E^3 that can touch one unit ball (without overlapping with each other)? This question caused a dispute between Isaac Newton and David Gregory. Newton conjectured that the answer was 12 while Gregory thought 13 was possible. It took 180 years before the question was answered: Hoppe [1874] proved that Newton was right.

For a convex body K, let $N(K)$ denote the maximum number of congruent copies of K that can touch K without overlapping with each other, call it the *Newton number* of K. The exact value of the Newton number of the d-dimensional ball B^d is known only for a few values of d. Obviously, $N(B^2) = 6$ and, as mentioned above, $N(B^3) = 12$. Considering the difficulty of the question disputed by Newton and Gregory, it is surprising that $N(B^d)$ could be computed for any $d > 3$ at all. Odlyzko and Sloane [1979] and, independently, Levenstein [1979] proved that $N(B^8) \leq 240$ and $N(B^{24}) \leq 196560$. Together with the fact that in the densest lattice packing of E^8 with congruent balls, each ball has 240 neighbors (Blichfeldt [1935]) and the fact that the Leech lattice in E^{24} generates a ball packing with the 196560-neighbor property (Leech [1964]), this implies that $N(B^8) = 240$ and $N(B^{24}) = 196560$. Moreover, the arrangements of balls producing these numbers are unique (see Bannai and Sloane [1981]).

Some results concerning the Newton number of convex bodies have been obtained in dimension $d = 2$ (see Böröczky [1971], Linhart [1973], Hortobágyi [1972] and L. Fejes Tóth [1967]). Recently, Wegner [1989] obtained a general bound for the Newton number for convex plane bodies which is sharp in many cases. In particular, his result includes the computation of the Newton number of the 30°-30°-120° isosceles triangle, which, as conjectured, turns out to be 21. Together with a previous result of Linhart [1977], this implies that in a congruent convex disk tiling in which each tile has its Newton number of neighbors, the maximum number of neighbors is 21.

Certain interesting problems arise when a restriction is imposed on the number of neighbors in a packing. If every member of a packing has exactly [at least] n neighbors, we call it an *n-neighbor* [*n⁺-neighbor*] *packing*. Osterreicher and Linhart [1981] consider the problem of existence of a finite n-neighbor packing in the plane with congruent smooth convex disks and that of the minimum cardinality of such a packing for $n = 3, 4$ and 5. (It is easy to observe that for $n \geq 6$ a finite n^+-neighbor packing with smooth convex disks does not exist). They show a 3-neighbor packing with 6 congruent disks, a 4-neighbor packing with 8 congruent disks and a 5-neighbor packing with 16 congruent disks, and they prove that the cardinality is minimal in each case.

In case a finite n-neighbor [n^+-neighbor] packing does not exist, one could ask about the infimum of the density of an n-neighbor [n^+-neighbor] packing with copies of, or translates of a convex disk. It is easily observed that there exists a 5-neighbor packing with translates of a parallelogram with density zero. It turns out that this property characterizes parallelograms (see L. Fejes Tóth [1973]). Let K be a convex disk which is not a parallelogram. Makai [1987] proved that every 5^+-neighbor packing with translates of K is of density greater than or equal to 3/7, and equality can occur only if K is a triangle (see Fig. 7.1).

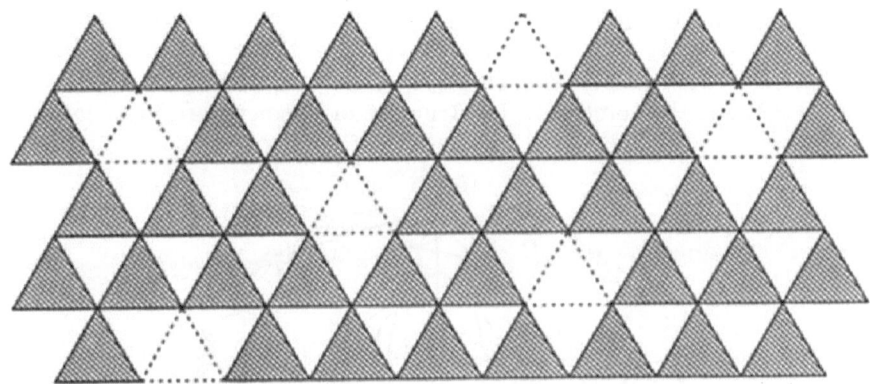

Fig. 7.1. Thinnest 5^+-neighbor packing with congruent convex disks.

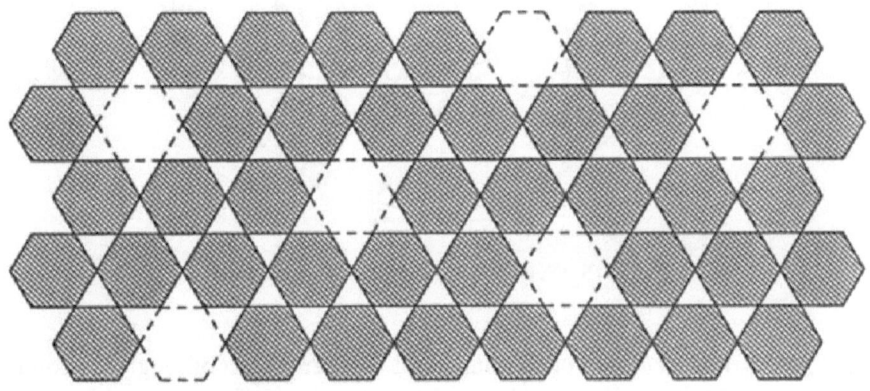

Fig. 7.2 Thinnest 5^+-neighbor packing with centrally symmetric congruent convex disks.

For any centrally symmetric K, Makai proved that every 5^+-neighbor packing with translates of K is of density greater than or equal to 9/14 and equality can hold only if K is an affine regular hexagon (see Fig. 7.2).

For the 6^+-neighbour packings, Makai obtained the sharp bounds of 1/2 (for the general K) and of 3/4 (for the centrally symmetric K), and again, the extreme values are produced only by the triangle and the affine regular hexagon, respectively.

In dimension 3 several results were obtained concerning neighbor-restricted packings with balls. Let $K(n)$ be the minimum cardinality of a finite n-neighbor packing with congruent balls. Obviously $K(1) = 2$ and $K(2) = 3$. The arrangements obtained by placing the centers of the balls in the vertices of a regular tetrahedron, a regular octahedron and a regular icosahedron show that $K(3) \leq 4$, $K(4) \leq 6$ and $K(5) \leq 12$, and it is not difficult to show that $K(3) = 4$ and $K(4) = 6$. The equality $K(5) = 12$ was shown by G. Fejes Tóth and Harborth [1987]. Wegner showed by an example that $K(6) \leq 240$. Figure 7.3 shows the combinatorial structure of Wegner's arrangement of the 240 balls.

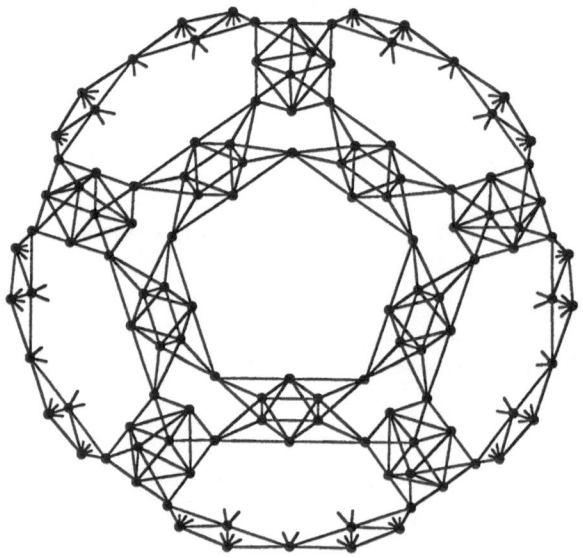

Fig. 7.3 Wegner's 6-neighbor ball packing.

The dots mark the centers of the balls and the bars (each of length 2) connect the centers of tangent balls. The design is made of "bridges" which follow the edges of a regular dodecahedron. The problem of finiteness of $K(7)$ and $K(8)$ remains open. For the remaining values of $n \leq 12$, $K(n) = \infty$ (Kertész [1989]). Thus, for n-neighbor packings of balls in E^3 with $n = 9$, 10, 11 or 12, one asks about the minimum density of such a packings. For $n = 9$, density zero is possible by arranging the balls in two adjacent hexagonal layers. For $n \geq 10$ any n^+-neighbor packing with congruent balls must have positive density, as G. Fejes Tóth [1981], Sachs [1986] and A. Bezdek and K. Bezdek [1988] proved. Each of them used an idea of L. Fejes Tóth who reduced the problem to proving that the maximum number of points on a closed unit hemisphere no two of which are closer than 1 to each other, is 9. The result of Kertész which was mentioned above was based on a similar idea: he proved that the maximum number of points on an open unit hemisphere no two of which are closer than 1 to each other, is 8. The minimum densities for the 10-neighbor and 11-neighbor packings with congruent balls are not known.

A packing with congruent copies of a convex body K is said to be *maximal* if it is an $N(K)$-neighbor packing, i.e. each copy of K has as many neighbors as it is possible for it to have in any arrangement. Observe that the maximum density lattice packing of congruent balls in E^3 is maximal. L. Fejes Tóth [1969, 1989] conjectures that any maximal packing of congruent balls in E^3 is essentially unique, i.e. it is composed of parallel layers of the same hexagonal structure as the layers in the densest lattice packing, and thus must have density $\pi/\sqrt{18}$. In general, in dimension $d > 3$, it is not known whether maximal ball packings exist at all. The surprising lattice packings in E^8 and E^{24} constitute remarkable exceptions: each of them is maximal and is uniquely determined by the property of maximality (see Bannai and Sloane [1981] and Conway and Sloane [1988], pp. 340–351).

Considering packings with balls of not necessarily equal sizes, define the *homogeneity* of such a packing as the infimum of the ratio between two radii of the balls. L. Fejes Tóth [1977] conjectured that a 6^+-neighbor circle packing must be either of homogeneity 1 or 0, which was confirmed by Bárány, Füredi and Pach [1984] who proved the stronger statement: In a 6^+-neighbor circle packing either all circles are congruent or arbitrarily small circles must occur. Their proof combines two ideas: a geometric one and a combinatorial one, each of interest of its own. The geometric part is a simple inequality on the radii of circles in a 6^+-neighbor packing: Let $\{C_i\}$ ($i = 0, 1, ..., n \geq 6$) be a packing with circular disks of radii $r(C_i)$, in which each C_i ($i = 0, 1, ..., n$) is a neighbor of C_0. Then

$$\frac{1}{n}\sum_{i=1}^{n}\frac{1}{r C_i)} \geq \frac{1}{r(C_0)}.$$

Now, assuming that a 6^+-neighbor circle packing contains two neighboring circles C' and C'' with $r(C') > r(C'')$, the function

$$f(C) = \frac{\frac{1}{r(C)} - \frac{1}{r(C')}}{\frac{1}{r(C)''} - \frac{1}{r(C)'}}$$

defined on the vertices of the nerve of the packing, by the above, turns out to be a convex function (a real valued function defined on the vertices of a graph is *convex* means that the value of the function of any vertex does not exceed the average of the values at the neighbors of the vertex). This leads to the combinatorial part: Under certain regularity conditions imposed on a graph (which are always satisfied by graphs obtained as nerves of plane circle packings), any convex function defined on the vertices of the graph is either constant or not bounded above. The conclusion applies to the function $f(C)$ yielding the existence of arbitrarily small circles in the packing.

It should be pointed out that the analogous statement for 12^{+}-neighbor packings with balls in E^3 does not hold. In fact, a 12-neighbor packing with 120 incongruent balls is obtained by a suitable chosen stereographic projection from the unit sphere S^3 in E^4 into E^3 of the 120 spheres inscribed in the faces of a regular spherical 120-cell on S^3. Nevertheless, a theorem analogous to the result of Bárány, Füredi and Pach may possibly hold in some dimensions, for example in dimensions 8 and 24, as suggested by the uniqueness results concerning the maximal packings with congruent balls in these dimensions.

8. Selected Topics in 3 Dimensions

In three dimensions, the packing and covering problems, as well as many other geometric problems, seem much more difficult than their plane versions, even in the special case in which only translates of a convex body are used. In particular, the problem of computing $\delta(B^3)$, i.e. the maximum density of any packing of E^3 with congruent balls, still remains open, despite the well-known, long-standing conjecture and the common belief in it, and despite the fact that the lattice version of the problem has been solved a long time ago (Gauss [1831]). The conjectured maximum density is $\pi/\sqrt{18} = 0.7404804\ldots$, the same as that in the densest lattice packing. Rogers [1958] obtained an upper bound for $\delta(B^d)$ which for $d = 3$ produces the inequality

$$\delta(B^3) \leq \sqrt{18}\{[\arccos(1/3)] - \pi/3\} = 0.7796355\ldots.$$

For many years, Rogers' bound was the best known and has been only recently improved by Lindsey [1986] to $0.7784\ldots$ and subsequently by Muder [1988] to

$$\delta(B^3) \leq \left(6\arccos\left(\frac{\sqrt{3}}{2}\sin\frac{\pi}{5}\right) - \frac{9\pi}{5}\right)\cot\frac{\pi}{5} = 0.7783683\ldots.$$

At the time this survey is being written, Muder's bound is the best known. The conjecture $\delta(B^3) = \pi/\sqrt{18}$ is supported by some other results as well. Dauenhauer and Zassenhaus [1987] proved that, in a certain sense, the densest lattice packing with balls is locally densest among all packings. A. Bezdek, W. Kuperberg and Makai [1991] proved that if a packing of E^3 with congruent balls consists of parallel strings of balls, then the density of the packing cannot exceed $\pi/\sqrt{18}$. A *string of balls* means a collection of congruent non-overlapping balls whose centers are collinear and such that each ball touches two other balls. The analogous problems for packings with strings which are not necessarily parallel and for packings with finite strings remain open. This leads naturally to the problem of packing E^3 with congruent circular cylinders. Generally, let $C(K, h)$ denote the (right) cylinder based on the convex disk K and with height $h > 0$, i.e. the Cartesian product $K \times [0, h]$, and let

$C(K, \infty)$ be the cylinder based on K, infinitely long in both directions, i.e. the Cartesian product $K \times (-\infty, \infty)$. We say that two cylinders are parallel if their generating segments [lines] are parallel. Obviously, if a packing of E^3 consists of translates of $C(K, h)$, then its density cannot exceed $\delta(K)$. For arbitrary packings with cylinders, the problem is less trivial. A. Bezdek and W. Kuperberg [1990] proved that the maximum density of any packing of E^3 with congruent copies of $C(B^2, \infty)$ is $\pi/\sqrt{12}$, attained in the parallel arrangement. In other words, $\delta(C(B^2, \infty)) = \pi/\sqrt{12}$. This was a first result of the kind; before that, no convex solid S in E^3 (bounded or not) with $\delta(S) \neq 1$ had its packing density $\delta(S)$ explicitly computed. Wilker [1987] conjectures that the circular cylinders of finite height have packing density $\pi/\sqrt{12}$ as well, i.e. that $\delta(C(B^2, h)) = \delta(B^2)$ for every $h > 0$.

One can conjecture that $\delta(C(K, \infty)) = \delta(K)$ for every convex disk K, and even that for every convex disk K there exists an h such that $\delta(C(K, h)) = \delta(K)$, but there are examples of convex disks K, even centrally symmetric ones, for which $\delta(C(K, h)) > \delta(K)$ for some h (see A. Bezdek and W. Kuperberg [1991]). Figure 8.1 illustrates a construction of such a packing with copies of $C(K, h)$ where K is an ellipse of axial ratio $k > \sqrt{3} + 1$ and h is sufficiently small.

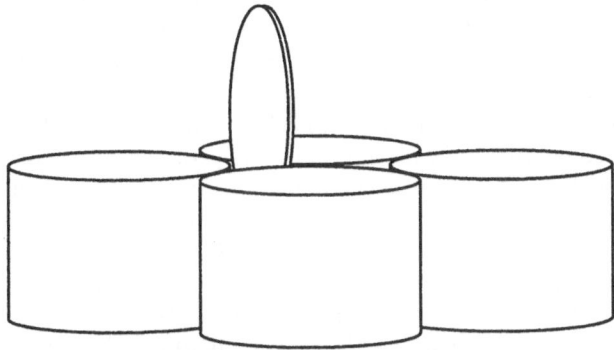

Fig. 8.1 Packing elliptical cylinders.

The idea of packing the elliptical cylinders in this manner led to a packing of E^3 with congruent ellipsoids and with density greater than $\pi/\sqrt{18}$, described in the same paper. Start with the densest lattice packing with balls in E^3 and observe that there exists a line (thus also an infinite cylinder of sufficiently small radius) which misses each of the balls. In fact, for every lattice packing of balls in E^3 there exist 3 independent directions in which a line can be drawn missing each of the balls, according to a theorem of Heppes [1960]. Because of the lattice structure of the family of balls, a family of mutually disjoint, congruent and parallel cylinders can be distributed in the gaps between the balls, so that the cylinders occupy a positive fraction of space. Now, in each cylinder a string of ellipsoids can be inscribed so that each ellipsoid matches the ball's volume and all ellipsoids are translates of each other. Obviously, this "mixed" packing with balls and ellipsoids is of density greater than $\pi/\sqrt{18}$ (the difference is the density of the family of ellipsoids). There exists an affinity $A : E^3 \rightarrow E^3$ turning this mixed packing into a packing with congruent ellipsoids. This construction shows that for certain ellipsoids E, $\delta(E) > \delta_L(E)$, which is in contrast to the situation in the plane. As we noted in Sect. 3, $\delta(K) = \delta_L(K)$ for all centrally symmetric convex disks K. Since the result of Heppes is valid in all dimensions $d \geq 3$ (see Ryškov and Horváth [1975]), the inequality $\delta(E) > \delta_L(E)$ is true for certain ellipsoids in every dimension $d \geq 3$.

Turning to a similar problem for coverings, G. Fejes Tóth and W. Kuperberg [1992] prove that for every strictly convex body K in dimension $d \geq 3$ there exist an affine equivalent K' of K such that $\vartheta(K') < \vartheta_L(K')$. The idea of the proof is similar to that applied in the ellipsoid packing scheme. We begin with a lattice covering of E^d with copies of K and shrink all copies of K homothetically until small holes (uncovered regions) appear. The holes can be covered by a discrete family of infinitely long, parallel congruent circular cylinders of a small diameter. Next, each of the cylinders is covered by a sequence of overlapping translates of a suitably affinely distorted copy of K, each of the same volume as K. If the holes are small enough, the resulting "mixed" covering is of smaller density than the original lattice covering. As before, there exists an affinity which turns the mixed covering into a covering with congruent copies of an affine equivalent of the body K.

We now turn to some ball packing problems in E^3 resembling the unsolved ball packing conjecture, yet solved completely in recent years. A packing is said to be totally separable if every pair of its members can be separated by a plane avoiding the interiors of the members of the packing (see G. Fejes Tóth and L. Fejes Tóth [1973]). Kertész [1988] proved that among all totally separable packings of E^3 with congruent balls, the cube-lattice packing, whose density is $\pi/6$, is the densest one. In fact, Kertész proved the following theorem about finite packings from which the density bound follows immediately: If a cube of volume V contains N unit balls forming a totally separable packing, then $V \geq 8N$. We present here the elegant proof of Kertész.

Begin with separating two of the N balls with a plane not intersecting any of the balls. The plane cuts the cube into two convex polyhedra. Whichever of the pieces contains more than one ball can be in turn cut into two pieces in a similar manner. At the end, the cube is cut into N pieces (convex polyhedra) each containing exactly on ball. For each such piece let v denote its volume, s its surface area and L the sum of the lengths of its edges. For any edge e of the piece, let α denote the dihedral angle at the edge. An inequality of Minkowski (see L. Fejes Tóth [1972] p. 287) implies that

$$s^2 \geq 3v \sum e \cot \frac{\alpha}{2},$$

where the sum is extended over all edges of the piece. Since the piece contains a unit sphere, it follows that $v \geq s/3$. Therefore

$$(*) \qquad\qquad 3v \geq \sum e \cot \frac{\alpha}{2} \,.$$

Let K be the number of pieces that meet along a common segment e of an edge and let $\alpha_1, \alpha_2, \ldots, \alpha_k$ be the respective dihedral angles. Observe that if e lies on an edge of the cube, then $k \geq 1$ and $\alpha_1 + \alpha_2 + \ldots + \alpha_k = \pi/2$; if e lies on a face of the cube or of a piece, but not on an edge of it, then $k \geq 2$ and $\alpha_1 + \alpha_2 + \ldots + \alpha_k = \pi$, and in the remaining case, $k \geq 4$ and $\alpha_1 + \alpha_2 + \ldots + \alpha_k = 2\pi$. In each of these cases, $\cot(\alpha_1/2) + \ldots + \cot(\alpha_k/2) \geq k$. Adding up the inequalities $(*)$ for all of the pieces, we get

$$3V \geq \sum L \,,$$

where the sum extends over the N pieces. Now, the conclusion of the theorem follows by a result of Besicovitch and Eggleston [1957] stating that if a convex polyhedron in E^3 contains a unit ball, then the sum of the lengths of its edges is at least 24 (as in a cube).

In a modified version of the ball packing problem, a different quantity measuring the tightness of the packing replaces density. Define the *closeness* of a packing with congruent balls as the ratio between the radius of a ball of the packing and the supremum of the radius of a ball disjoint with the packing. Analogously, the *looseness* of a covering with congruent balls is defined as the ratio between the radius of the balls of the covering and the supremum of the radius of a ball contained in the intersection of two balls of the covering. Confirming a conjecture of L. Fejes Tóth [1976], Böröczky [1986] proved that the closest packing with unit balls in E^3 is unique and is obtained by placing the centers of the balls in the vertices of a cubic lattice of edge-length $4/\sqrt{3}$ and in the centers of the cubes. The closeness so obtained equals to $(\sqrt{5/3}-1)^{-1} =$ 2.3094 .... Since the problems of the loosest covering and the closest packing with congruent balls are equivalent to each other, Böröczky's result solves both of them.

References

Books and Survey Papers

E.P. Baranovskii

[1969] Packings, coverings, partitionings and certain other distributions in spaces of constant curvature (Russian). Itogi Nauki – Ser. Mat. (Algebra, Topologiya, Geometriya) 185–225 = Progress Math. 9 (1971) 209–253

J.W.S. Cassels

[1971] An introduction to the geometry of numbers, 2nd edn. Springer, Berlin Heidelberg New York

J.H. Conway and N.J.A. Sloane
[1988] Sphere packings, lattices and groups. Springer, New York Berlin Heidelberg London Paris Tokyo

P. Erdős, P.M. Gruber and J. Hammer
[1989] Lattice points. Longman Sci. Techn., Harlow

L. Fejes Tóth
[1964] Regular figures. Pergamon Press, Oxford New York
[1972] Lagerungen in der Ebene, auf der Kugel und im Raum, 2nd edn. Springer, Berlin New York
[1984] Density bounds for packing and covering with convex discs. Expositiones Mathematicae 2, 131–153

G. Fejes Tóth
[1983] New results in the theory of packing and covering. In: Convexity and its Applications, ed. by P.M. Gruber and J.M. Wills. Birkhäuser, Basel Boston Stuttgart

A. Florian
[1987] Packing and covering with convex discs. Intuitive Geometry (Siófok, 1985), Colloq. Math. Soc. János Bolyai, vol. 48. North-Holland, Amsterdam New York, pp. 191–207

P. Gritzmann and J. M. Wills
[1986] Finite packing and covering. Stud. Sci. Math. Hungar. 21, 149–162

H. Groemer
[1987] Coverings and packings by sequences of convex sets. In: Discrete Geometry and Convexity, ed. by J.E. Goodman, E. Lutwak, J. Malkewitch and R. Pollack. Annals of the New York Academy of Sciences, vol. 440

P.M. Gruber
[1979] Geometry of numbers. Contributions to Geometry, Proc. Symp. Siegen, Birkhäuser, Basel Boston Stuttgart

P.M. Gruber and C.G. Lekkerkerker
[1987] Geometry of numbers. North-Holland, Amsterdam

B. Grünbaum and G.C. Shephard
[1986] Tilings and patterns. Freeman & Co., New York

W. Moser and J. Pach
[1986] 100 research problems in discrete geometry. Mimeograph

J. Pach and P.K. Agarwal
[1991] Combinatorial Geometry. Courant Lecture Notes Series, New York University, New York

C.A. Rogers
[1964] Packing and covering. Cambridge University Press, Cambridge

T.L. Saaty and J.M. Alexander
[1975] Optimization and the geometry of numbers: packing and covering. SIAM Review 17, 475–519

J.M. Wills
[1990] Kugellagerungen und Konvexgeometrie. Jber. Deutsch. Math. Verein. 92, 21–46

Research Papers

E. Bannai and N.J.A. Sloane
[1981] Uniqueness of certain spherical codes. Canad. J. Math. **33**, 437–449

I. Bárány, Z. Füredi and J. Pach
[1984] Discrete convex functions and proof of the six circle conjecture of Fejes Tóth. Can. J. Math. **36**, 569–576

R.S. Besicovitch and E.G. Eggleston
[1957] The total length of the edges of a polyhedron. Quart. J. Math. Oxford Ser. (2) **8**, 172–190

A. Bezdek and K. Bezdek
[1988] A note on the ten-neighbour packings of equal balls. Beiträge zur Algebra und Geometrie **27**, 49–53

A. Bezdek and W. Kuperberg
[1990] Maximum density space packing with congruent circular cylinders of infinite length. Mathematika **37**, 74–80
[1991] Packing Euclidean space with congruent cylinders and with congruent ellipsoids. Applied Geometry and Discrete Mathematics: The Victor Klee Festschrift, eds. P Gritzmann and B. Sturmfels, DIMACS Series on Discrete Mathematics and Computer Science, Amer. Math. Soc. & ACM, New York, pp. 71–80

A. Bezdek, W. Kuperberg and E. Makai Jr.
[1991] Maximum density space packing with parallel strings of spheres. Discrete Comput. Geom. **6**, 277–283

H.F. Blichfeldt
[1929] The minimum value of quadratic forms and the closest packing of spheres. Math. Ann. **101**, 605–608
[1935] The minimum values of positive quadratic forms in six, seven and eight variables. Math. Zeit. **39**, 1–15

U. Bolle
[1989] On the density of multiple packings and coverings of convex discs. Studia Sci. Math. Hungar. **24**, 119-126

K. Böröczky
[1971] Über die Newtonsche Zahl regulärer Vielecke. Period. Math. Hungar. 1, 113-119
[1986] Closest packing and loosest covering of the space with balls. Stud. Sci. Math. Hungar. **21**, 79–89

G.D. Chakerian and L.H. Lange
[1971] Geometric extremum problems. Math. Mag. **44**, 57–69

M.J. Cohn
[1976] Multiple lattice covering of space. Proc. London Math. Soc. (3) **32**, 117–132

R. Courant
[1965] The least dense lattice packing of two-dimensional convex bodies. Comm. Pure Appl. Math. **18**, 339–343

M.H. Dauenhauer and H.J. Zassenhaus
[1987] Local optimality of the critical lattice sphere-packing of regular tetrahedra.
Discrete Math. **64**, 129–146

D.P. Dobkin and L. Snyder
[1979] On a general method for maximizing and minimizing among certain geometric problems. Proc. of the 20th IEEE Symp. on Foundations of Computer Science, 9–17

C. H. Dowker
[1944] On minimum circumscribed polygons. Bull. Amer. Math. Soc. **50**, 120–122

V.C. Dumir and R.J. Hans-Gill
[1972a] Lattice double coverings in the plane. Indian J. Pure Appl. Math. **3**, 466–480
[1972b] Lattice double packings in the plane. Indian J. Pure Appl. Math. **3**, 481–487

H. Edelsbrunner
[1987] Algorithms in Combinatorial Geometry. Springer, Berlin Heidelberg New York

V. Ennola
[1961] On the lattice constant of a symmetric convex domain. J. London Math. Soc. **36**, 135–138

P. Erdős and C.A. Rogers
[1962] Covering space with convex bodies. Acta Arith. **7**, 281–285

I. Fáry
[1950] Sur la densité des réseaux de domaines convexes. Bull. Soc. Math. France **78**, 152–161

G. Fejes Tóth
[1976] Multiple packing and covering of the plane with circles. Acta Math. Acad. Sci. Hungar. **27**, 135–140
[1979] Multiple packing and covering of spheres. Acta Math. Acad. Sci. Hungar. **34**, 165–176
[1981] Ten-neighbour packing of equal balls. Period. Math. Hungar. **12**, 125–127
[1983] Multiple lattice packings of symmetric convex domains in the plane. J. London Math. Soc. (2) **29**, 556–561

G. Fejes Tóth and L. Fejes Tóth
[1973] On totally separable domains. Acta Math. Acad. Sci. Hungar. **24**, 229–232

G. Fejes Tóth and H. Harborth
[1987] Kugelpackungen mit vorgegebenen Nachbarnzahlen. Studia Sci. Math. Hungar. **22**, 79-82

G. Fejes Tóth and W. Kuperberg
[1992] Thin non-lattice coverings with congruent strictly convex bodies (in preparation)

L. Fejes Tóth
[1950] Some packing and covering theorems. Acta Sci. Math. Szeged **12/A**, 62–67
[1967] On the number of equal discs that can touch another of the same kind. Studia Sci. Math. Hungar. **2**, 363–367
[1969] Remarks on a theorem of R.M. Robinson. Studia Sci. Math. Hungar. **4**, 441–445
[1973] Five-neighbour packing of convex discs. Period. Math. Hungar. **4**, 221–229
[1976] Close packing and loose covering with balls. Publ. Math. Debrecen **23**, 323–326

[1977] Research problem no. 21. Period. Math. Hungar. **8**, 103–104
[1989] Research problem no. 44. Period. Math. Hungar. **20**, 89–91

L. Few
[1964] Multiple packing of spheres. J. London Math. Soc. **39**, 51–54
[1968] Double packing of spheres: A new upper bound, Mathematika **15**, 88–92

A. Florian
[1978] Mehrfache Packung konvexer Körper. Österreich. Akad. Wiss. Math.- Natur. Kl. Sitzungsber. II., 238–247

C.F. Gauss
[1831] Untersuchungen über die Eigenschaften der positiven ternären quadratischen Formen von Ludwig August Seber. Göttingische gelehrte Anzeigen, Juli 9 = J. reine angew. Math. **20** (1940), 312–320 = Werke II, 188–196

H. Groemer
[1963] Existenzsätze für Lagerungen im Euklidischen Raum. Math. Zeitschr. **81**, 260–278
[1968] Existenzsätze für Lagerungen in metrischen Räumen, Monatsh. Math. **72**, 325–334

A. Heppes
[1960] Ein Satz über gitterförmige Kugelpackungen. Ann. Univ. Sci Budapest. Eötvös, Sect Math. **3–4**, 89–90

R. Hoppe
[1874] Bemerkung der Redaktion. Archiv Math. Physik (Grunert) **56**, 307–312

I. Hortobágyi
[1972] The Newton number of convex plane regions (Hungarian). Mat. Lapok **23**, 313–317

R. Kershner
[1939] The number of circles covering a set. Amer. J. Math. **61**, 665–671

G. Kertész
[1988] On totally separable packings of equal balls. Acta Math. Hungar. **51**, 363–364
[1989] Private communication

G. Kuperberg and W. Kuperberg
[1990] Double-lattice packings of convex bodies in the plane. Discrete Comput. Geom. **5**, 389–397

W. Kuperberg
[1987a] On packing the plane with congruent copies of a convex body. Intuitive Geometry (Siófok, 1985), Colloq. Math. Soc. János Bolyai, vol. 48. North-Holland, Amsterdam New York, pp. 317–329
[1987b] An inequality linking packing and covering densities of plane convex bodies. Geom. Dedicata **23**, 59–66
[1989] Covering the plane with congruent copies of a convex body. Bull. London Math. Soc. **21**, 82–86

J.L. Lagrange
[1773] Recherches d'arithmetique. Nouv. Mem. Acad. Roy. Sc. Belle Letteres, Berlin 1773, 265–312 = Oeuvres III, 693–758

J. Leech
[1964] Some sphere packings in higher space. Canad. J. Math. **16**, 657–682

V.I. Levenstein
[1979] On bounds for packings in n-dimensional Euclidean space. Dokl. Acad. Nauk
    SSSR **245**, 1299–1303 = Soviet Mathematics, Doklady **20**, 417–421

J.H. Lindsey
[1986] Sphere packing in R^3. Mathematika **33**, 137–147

J. Linhart
[1973] Die Newtonsche Zahl von regelmäßigen Fünfecken. Period. Math. Hungar. **4**,
    315–328
[1977] Scheibenpackungen mit nach unten beschränkter Nachbarnzahlen. Studia Sci.
    Math. Hungar. **12**, 281–293
[1983] Eine Methode zur Berechnung der Dichte einer dichtesten gitterförmigen *k*-
    fachen Kreispackung. Arbeitsber. Math. Inst. Univ. Salzburg, 1983

K. Mahler
[1946] The theorem of Minkowski-Hlawka. Duke Math J. **13**, 611–621
[1947] On the minimum determinant and the circumscribed hexagons of a convex
    domain. Proc. K. Ned. Acad. Wet. Amsterdam **50**, 692–703

E. Makai Jr.
[1987] Five-neighbour packing of convex plates. Intuitive Geometry (Siófok, 1985),
    Colloq. Math. Soc. János Bolyai, vol. 48, North-Holland, Amsterdam New
    York, pp. 373–381

P. Mani-Levitska and J. Pach
[1990] Decomposition problems for multiple coverings of unit balls. Preprint

H. Minkowski
[1904] Dichteste gitterförmige Lagerung kongruenter Körper. Nachr. Ges. Wiss.
    Göttingen, 311–355

D.M. Mount and R. Silverman
[1990] Packing and covering the plane with translates of a convex polygon. J. Algo-
    rithms **11**, 564–580

D.J. Muder
[1988] Putting the best face on a Voronoi polyhedron. Proc. London Math. Soc. (3)
    56, 329–348

A.M. Odlyzko and N.J.A. Sloane
[1979] New bounds on the number of unit spheres that can touch a unit sphere in n
    dimensions. J. Comb. Theory Ser. A **26**, 210–214

F. Österreicher and J. Linhart
[1981] Packungen kongruenter Stäbchen mit konstanter Nachbarnzahl. Elem. Math.
    37, 5–16

J.Pach
[1980] Decomposition of multiple packing and covering, Diskrete Geometrie. 2. Kol-
    loq. Inst. Math. Univ. Salzburg, 169–178.
[1986] Covering the plane with convex polygons. Discrete Comput. Geom. **1**, 73–81

K. Reinhardt
[1934] Über die dichteste gitterförmige Lagerung kongruenter Bereiche in der Ebene
    und eine besondere Art konvexer Kurven. Abh. math. Sem. hansischer Univ.
    10, 216–230

C. A. Rogers
[1951] The closest packing of convex two-dimensional domains. Acta Math. 86, 309–321
[1958] The packing of equal spheres. Proc. London Math. Soc. (3) 8, 609–620

S.S. Ryškov and J. Horváth
[1975] Estimation of the radius of a cylinder that can be imbedded in every lattice packing of n-dimensional unit balls (Russian). Mat. Zametki 17, 123–128 = Math. Notes 17, 72–75

H. Sachs
[1986] No more than nine unit balls can touch a closed unit hemisphere. Studia Sci. Math. Hungar. 21, 203–206

E. Sas
[1939] Über eine Extremumeigenschaft der Ellipsen. Compositio Math. 6, 468–470

W.M. Schmidt
[1961] Zur Lagerung kongruenter Körper im Raum. Monatsh. Math. 65, 154–158

M.I. Shamos and D. Hoey
[1975] Closest point problems. Proc. 16th Ann. IEEE Sympos. Found. Comput. Sci., 151–162

P. Tammela
[1970] An estimate of the critical determinant of a two-dimensional convex symmetric domain (Russian). Izv. Vyss. Ucebn. Zaved. Mat. 12 (103), 103–107

A.H. Temesvári
[1988] Eine Methode zur Bestimmung der dünnsten gitterförmigen k-fachen Kreisüberdeckungen. Studia Sci. Math. Hungar. 23, 23–35

A.H. Temesvári, J. Horváth and N.N. Yakovlev
[1987] A method for finding the densest lattice k-fold packing of circles (Russian). Math. Zametki 41, 625–636, 764 = Math. Notes 41, 349–355

A. Thue
[1910] Über die dichteste Zusammenstellung von kongruenten Kreisen in der Ebene. Norske Vid. Selsk. Skr. 1, 1–9

G. Wegner
[1989] Relative Newton numbers. Preprint

J.B. Wilker
[1987] Problem II. Intuitive Geometry (Siófok, 1985), Colloq. Math. Soc. János Bolyai, vol. 48. North-Holland, Amsterdam New York, p. 700

Chapter XI. Recent Developments in Combinatorial Geometry

W. Moser and J. Pach

Dedicated to Paul Erdős
on his 80th birthday

Over a span of fifty years Paul Erdős has written many articles with this or a similar title. His countless results, which were obtained by the application of combinatorial and counting (random) methods, and the many deep problems raised and popularized in these papers, generated much research in combinatorics and graph theory. They played an important role in the emergence of a number of new areas in mathematics. One of these is combinatorial geometry, the study of extremal problems about finite arrangements of points, lines, circles, etc.

In this chapter we survey some of the central problems in Erdős-type combinatorial geometry, particularly where significant progress has been made recently, or the attack has been persistent, or the problem deserves more attention.

This chapter is organized as follows. Sections 1 and 2 deal mainly with metric problems. In Section 3 we consider several recent results on geometric graphs and hypergraphs. In Section 4 we review some recent developments in a new area of research on the structure of arrangements of lines in space.

1. The Distribution of Distances

Almost fifty years ago Erdős (1946, 1960) raised the following question. What is the minimum number $f_d(n)$ of distinct distances determined by a set of n points in d-dimensional Euclidean space E^d? He showed that

$$cn^{1/2} < f_2(n) < \frac{c'n}{\sqrt{\log n}},$$

$$c_d n^{1/d} < f_d(n) < c'_d n^{2/d}, \qquad d \geq 3,$$

and remarked: "Though I have sought to improve this result for many years, I have not been able to do so." The upper bounds are realized by finite portions of the square lattice and the higher dimensional cubic lattices. The lower bound on $f_2(n)$ was subsequently improved by Leo Moser (1952) to $cn^{2/3}$, by Fan Chung (1984) to $cn^{5/7}$ and by J. Beck (1983) to $cn^{\frac{58}{81}-\epsilon}$ for any $\epsilon > 0$. In fact, Beck established the stronger statement that there always exists a point from which there are at least $cn^{\frac{58}{81}-\epsilon}$ distinct distances. According to a recent result of Chung, Szemerédi and Trotter (1992), there exists a constant c such that

$$f_2(n) \geq n^{4/5}/(\log n)^c .$$

They note that although one could perhaps show $f_2(n) \geq cn^{4/5}$ " ... we find little reason to attempt such improvements since they would still leave us far from the conjectured lower bound which we suspect is correct." In fact, Erdős conjectured that

$$f_2(n) > cn/\sqrt{\log n} ,$$

and has repeatedly offered \$500 for a proof or disproof.

In the case when the points are in general position, i.e., no three are on a line and no four on a circle, we get a radically different problem. Clearly, the minimum number of distinct distances determined by n points in the plane in general position is $f_2^{gen}(n) \geq \frac{1}{3}(n-1)$. On the other hand, Erdős, Fűredi, Pach and Ruzsa (1993) have recently proved that

$$f_2^{gen}(n) < n^{1+c/\sqrt{\log n}}$$

for a suitable constant c.

A closely related problem discussed in Erdős (1946) is the following. Determine (or estimate) $F_d(n)$, the maximum number of times that unit distance can occur among n points in E^d. Evidently, $f_d(n)F_d(n) \geq \binom{n}{2}$. Erdős (1946, 1960) proved

$$n^{1+c/\log\log n} < F_2(n) < c'n^{3/2} ,$$

$$c_3 n^{4/3} \log\log n < F_3(n) < c_3' n^{5/3} ,$$

$$F_d(n) = \frac{1}{2}n^2 \left(1 - \frac{1}{\lfloor d/2 \rfloor} + o(1)\right) , \quad d \geq 4.$$

The lower bounds for $F_2(n)$ and $F_3(n)$ are given by the square lattice and the cubic lattice, respectively. For $d \geq 4$ the lower bound follows from the following well-known construction due to H. Lenz: take $\lfloor d/2 \rfloor$ pairwise perpendicular circles of radius $1/\sqrt{2}$ around the origin, and put n points on them as evenly distributed as possible. The upper bounds can be derived from some classical theorems in extremal graph theory (Kővári, Sós and Turán 1954; Erdős and Stone 1946).

The upper bound for $F_2(n)$ given by Erdős was improved by Józsa and Szemerédi (1975) and Beck and Spencer (1984). The currently best known bound

$$F_2(n) \leq cn^{4/3} ,$$

is due to Spencer, Szemerédi and Trotter (1984). A very nice alternative proof based on random sampling techniques was given by Clarkson, Edelsbrunner, Guibas, Sharir and Welzl (1988). The upper bound on $F_3(n)$ was first improved by Beck (1983) to $n^{\frac{22}{24}+o(1)}$, and then by Chung (1989) to $cn^{8/5}$. In fact, the arguments in Clarkson et al yield

$$F_3(n) < n^{3/2}\beta(n),$$

where $\beta(n)$ is an extremely slowly growing function depending on the inverse of Ackermann's function. Note that this immediately implies that

$$n^{\frac{1}{2}-o(1)} \leq \binom{n}{2} \Big/ F_3(n) \leq f_3(n).$$

Erdős (1967) started investigating the error term of $F_d(n)$, $d \geq 4$. He proved that

$$F_d(n) = \frac{1}{2}n^2\left(1 - \frac{1}{d/2}\right) + n + O(1), \qquad \text{if } d \text{ is even,}$$

but he had " ... not been able to disprove that

$$F_d(n) = \frac{1}{2}n^2\left(1 - \frac{1}{\lfloor d/2 \rfloor}\right) + O(n) \quad \text{for every } d.''$$

This gap was filled in a recent paper of Erdős and Pach (1990), where it was shown that

$$\frac{1}{2}n^2\left(1 - \frac{1}{\lfloor d/2 \rfloor}\right) + c_d n^{4/3} < F_d(n) < \frac{1}{2}n^2\left(1 - \frac{1}{\lfloor d/2 \rfloor}\right) + c'_d n^{4/3}, \text{if } d \text{ is odd,}$$

where c_d and c'_d are suitable positive constants. The proof of this fact uses the following result of Clarkson et al (1988) and Erdős, Hickerson and Pach (1989), which settles a problem posed by Leo Moser. Let $F_{\text{sphere}}(n, \alpha)$ denote the maximum number of times that distance α can occur among n points on the unit sphere S^2. Then

$$F_{\text{sphere}}(n, \alpha) < cn^{4/3} \quad \text{for every } 0 < \alpha < 2,$$

and the order of magnitude of this bound cannot be improved if $\alpha = \sqrt{2}$. For all distances $\alpha > 2$ the currently known best lower bound is

$$F_{\text{sphere}}(n, \alpha) \geq cn \log^* n,$$

where $\log^*$ stands for the iterated logarithm function. It is an intriguing open problem to improve this bound for any $\alpha \neq \sqrt{2}$.

A modified version of the unit distance problem was discussed by Avis, Erdős and Pach (1988). Given a set $P = \{p_1, p_2, \cdots, p_n\}$ of n points in E^d, and positive real numbers $\alpha_1, \alpha_2, \cdots, \alpha_n$, let $g(p_i, \alpha_i)$ denote the number of points in P whose distance from p_i is α_i. Let

$$G_d(n) = \max_{\substack{p_1, \cdots, p_n \\ \alpha_1, \cdots, \alpha_n}} \sum_{i=1}^{n} g(p_i, \alpha_i).$$

Choosing every α_i to be 1, we obtain

$$G_d(n) \geq 2F_d(n).$$

It follows from the results of Clarkson et al (1988) that $G_2(n) \leq cn^{7/5}$. On the other hand, Avis, Erdős and Pach (1988) proved:

$$\frac{1}{4}n^2 + \frac{3}{2}n \leq G_3(n) < \frac{1}{4}n^2 + cn^{2-\delta} \quad \text{for some } c, \delta > 0;$$

$$G_d(n) = n^2 \left(1 - \frac{1}{\lfloor d/2 \rfloor} + o(1)\right) \quad \text{for any even } d \geq 4.$$

Erdős and Pach (1990) extended this result to any integer $d \geq 4$.

There are quite a few recent results about the distribution of the smallest and largest distances determined by a point set. Let $m_i(n)$ and $M_i(n)$ denote the maximum number of times that the ith smallest resp. the ith largest distance occurs among n points in the plane. H. Harborth (1974) showed that $m_1(n) = \lfloor 3n - \sqrt{12n - 3} \rfloor$ for every n, where equality is attained for certain subsets of the regular triangular lattice. These subsets have been completely characterized by Kupitz (1990a). Very recently Brass (1992) has shown that

$$m_2(n) = \left(\frac{24}{7} + o(1)\right) n.$$

As to the large distances, it has been known for a long time that $M_1(n) = n$ (Hopf and Pannwitz 1934). More recently, Vesztergombi (1985, 1987) proved that $M_2(n) \leq 3n/2$ with equality if n is even.

In the case when our points are the vertices of a convex n-gon, she proved that the second largest distance occurs at most $M_2^{\text{conv}}(n) \leq 4n/3$ times. Furthermore, $M_i^{\text{conv}}(n) \leq 2in$ holds for every i. In fact, Erdős and Leo Moser conjectured a long time ago that there exists an absolute constant c such that no distance can occur more than cn times among the vertices of a convex n-gon. In other words, $M_i^{\text{conv}}(n) \leq cn$ for every i. The only nontrivial result in this direction is due to Füredi (1990), who proved

$$M_i^{\text{conv}}(n) \leq cn \log n \quad \text{for every } i.$$

Note that the Erdős-Moser conjecture would follow immediately from a conjecture of Erdős which states that there exists an integer k such that every convex n-gon has a vertex from which no k other vertices are equidistant. L. Danzer showed that if such a k exists it must be at least 4 (Erdős 1987). On the other hand, a recent construction of Edelsbrunner and Hajnal (1991) shows that

$$\max_i M_i^{\text{conv}}(n) \geq 2n - 7,$$

which is considerably better than the lower bound $\frac{5}{3}(n-1)$ given by Erdős and Moser (1959).

Erdős, Lovász and Vesztergombi (1989) showed that Erdős' conjecture is valid "in average" for the "large" distances. More precisely, if S is a set of n points forming a convex polygon and i is a natural number, then let $G_{\leq i}(S)$ be the graph with vertex set S and pairs of points joined by an edge if and only if their distance is one of the i largest distances determined by S. Erdős et al proved that $G_{\leq i}(S)$ has a vertex of degree at most $3i - 1$. Therefore the chromatic number of $G_{\leq i}(S)$ is at most $3i$, and it has at most $3in$ edges. Furthermore, if $|S| = n$ is large enough ($n > ci^2$ for some $c > 0$), then the chromatic number of $G_{\leq i}(S)$ drops to ≤ 7. A similar phenomenon in higher dimensions is discussed in Erdős, Lovász and Vesztergombi (1987).

A slightly different approach to the same problem was taken by Vesztergombi (1985). Given a set S of n points in the plane forming a convex n-gon, let $M_i^{\mathrm{conv}}(S)$ denote the number of occurrences of ith largest distance in S. Then all linear inequalities involving $M_1^{\mathrm{conv}}(S)$, $M_2^{\mathrm{conv}}(S)$ and n, which are true for every S, can be deduced from the following system:

$$M_1^{\mathrm{conv}}(S) \leq n,$$

$$M_1^{\mathrm{conv}}(S) + 2M_2^{\mathrm{conv}}(S) \leq 3n,$$

$$M_2^{\mathrm{conv}}(S) - M_1^{\mathrm{conv}}(S) \leq n.$$

Another problem similar to the question of largest distances was studied by Avis (1984), Avis, Erdős and Pach (1988) and Edelsbrunner and Skiena (1989). Given a set of points $p_1, p_2, \cdots, p_n \in E^d$, we say that p_j is a *farthest neighbour* of p_i if

$$|p_i - p_j| = \max_{1 \leq k \leq n} |p_i - p_k|.$$

The number of farthest neighbours of p_i will be denoted by $\varphi(p_i)$. Let

$$\Phi_d(n) = \max_{p_1, \cdots, p_n} \sum_{i=1}^{n} \varphi(p_i).$$

Obviously, $\Phi_d(n) \leq G_d(n)$. It is known that

$$\Phi_2(n) = \begin{cases} 3n - 3 & \text{if } n \text{ is even,} \\ 3n - 4 & \text{if } n \text{ is odd,} \end{cases}$$

$$\frac{n^2}{4} + \frac{3n}{2} < \Phi_3(n) < \frac{n^2}{4} + \frac{3n}{2} + 255,$$

$$2F_d(n) \leq \Phi_d(n) \quad \text{for } d \geq 4.$$

No upper bounds better than those known for $G_d(n)$ have been established for $\Phi_d(n), d \geq 4$.

It is interesting to note in this context that the number of those pairs of points whose distance is *almost* maximal can be much larger than $\Phi_d(n)$.

For instance, in the plane, let $\{p_1, \ldots, p_n\}$ be a set of points with minimum distance 1 and maximum distance (diameter) D. Erdős, Makai, Pach and Spencer (1991) have recently shown that the number of pairs $p_i p_j$ $(i < j)$ whose distance is between $D-1$ and D is at most $cn^{3/2}$. This can be attained, for example, if $D = n$ and the points are uniformly distributed along the boundary of a circle of diameter D, so that no two points are closer than 1 to each other. They also prove that, if n is sufficiently large, then the maximum number of pairs $p_i p_j$ $(i < j)$ whose distances are in the same interval of length 1 is $\lfloor n^2/4 \rfloor$ and this bound is tight.

As Erdős, Hickerson and Pach (1989) have pointed out, a serious difficulty in the way of research in this field is that the extremal configurations for almost all problems of this kind have many symmetries. However, apart from the integer grid and the regular n-gon, we know very few nontrivial symmetric constructions. Therefore, as they put it, "Fighting against these problems is a little bit like shadow boxing. You do not know where the enemy is." For the same reason, until very recently there was hardly any nontrivial extremal problem in this area for which the optimal configuration was conjectured to be a finite piece of the integer grid (or some other lattice) and one was actually able to prove this. In fact, in most cases there was very little evidence supporting these conjectures. One of the very few exceptions is the following result of Pach and Sharir (1992): For every $0 < \alpha < \pi$ there is a constant c_α such that for any system of n points in the plane the number of triples determining angle α is at most $c_\alpha n^2 \log n$. This bound is asymptotically tight (e.g., for every α with $\tan \alpha$ rational) and the (asymptotically) extremal configuration is a subset of the integer grid. This is an improvement over the weaker bounds obtained by Conway, Croft, Erdős and Guy (1979), Croft (1967) and de Berg and van Kreveld (1989).

Many similar problems were raised by Erdős and Purdy (1971, 1975, 1976). One of them, which has been solved very recently, is the following. Let d, k be fixed, and let n tend to infinity. Let $h = h_d(k)$ be the largest integer for which it is true that almost all k-element subsets of any n-element set $P \subseteq E^d$ determine at least h distinct distances. More precisely, letting $f(P)$ denote the number of different distances determined by the pairs of points in P, $h = h_d(k)$ is the largest integer for which

$$\lim_{n \to \infty} \min_{\substack{P \subseteq E^d \\ |P| = n}} \left| \{ P_k \subseteq P : |P_k| = k \text{ and } f(P_k) \geq h \} \right| \Big/ \binom{n}{k} = 1.$$

It is not difficult to see that $h_1(k) = h_2(k) = \binom{k}{2}$ for every k. Avis, Erdős and Pach (1991) proved that $h_3(k) = \lfloor k^2/4 \rfloor$, and they also determined the exact value of $h_d(k)$ for all $d \geq 4$ and k:

$$h_d(k) = \frac{1}{2} \lfloor k/\lfloor d/2 \rfloor \rfloor \, (k + r - \lfloor d/2 \rfloor) + 1 + \begin{cases} \lfloor k/\lfloor d/2 \rfloor \rfloor & \text{if } d \text{ is odd,} \\ 0 & \text{if } d \text{ is even.} \end{cases}$$

In the extremal configurations most points are distributed on $\lfloor d/2 \rfloor$ pairwise perpendicular circles and, if d is odd, on an additional line perpendicular to all of them.

Erdős (1967) suggested that it might be interesting to investigate the distribution of distances among n points, restricting attention to some large subset of the $\binom{n}{2}$ pairs. For example, divide the set into two equal parts, and consider only those pairs whose points belong to different parts. The only result of this type is due to Edelsbrunner and Sharir (1990). Given n red points and m blue points in $\mathbf{R}^3$ such that the minimum distance between points of different colours is 1, they proved that the maximum number of red-blue unit-distance pairs is $O(m^{2/3}n^{2/3} + m + n)$. However, no superlinear lower bound is known.

It is clear that many of the problems discussed above are closely related to questions concerning the incidence structure of systems of points and curves (or surfaces). For example, estimating $F_2(n)$, the maximum number of times the unit distance can occur among n points in the plane, is obviously equivalent to finding good upper bounds for the maximum number of incidences between n points and n unit circles in the plane. (We say that a point p is *incident* with a circle C if $p \in C$.) The investigations in this direction were stimulated by the spectacular development of projective geometry in the last century, and have led, for example, to the discovery of finite geometries. One of the most elementary classical problems that remained unsolved was raised by Sylvester (1893). Given a set P of n noncollinear points in the plane, is it true that they always determine a simple connecting line, i.e., a line which passes through exactly two elements of P? Forty years later this question was settled in the affirmative (see Erdős (1944)) by T. Gallai (= Grünwald). G. Dirac (1951) conjectured that if n is sufficiently large then the number of simple connecting lines determined by n noncollinear points is at least $\lfloor n/2 \rfloor$, and Böröczky (see Crowe and McKee (1968)) showed that this bound can be attained. In his 200 page thesis Hansen (1981) claimed to have proved Dirac's conjecture, but no one managed to verify his argument. Now, Csima and Sawyer (1993) have found a serious flaw in Hansen's proof, and have shown that any set of n noncollinear points determine at least $6n/13$ $(n > 7)$ simple connecting lines, thus superceding the previously best known lower bound $3n/7$ established by Kelly and Moser (1958).

2. Graph Dimensions

Construct an infinite graph G^d on the vertex set $V(G^d) = E^d$ by connecting two points with an edge if and only if their distance is one. Erdős, Harary and Tutte (1965) defined the (Euclidean) *dimension* of a graph G as the minimum d such that G is a subgraph of G^d. In other words, the dimension of G (denoted by dim G) is the dimension of the smallest Euclidean space into which G can be embedded with every edge having length 1.

There are many results concerning the chromatic number of G^d. In particular, Hadwiger (1961), Woodall (1973) and L. Moser and W. Moser (1961) showed that

$$4 \leq \chi(G^2) \leq 7.$$

Answering a question of Erdős, Wormald (1979) found a finite triangle-free 2-dimensional graph whose chromatic number is 4. In higher dimensions, Frankl and Wilson (1981) and Larman and Rogers (1972) proved

$$(1 + o(1)) \left(\frac{6}{5}\right)^d < \chi(G^d) < (3 + o(1))^d.$$

It is easy to see that $\dim G \leq 2\chi(G)$ holds for every graph G. Erdős and Simonovits (1980) proved that $\dim G \leq \Delta(G) + 2$, where $\Delta(G)$ denotes the maximum degree of a vertex in G. They also introduced a slightly different notion of dimension of a graph. The *faithful dimension* of G, denoted by Dim G, is defined as the smallest integer d such that G is an *induced* subgraph of G^d, i.e., G can be embedded into E^d so that two vertices are adjacent if and only if their images are at unit distance from each other. Of course dim $G \leq$ Dim G for every graph. Erdős and M. Simonovits showed that while dim G depends on the chromatic number, Dim G is more closely related to the maximum degree. More precisely,

$$\text{Dim } G \leq 2\Delta(G) + 1.$$

As a matter of fact, they conjectured that Dim $G \leq \Delta(G)$ with equality, for example, for every complete graph. The difference between the dimension and the faithful dimension of a graph is well demonstrated by the following example. Let $K'_{m,m}$ denote a graph which can be obtained from $K_{m,m}$ (the complete bipartite graph with m vertices in each of its classes) by deleting m independent edges. Then $\dim K'_{m,m} \leq 4$ but Dim $K'_{m,m} \geq m - 2$. Some weaker results were presented by Maehara (1989). The faithful dimensions of all complete k-partite graphs were recently determined by Buckley and Harary (1988) and Maehara (1988).

When we want to establish upper bounds for the faithful dimension of a graph, in many cases we construct an embedding of the vertices on the surface of a d-dimensional ball of radius $1/\sqrt{2}$ such that two points are at distance 1 if and only if the corresponding vertices are adjacent. Two such points can be seen from the center of the ball at angle $\pi/2$. In other words, we wish to determine the smallest dimension $\overline{\text{Dim}} \, G = d$ such that one can assign d-dimensional vectors to the vertices of G with the property that two vectors are orthogonal if and only if the corresponding vertices are adjacent. Evidently, $\overline{\text{Dim}} \, G \geq$ Dim G. This notion of dimension was introduced by Lovász (1979a). Rosenfeld (1991) showed that if G is a graph with n vertices such that no 3 of them are independent, then $\overline{\text{Dim}} \, G \geq \lceil n/2 \rceil$. A set of n vectors in $\mathbf{R}^d$ is called *almost orthogonal* if among any 3 elements there is at least one orthogonal pair. By Rosenfeld's theorem, any almost orthogonal set of vectors in $\mathbf{R}^d$ has at most $2d$ elements. The union of two orthonormal bases of $\mathbf{R}^d$ demonstrates that this bound is tight. Another interesting result for almost orthogonal vectors was obtained by Kashin and Konyagin (1983). Answering a question of Lovász (1979b) they proved that

$$c\frac{d^{2/3}}{\sqrt{\log d}} \le \max \left\| \sum_{i=1}^{d} u_i \right\| \le 2^{1/3} d^{2/3},$$

where the maximum is taken over all almost orthogonal sets of d vectors in $\mathbf{R}^d$. Some weaker bounds were given in Konyagin (1981).

The investigation of Condim G, the *contact dimension* of G, was initiated by Pach (1980). This is defined as the smallest integer d such that G can be embedded into E^d so that any two vertices are at least unit distance apart, with equality if and only if they are adjacent. In other words, the vertices of G can be represented by nonoverlapping equal balls in E^d, two of them touching each other if and only if the corresponding vertices are connected by an edge of G. Obviously,

$$\text{Dim } G \le \text{ Condim } G \le |V(G)| - 1$$

holds for every graph G. Maehara (1985) determined the contact dimension of all complete k-partite graphs where the size of none of the classes exceeds k, and gave estimates for many other graphs. Frankl and Maehara (1986) proved that

$$cn/\log n < \text{Condim } Q_n < c'n/\log n,$$

where Q_n denotes the graph formed by the edges of the n-dimensional cube. They also showed (Frankl and Maehara 1988) that

$$\text{Condim } T < 7.3 \log |V(T)| \quad \text{for every tree } T.$$

Apart from this, very few general bounds involving the contact dimension are known.

The *sphericity* of a graph G is defined to be the minimum integer d such that G can be embedded into E^d so that two vertices are adjacent if and only if their distance is at most 1. In other words, the sphericity of G is the *smallest* dimension in which the vertices can be represented by equal (closed) balls so that two balls have a point in common if and only if they are connected by an edge. This notion first appeared in Gutman (1977), then in Havel (1982) and Maehara (1984a). More precisely, graphs of sphericity one, that is those that represent intersection patterns of systems of unit intervals on a line, had been considered much earlier by Roberts (1969a). Some other higher dimensional generalizations (cubicity, boxicity, gridicity) were discussed in Roberts (1969b), Trotter (1979), Trotter and Harary (1979), Cozzens and Roberts (1983), Schneierman and West (1983), and Bellantoni, Ben-Arroyo Hartman, Przytycka and Whitesides (1990). Let Sphdim G denote the sphericity of G.

Clearly, Sphdim $G \leq$ Condim G for every G. Maehara (1984a) showed that if G is not a complete graph then

$$\text{Sphdim } G \leq |V(G)| - \omega(G),$$

where $\omega(G)$ is the size of the largest complete subgraph in G. Reiterman, Rödl and Šiňajova (1989b) established

$$\text{Sphdim } G \leq c\Delta^3(G) \log |V(G)|.$$

The sphericity of some other special classes of graphs is estimated in Maehara (1984b) and Frankl and Maehara (1989).

Finally, we would like to recall one further notion of dimension introduced by Paturi and Simon (1984) and Reiterman, Rödl and Šiňajová (1989b). Let Vecdim G stand for the minimum integer d such that to each vertex of G one can assign a d-dimensional vector so that the scalar product of two vectors is at least t (for some real number t) if and only if the corresponding vertices are adjacent. It is not hard to see that

$$\text{Vecdim } G \leq \text{Sphdim } G + 1$$

for every graph G. Reiterman, Rödl and Šiňajová (1989a) established the following nice general upper bound:

$$\text{Vecdim } G \leq |V(G)| - \max(\alpha(G), \chi(G)),$$

where $\alpha(G)$ and $\chi(G)$ denote the size of the maximal independent set and the chromatic number of G, respectively. From this one immediately obtains that Vecdim $G \leq n - \sqrt{n}$ for every graph with n vertices. They also proved that Vecdim $G \geq n/15$ for almost all graphs of n vertices. Clearly, one can pose many similar questions. For instance, we might want to represent the vertices of a graph G by not necessarily equal nonoverlapping balls so that two balls touch each other if and only if the corresponding vertices are adjacent in G. The smallest dimension of a Euclidean space in which this is possible would define another kind of dimension of a graph. But we do not have to insist on balls; in fact, the same question can be asked if we want to represent G by low dimensional convex bodies (which might or might not be congruent, homothetic or translated copies of each other), and we can also drop the condition that these bodies do not overlap. We mention only one result of this kind. It was shown by P. Koebe (1936) and rediscovered by Andreev (1970a, b) that any triangulated planar graph can be represented with nonoverlapping circular disks in the plane such that two of them touch each other if and only if the corresponding vertices are adjacent. Furthermore, Thurston (1985) has proved that this representation is essentially unique. (For an algorithmic approach to the same problem, see Y. Colin de Verdière (1989, 1990).) A particularly nice reformulation of Koebe's result is the following. For every triangulated planar graph G there exists a convex polytope in 3-space whose edge-structure is isomorphic to G and every edge is tangent to the unit ball.

This is in sharp contrast to a well-known theorem of Steinitz (1928) (see also Grünbaum (1963, 1967)) stating that there are infinitely many triangulated planar graphs for which there exist no 3-dimensional convex polytopes with the same edge-structure and such that all vertices lie on the unit sphere. Some analogous results in higher dimensions were obtained by E. Schulte (1987).

3. Geometric Graphs

One of the classical results in graph theory is Turán's theorem (Turán 1954; B. Bollobás 1978) according to which the maximum number of edges of a graph with n vertices containing no complete subgraph of $k + 1$ vertices is

$$T(k, n) = \frac{k-1}{2k}(n^2 - r^2) + \binom{r}{2},$$

where r is the remainder of n upon division by k.

As far as we know, Erdős was the first to suggest that similar questions can be raised for *geometric graphs*, i.e., for graphs whose vertices are embedded in the plane and whose edges are straight line segments. We will always assume that the vertices of a geometric graph are in general position, i.e., there are no 3 vertices along the same line. Given a class $\mathcal{H}$ of so called *forbidden* geometric subgraphs, let $t(\mathcal{H}, n)$ denote the maximum number of edges a geometric graph with n vertices can have without containing a subgraph isomorphic to a member of $\mathcal{H}$. Let $t_c(\mathcal{H}, n)$ be defined similarly, except that now the maximum is taken over all geometric graphs whose n vertices form a convex polygon. Two edges of a geometric graph are called *independent* if the corresponding closed segments are disjoint.

In particular, Erdős asked what $t(\mathcal{I}_{k+1}, n)$ is, where $\mathcal{I}_{k+1}$ is the family consisting of all systems of $k + 1$ pairwise independent edges. For $k = 1$, Erdős noticed that

$$t(\mathcal{I}_2, n) = n.$$

(This also follows from the argument of Hopf and Pannwitz (1934)). Alon and Erdős (1989) showed that

$$t(\mathcal{I}_3, n) \leq 6n - 5,$$

and this has recently been improved by O'Donnell and Perles (1990) to

$$\left\lfloor \frac{5}{2}(n-1) \right\rfloor \leq t(\mathcal{I}_3, n) \leq 3.6n + c.$$

For larger (fixed) values of k it is not even clear whether or not $t(\mathcal{I}_{k+1}, n)$ is linear in n as n tends to infinity.[*] Of course, the same question can be asked

---

[*]   **Added in proof.** This has been proved very recently by Pach and Törőcsik (1992).

for geometric graphs whose vertices are in convex position. As Kupitz (1984) observed, it is easy to see that in this case

$$t_c(\mathcal{I}_{k+1}, n) = kn \qquad \text{for every } k \text{ and } n \geq 2k.$$

The "dual" problem, i.e., when the forbidden configuration is $\mathcal{C}_{k+1}$, a system of $k+1$ pairwise *crossing* edges, was raised by Bernd Gärtner (1989). It follows from Euler's theorem that

$$t(\mathcal{C}_2, n) = 3n - 6 \qquad \text{for all } n \geq 3,$$

and there are no good estimates known for $t(\mathcal{C}_{k+1}, n)$, $k \geq 2$. Recently, M. Perles conjectured, and Capoyleas and Pach (1991) proved, that

$$t_c(\mathcal{C}_{k+1}, n) = \begin{cases} \binom{n}{2} & \text{if } n \leq 2k+1, \\ 2kn - \binom{2k+1}{2} & \text{if } n \geq 2k+1. \end{cases}$$

For $k = 2$, the conjecture of Perles was also verified independently by I. Z. Ruzsa (personal communication). Note that $t_c(\mathcal{C}_{k+1}, n) \leq k^3 2^k n$ follows from a theorem of Gyárfás (1985) and Kostochka (1988), who gave an upper bound on the chromatic number of a graph whose vertices correspond to chords of a circle, two vertices being joined by an edge if and only if the corresponding chords cross each other. Kostochka also gave a construction showing that the chromatic number of these graphs can be superlinear in k. It is very easy to see that, if $\mathcal{H}$ is a forbidden planar configuration (geometric graph) containing a cycle, then $t_c(\mathcal{H}, n)$ is quadratic in n. If $\mathcal{H}$ has no cycle (i.e., if $\mathcal{H}$ is a *forest*) then $t_c(\mathcal{H}, n)$ is linear in n if and only if $\mathcal{H} \not\supseteq Y$, where Y is the union of 3 paths of lengths 2 meeting at a point. If $\mathcal{H} \not\supseteq Y$, then it is called a *caterpillar*. The following beautiful result was established by Perles (1990). For any fixed connected caterpillar cp_k with k vertices,

$$t_c(cp_k, n) = \left\lfloor \frac{n(k-2)}{2} \right\rfloor \qquad \text{for all } n > k.$$

If the forbidden configuration $\mathcal{H}$ is an outerplanar graph, i.e., a noninter-secting cycle of length $k + 1$ with some noncrossing internal diagonals, then

$$t(\mathcal{H}, n) = T(k, n) \qquad \text{for every } n \geq k + 1,$$

where $T(k, n)$ is the value occurring in Turán's theorem quoted at the be-ginning of this section. This was shown independently by Gritzmann, Mohar, Pach and Pollack (1989) and Perles (1990) (see also de Fraysseix, Pach and Pollack (1990)).

It appears to be hard to generalize any of the above results to higher dimensions. However, the following well-known fact has a higher dimensional analogue. Given n red and n blue points in general position in the plane, there is a matching between them so that the corresponding segments are pairwise disjoint. Akiyama and Alon (1989) proved that, if we have d equal sets of points of d different colours in general position in d-space, then we can partition their union into multicoloured $(d-1)$-dimensional simplices so that no two of them have a point in common. Combining this result with a combinatorial lemma of Erdős (1965), we obtain the following theorem. Let d and k be positive integers. Then there exists $\epsilon_{d,k} = \epsilon > 0$ such that, given any set of n points in $\mathbf{R}^d$ in general position and any family of at least $n^{d-\epsilon}$ $(d-1)$-dimensional simplices spanned by them, we can always choose k simplices from this family which are pairwise disjoint (provided that n is sufficiently large). A natural way of associating a geometric graph with a set S of n points in the plane was studied by Lovász (1971) and Erdős, Lovász, Simmons and Straus (1973). Fix an integer k, and join two points $p, q \in S$ by a directed segment if and only if there are exactly k points in the open halfplane to the right of $\overrightarrow{pq}$. Let $e_k(n)$ denote the maximum number of edges of this graph over all n−element point sets S in general position. Almost exactly the same problem in a slightly different setting was raised by Edelsbrunner and Welzl (1985). It was shown that there are positive constants c_1 and c_2 such that

$$c_1 n \log n < e_k(n) < c_2 n \sqrt{k}.$$

The upper bound has been recently improved by Pach, Steiger and Szemerédi (1989) to

$$e_k(n) < \frac{cn\sqrt{k}}{\log^* k},$$

where $\log^*$ denotes the iterated logarithm function. Some related results were obtained by Alon and Győri (1986), Goodman and Pollack (1984) and for small values of k by Edelsbrunner and Stöckl (1986).

The same question can be asked in d-space. Let S be a set of n points in E^d in general position. What is the maximum number of $(d-1)$-element subsets of S with the property that there are exactly k points of S on one side of the hyperplane induced by them? Let $e_k^d(n)$ denote this number. For $d \geq 3$ it is quite difficult to give any nontrivial estimate for $e_k^d(n)$ even in the special case $k = \lfloor n/2 \rfloor$. It is not hard to see (Edelsbrunner 1987) that

$$e_{\lfloor n/2 \rfloor}^d(n) > c_d n^{d-1} \log n,$$

for some constant $c_d > 0$. In 3-dimensional space, Bárány, Füredi and Lovász (1989) showed that $e_{\lfloor n/2 \rfloor}^3(n) \leq cn^{3-\epsilon}$ for some small $\epsilon > 0$. This has been recently improved by Aronov, Chazelle, Edelsbrunner, Guibas, Sharir and Wenger (1991) to

$$e_{\lfloor n/2 \rfloor}^3(n) < cn^{8/3} \log^{5/3} n.$$

Their arguments were extended by Živaljević and Vrećica (1991) to prove that $e_{\lfloor n/2 \rfloor}^d(n) = o(n^d)$ for any $d \geq 3$. Some other bounds can be found in Clarkson and Shor (1989).

Alon, Füredi and Katchalski (1985) have studied the following analogue of the above problem. Given a set of S of n distinct points in E^d, we say that a pair $p, q \in S$ is *separable* if one can find a closed (rectangular) box, with sides parallel to the axes, so that it contains p and q but no other points of S. They showed that the maximum number of separated pairs in a set of n points in general position

$$b^d(n) = \left(1 - \frac{1}{2^{2^{d-1}-1}}\right) \frac{n^2}{2} + o(n^2).$$

Similarly, one can attempt to estimate $b_k^d(n)$, the maximum number of k-tuples $(k > 2)$ of an n element set in d-space that can be separated from the remaining $n - k$ points by a box. One can also drop the condition that the sides of the boxes are parallel to the axes or consider the same problem for separation by cubes, balls, congruent copies of a fixed convex body, etc. These problems were posed by Gamble and Katchalski (1986), who established some preliminary results. In particular, they proved the surprising result that there exists a convex set C and an n element point set S in 3-space such that any pair of points in S can be separated from the others by a homothetic copy of C.

A celebrated theorem of T. Gallai mentioned earlier (see e.g., Kelly (1986), Borwein and Moser (1990)) states that for any noncollinear point set S in the plane one can find a straight line passing through *exactly* two elements of S. Given any two natural numbers k and ℓ, let $\varphi_d(k, \ell)$ denote the largest n for which there is a full-dimensional set S of n points in $\mathbf{R}^d$ such that no oriented hyperplane spanned by S has at least k points on its right-hand side and at least ℓ points on its left-hand side. Kupitz (1990b) showed that if d is odd then

$$\varphi_d(0, k) = \frac{1}{2}(d + 1)k \qquad \text{for all } k \geq 2.$$

For even d, Kupitz and Perles (1990) established

$$\varphi_d(0, k) = k + \left\lceil \log_2 \left(\frac{16}{45}\left(k + 2 + \log_2\left(\frac{16}{45}\right)\right)\right)\right\rceil \qquad \text{for all } k \geq 2.$$

Alon and Perles (1990) proved that for $k \to \infty$

$$(2 + o(1))k \leq \varphi_2(k, k) \leq 2k + O(\log k),$$

and Kupitz (1990c) improved some of these bounds by showing

$$2k + 4 \leq \varphi_2(k, k) \leq 3k - 2 \qquad \text{for all } k \geq 6.$$

Kupitz (1990d) has also obtained some similar results in higher dimensions. In particular, he proved that

$$\varphi_3(k, k) = 4k \qquad \text{for every } k.$$

4. Arrangements of Lines in Space

In contrast to the vast literature about planar arrangements of lines, there have been very few results on systems of lines in 3-space. The recent developments in this field were partly motivated by hidden surface removal problems in computational geometry. An arrangement of lines or line segments in the plane is called *simple* if no two are parallel and no three are concurrent. If such an arrangement W is equipped with a binary relation that specifies for any pair of crossing elements $a, b \in W$ which one lies "above" the other, then W is said to be a *weaving*. A system R of lines or line segments (rods) in 3-space is a *realization* of a weaving W if its orthogonal projection onto the x, y-plane is W and the relative position of the elements of R respect the specifications. That is, if $a', b' \in W$ are the projections of $a, b \in R$ respectively, and b' is "above" a' then there exists a vertical ray pointing in the direction of the positive z-axis whose starting point is on a and which passes through b. A weaving is called *realizable* if it has at least one realization R. A weaving W is *bipartite* of size $m \times n$ if it splits into two ordered sets $W = H \cup V$, $H = \{h_1, \ldots, h_m\}$, $V = \{v_1, \ldots, v_n\}$ and

(i) h_1, v_1, h_m, v_n enclose a convex quadrilateral Q (in this order),

(ii) all elements of H (resp. V) intersect the elements of V (resp. H) in the same order, and all these intersections occcur within Q

and

(iii) no two elements of H (resp. V) cross each other.

A weaving (resp. bipartite weaving) W is called *perfect* if along each line of W the lines intersecting it are alternately "above" and "below".

It was pointed out by Pach, Pollack and Welzl (1990) that if n is sufficiently large then almost all weavings W of n lines are nonrealizable. This remains true even if we relax the condition that the projection of R must coincide with W, and we require only that it defines a planar map *combinatorially* equivalent to W. In spite of this, for a given large n, it seems to be difficult to exhibit a weaving of n lines which cannot be realized even in this weak sense, but if we remove any of its lines then it becomes realizable. Pach, Pollack and Welzl showed that

(i) no perfect weaving of $n \geq 4$ lines can be realized

and

(ii) no perfect bipartite $m \times n$ weaving of line segments can be realized provided that $\min\{m, n\} \geq 4$. (Otherwise this statement is not true.)

Given a set of k elements $a_1, \ldots, a_k$ in a weaving W, we say that they form a *cycle* if a_{i+1} is "above" a_i for every $1 \leq i < k$ and a_1 is "above" a_k. If, in addition, $a_1, \ldots, a_k$ (in this order) enclose a cell in the arrangement W, then they are said to form an *elementary cycle*. Note that in a perfect weaving the sides of each cell determine an elementary cycle. Chazelle, Edelsbrunner, Guibas, Pollack, Seidel, Sharir and Snoeyink (1992) proved that $\gamma(n)$, the maximum number of elementary cycles in a realizable $n \times n$ bipartite weaving, satisfies

$$c_1 n^{4/3} \le \gamma(n) \le c_2 n^{3/2}$$

for some positive constants c_1, c_2. It is easy to see that a (not necessarily bipartite) realizable weaving of n lines can have as many as $c n^{3/2}$ elementary cycles. However, in this case no nontrivial (subquadratic) upper bound is known.

One can eliminate all cycles in a weaving by cutting some of its rods into smaller pieces. To *cut* a rod means to remove one of its points and thus to decompose it into two rods. Of course, n^2 cuts are sufficient to "kill" all cycles in any weaving, because we can cut every rod at every crossing. Let $k(n)$ denote the maximum number of cuts necessary to eliminate all cycles (not just the elementary ones!) in a realizable $n \times n$ weaving. Since we have to kill each elementary cycle, the last result immediatley implies that $k(n) \ge \frac{c_1}{4} n^{4/3}$. On the other hand, Chazelle et al have shown that

$$k(n) \le c n^{9/5}.$$

In a recent paper Sharir (1991) proved the following related result. Given n lines in 3-dimensional space, for every $\epsilon > 0$ there are at most $c_\epsilon n^{8/5+\epsilon}$ points p with the property that we can find 3 non-coplanar lines in our system passing through p. An easy construction shows that the number of such points can be as large as $c' n^{3/2}$. Consider all axis-parallel lines in $\mathbf{R}^3$ passing through at least one point (x, y, z), where $1 \le x, y, z \le \sqrt{n}$ are integers.

Acknowledgement. The work of the second author has been supported by NSF Grant CCR–89–01484 and Hungarian Science Foundation Grant OTKA–1814.

References

Akiyama, J. and N. Alon (1989): Disjoint simplices and geometric hypergraphs. Combinatorial Mathematics. Proc. Third Internat. Conference (G. Bloom, R. Graham, J. Malkevitch, eds.). Ann. New York Acad. Sci. **555**, 1–3

Alon, N. and P. Erdős (1989): Disjoint edges in geometric graphs. Discr. Comput. Geom. **4**, 287–290

Alon, N., Z. Füredi and M. Katchalski (1985): Separating pairs of points by standard boxes. Europ. J. Combinatorics **6**, 205–210 [MR 87b: 05011]

Alon, N. and E. Győri (1986): The number of small semispaces of a finite set of points in the plane. J. Combinat. Th. Ser. A **41**, 154-157 [MR 87f: 52014]

Alon, N. and M. Perles (1990): Personal communication

Andreev, E. M (1970a): On convex polyhedra in Lobacevskii spaces. Math. Sbornik USSR, Nov. Ser. **81**, 445–478

Andreev, E. M (1970b): On convex polyhedra of finite volume in Lobacevskii spaces. Math. Sbornik USSR, Nov. Ser. **83**, 256–260

Aronov, B., B. Chazelle, H. Edelsbrunner, L. Guibas, M. Sharir and R. Wenger (1991): Points and triangles in the plane and halving planes in space. Discr. Computat. Geom., to appear

Avis, D (1984): The number of farthest neighbour pairs of a finite planar set. Amer. Math. Monthly **91**, 417–420. [MR 85j: 52012 Zbl 571.51009]

Avis, D., P. Erdős and J. Pach (1988): Repeated distances in space. Graphs and Combinatorics **4**, 207–217. [Zbl 656.05039]

Avis, D., P. Erdős and J. Pach (1991): Distinct distances determined by subsets of a pointset in space. Computat. Geometry, Theory and Applications **1**, 1–11

Bárány, I., Z. Fűredi and L. Lovász (1989): On the number of halving planes. Proc. 5th Ann. Sympos. Computat. Geom. 1989: 140–144. To appear also in Combinatorica **10** (1990)

Beck, J (1983): On the lattice property of the plane and some problems of Dirac, Motzkin and Erdős in combinatorial geometry. Combinatorica **3**, 281–297. [Zbl 533.52004]

Beck, J. and J. Spencer 1984. Unit distances. J. Combinat. Theory Ser A **37**, 231–238. [Zbl 551.05005]

Bellantoni, S., I. Ben-Arroyo Hartman, T. Przytycka and S. Whitesides. 1990. Graphs and boxicity. Preprint

Berg, M. T. de and M. J. van Kreveld (1989): Finding squares and rectangles in sets of points. Univ. of Utrecht, Dept. Computer Science Technical Report RUU-CS-89-10

Bollobás, B (1978): Extremal Graph Theory. Academic Press, London, New York

Borwein, P. and W. O. J. Moser (1990): A survey of Sylvester's problem and its generalizations. Aequationes Math. **40**, 111–135

Brass, P (1992): The maximum number of second smallest distances in finite planar graphs. Discr. Computat. Geom. **7**, 371–379

Buckley, F. and F. Harary (1988): On the Euclidean dimension of a wheel. Graphs and Combinatorics **4**, 23–30. [Zbl 642.05018]

Capoyleas, V. and J. Pach (1991): A Turán-type theorem on chords of a convex polygon. J. Combinat. Theory, ser. B, to appear

Chazelle, B., E. Edelsbrunner, L. Guibas, R. Pollack, R. Seidel, M. Sharir and J. Snoeyink (1992): Counting and cutting cycles of lines and rods in space. Computational Geom., Theory and Appl. **1**, 305–323

Chung, F. R. K (1984): The number of different distances determined by n points in the plane. J. Combinat. Theory Ser A **36**, 342–354. [Zbl 536.05003]

Chung, Fan (1989): Sphere-and-point incidence relations in high dimensions with applications to unit distances and furthest neighbour pairs. Discr. Computat. Geometry **4**, 183–190. [Zbl 662.52005]

Chung, F. R. K., E. Szemerédi and W. T. Trotter (1992): The number of different distances determined by a set of points in the Euclidean plane. Discr. Computat. Geom. **7**, 1–11

Clarkson, K. L., H. Edelsbrunner, L. J. Guibas, M. Sharir and E. Welzl (1988): Combinatorial complexity bounds for arrangements of curves and surfaces. IEEE Proc. 29th Symposium on Foundations of Computer Science: 568–579. Also in Discr. Computat. Geom. **5**, (1990) 99–160

Clarkson, K. L. and P. W. Shor (1989): Applications of random sampling in computational geometry, II. Discr. Computat. Geom. **4**, 387–421

Colin de Verdière, Y (1989): Empilements de cercles: Convergence d'une méthode de point fixe . Forum Math. **1**, 395–402

Colin de Verdière, Y (1990): Un principe variationnel pour les empilements de cercles. Prépublication de l'Institut Fourier, Laboratoire de Math. No. 147

Conway, J. H., H. T. Croft, P. Erdős and M. J. T. Guy (1979): On the distribution of values of angles by coplanar points. J. London Math. Soc.(2) **19**, 137–143. [MR 80h: 51021]

Cozzens, M. B. and F. S. Roberts (1983): Computing the boxicity of a graph by covering its complement by cointerval graphs. Discr. Appl. Math. **6**, 217–228. [MR 85d: 05142]

Croft, H. T (1967): Some geometrical thoughts II. Math. Gazette **51**, 125–129

Crowe, D. W. and T. A. Mckee (1968): Sylvester's problem on collinear points. Math. Mag. **41**, 30–34. [MR 38 #3761]

Csima, J. and E. T. Sawyer (1993): A short proof that there exist $6n/13$ ordinary points. Discr. Computat. Geom., to appear

Dirac, G. A (1951): Collinearity properties of sets of points. Quart. J. Math. **2**, 221–227. [MR 13 p. 270]

Edelsbrunner, H (1987): Algorithms in Combinatorial Geometry. Springer, Heidelberg. [MR 89a: 68205]

Edelsbrunner, H. and P. Hajnal (1991): A lower bound on the number of unit distances between the points of a convex polygon. J. Combinat. Theory, Ser. A **56**, 312–316

Edelsbrunner, H. and M. Sharir (1990): A hyperplane incidence problem with applications to counting distances. In: Algorithms (Proc. Internat. Symp. SIGAL '90, Tokyo. T. Asano et al, eds.) Lecture Notes in Comp. Sci. **450**, 419–428. Springer

Edelsbrunner, H and S. Skiena (1989): On the number of furthest neighbour pairs in a point set. Amer. Math. Monthly **96**, 614–618

Edelsbrunner, H. and G. Stöckl (1986): The number of extreme pairs of finite point-sets in Euclidean spaces. J. Combinat. Theory Ser A **43**, 344–349. [MR 87k: 52025. Zbl 611.51002]

Edelsbrunner, H. and A. Welzl (1985): On the line-separations of a finite set in the plane. J. Combinat. Theory Ser A **38**, 15–29. [MR 86m: 52012. Zbl 616.52003]

Erdős, P (1944): Solution of Problem 4065. Amer. Math. Monthly **51**, 169–171

Erdős, P (1946): On sets of distances of n points. Amer. Math. Monthly **53**, 248–250. [MR 7 – 471] (This paper is also in: Erdős, P. 1973. The Art of Counting. MIT Press. Cambridge, Mass.) [MR 7 – 471]

Erdős P (1960): On sets of distances of n points in Euclidean space. Publ. Math. Inst. Hungar. Acad. Sci. **5**, 165–169. [MR 25 #4420]

Erdős, P (1965): On extremal problems for graphs and generalized graphs. Israel J. Math. **2**, 183–190

Erdős P (1967): On some applications of graph theory to geometry. Canad. J. Math. **19**, 968–971. [MR 36 #2520]

Erdős, P (1987): Some combinatorial and metric problems in geometry. Intuitive Geometry: 167–177. Colloq. Math. Soc. J. Bolyai **48**, North-Holland. [Zbl 625.52008]

Erdős, P., Z. Füredi, J. Pach and I. Z. Ruzsa (1993): The grid revisited. Discrete Math. **110**, to appear

Erdős, P., F. Harary and W. T. Tutte (1965): On the dimension of a graph, Mathematika **12**, 118–122. [MR 32 #5537]

Erdős, P., D. Hickerson and J. Pach (1989): A problem of Leo Moser about repeated distances on the sphere. Amer. Math. Monthly **96**, 569–575

Erdős, P., L. Lovász, A. Simmons and E. G. Straus (1973): Dissection graphs of planar point sets. A Survey of Combinatorial Theory (J. N. Shrivastava et al, eds.): 139–149. North Holland. [MR 51 #241]

Erdős, P., L. Lovász and K. Vesztergombi (1987): The chromatic number of the graph of large distances. Colloquia Math. Soc. J. Bolyai 52, Combinatorics, Eger (Hungary): 547–551, North-Holland

Erdős, P., L. Lovász and K. Vesztergombi (1989): On the graph of large distances. Discr. Computat. Geom. **4**, 541–549

Erdős, P., E. Makai, J. Pach and J. Spencer (1991): Gaps in difference sets and the graph of nearly equal distances. In: Applied Geometry and Discrete Mathematics, The Victor Klee Festschrift (P. Gritzmann, B. Sturmfels, eds.), DIMACS Series, Vol. 4, AMS-ACM, 265–273

Erdős, P.and L. Moser (1959): Problem 11. Canad Math. Bull. 2 43

Erdős, P. and J. Pach (1990): Variations on the theme of repeated distances. Combinatorica 10, 261–269

Erdős, P. and G. Purdy (1971): Some extremal problems in geometry. J. Combinat. Theory A 10, 246–252. [MR 43 #1045]

Erdős, P. and G. Purdy (1975): Some extremal problems in geometry III. Proceedings of the Sixth Southeastern Conference on Combinatorics, Graph Theory and Computing (Florida Atlantic Univ., Boca Raton, Fla., 1975): 291–308. Congressus Numerantium XIV, Utilitas Math., Winnipeg, Man. [MR 52 #13650]

Erdős, P. and G. Purdy (1976): Some extremal problems in geometry. IV. Proceedings of the Seventh Southeastern Conference on Combinatorics, Graph Theory and Computing (Florida Atlantic Univ., Boca Raton, Fla., 1976): 307–322. Congressus Numerantium XVII, Utilitas Math., Winnipeg, Man. [MR 55 #10292]

Erdős, P. and M. Simonovits (1980): On the chromatic number of geometric graphs. Ars Combinatoria 9, 229–246. [Zbl 466.05031]

Erdős, P. and A. H. Stone (1946): On the structure of linear graphs. Bull. Amer. Math. Soc. 52, 1087–1091. [MR 8 – 333]

Frankl, P. and H. Maehara (1986): Embedding the n-cube in lower dimensions. Europ. J. Combinat. 7, 221–225. [Zbl 627.05038]

Frankl, P. and H. Maehara (1988): On the contact dimension of graphs. Discr. Computat. Geom. 3, 89–96. [Zbl 625.05048]

Frankl, P. and H. Maehara (1988a): The Johnson-Lindenstrauss lemma and the sphericity of some graphs. J. Combinat. Theory Ser. B 44, 355–362. [MR 89e: 05078]

Frankl, P. and R. Wilson (1981): Intersection theorems with geometric consequences. Combinatorica 1, 357–368. [Zbl 498.05046]

Fraysseix, H. de, J. Pach and R. Pollack (1990): Drawing a planar graph on a grid. Combinatorica 10, 41–51

Füredi, Z. (1990): The maximum number of unit distances in a convex n–gon. J. Combinat. Theory ser. A 55, 316–320

Gärtner, B. (1989): in a letter from Emo Welzl to János Pach.

Gamble, A. B. and M. Katchalski (1986): Separating pairs of points by convex sets. Manuscript

Goodman, J. E. and R. Pollack (1984): On the number of k-subsets of a set of n points in the plane. J. Combinat. Theory Ser. A. 36, 101-104. [MR 85d: 05015]

Gritzmann, P., B. Mohar, J. Pach and R. Pollack (1989): Problem E 3341. Amer. Math. Monthly 96, 642

Grünbaum, B. (1963): On Steinitz's theorem about noninscribable polyhedra. Proc. Ned. Akad. Wetenschap Ser. A 66, 452–455

Grünbaum, B. (1967): Convex Polytopes. Interscience, New York

Gutman, I. (1977): A definition of dimensionality and distance for graphs. Geometric Representation of Relational Data: 713–723. (J. C. Lingoes, ed.) Mathesis Press, Michigan

Gyárfás, A. (1985): On the chromatic number of multiple intervals graphs. Discr. Math. 55, 161–166. [MR 86k: 05052] (Corregendum, ibid 55, 193–204. [MR 87k: 05079])

Hadwiger, H. (1961): Ungelöste Probleme No. 40. Elemente der Math. 16, 103–104

Hansen, S. (1981): Contributions to the Sylvester-Gallai Theory. Ph.D.Thesis. Copenhagen

Harborth, H. (1974): Solution to problem 664a. Elem. Math 29, 14–15

Havel, T. H. (1982): The combinatorial distance geometry approach to the calculation of molecular conformation. Ph. D. Thesis, Group in Biophysics, University of California, Berkeley

Hopf, H. and E. Pannwitz (1934): Aufgabe No. 167. Jber. Deutsch. Math. Verein. **43**, 114

Józsa, S. and E. Szemerédi (1975): On the number of unit distances on the plane. Infinite and finite sets (Colloq., Keszthely, 1973; dedicated to P. Erdős on his 60th birthday), Vol. II: 939–950. Colloq. Math. Soc. János Bolyai, Vol. 10, North-Holland, Amsterdam. [MR 52 #10624]

Kashin, B. S. and S. V. Konyagin (1983): On systems of vectors in a Hilbert space. Proc. Steklov Inst. Math. (AMS Translation) Issue 3: 67–70

Kelly, L. M. (1986): A resolution of the Sylvester-Gallai problem of J.-P. Serre. Discr. Computat. Geom. **1**, 101–104

Kelly, L. M. and W. Moser (1958): On the number of ordinary lines determined by n points. Canad. J. Math. **10**, 210–219 [MR20 #3494]

Koebe, K. (1936): Kontaktprobleme der konformen Abbildung. Berichte über die Verhandlungen der Sächsischen Akad. d. Wissenschaften, Math.-Physikalische Klasse **88**, 141–164

Konyagin, S. V. (1981): Systems of vectors in Euclidean space and an extremal problem for polynomials. Mathematicheskie Zametki **29/1**, 63–74

Kővári, T., V. T. Sós and P. Turán (1954): On a problem of K. Zarankiewicz. Colloq. Math. **3**, 50–57. [MR 16 – 456]

Kostochka, A. V. (1988): On upper bounds on the chromatic number off graphs (Russian). Trudy Inst. Math. Akad. Nauk SSSR, Sibirskoie Otdel. **10**, 204–226

Kupitz, Y. (1984): On pairs of disjoint segments in convex position in the plane. Annals Discr. Math. **20**, 203–208

Kupitz, Y. (1990a): On the maximal number of appearances of the minimal distance among n points in the plane. Manuscript

Kupitz, Y. (1990b): k-supporting hyperplanes of a finite set in d-space. Manuscript

Kupitz, Y. (1990c): k-bisectors of a planar set. Combinatorica, to appear

Kupitz, Y. (1990d): Separation of a finite set in d-space by spanned hyperplanes. Manuscript

Kupitz, Y. and M. Perles (1990): The maximal cardinality of a k-thin set in the plane. Manuscript

Larman, D. G. and C. A. Rogers (1972): The realization of distances within sets in Euclidean space. Mathematika **19**, 1–24. [MR 47 #7601]

Lovász, L. (1971): On the number of halving lines. Ann. Univ. Sci. Budapest Eőtvős, Sect. Math. **14**, 107–108

Lovász, L. (1979a): On the Shannon capacity of a graph. IEEE Trans. Infom. Theory **25**, 1–7. [MR 81g: 05095]

Lovász, L. (1979b): Combinatorial Problems and Exercises. Akadémiai Kiadó – North-Holland

Maehara, H. (1984a): Space graphs and sphericity. Discr. Appl. Math. **7**, 55–64. [MR 85i: 05092]

Maehara, H. (1984b): On the sphericity for the join of many graphs, Discr. Math. **49**, 311–313. [MR 85i: 05093]

Maehara, H. (1985): Contact patterns of equal nonoverlapping spheres. Graphs and Combinatorics **1**, 271–282. [Zbl 581.05022]

Maehara, H. (1988): On the Euclidean dimension of a complete multipartite graph. Discr. Math. **72**, 285–289

Maehara, H. (1989): Note on induced subgraphs of the unit distance graph E^n. Discr. Computat. Geometry **4**, 15–18. [MR 89k: 05035]

Moser, L. (1952): On different distances determined by n points. Amer. Math. Monthly **59**, 85–91. [MR 13 – 768]

Moser, L. and W. Moser (1961): Solution P. 10. Canad. Math. Bull. **4**, 187–189

O'Donnel, P. and M. Perles (1990): Personal communication

Pach, J. (1980): Decomposition of multiple packing and covering. Diskrete Geometrie, 2 Kolloq. Math. Inst. Univ. Salzburg: 169–178

Pach, J, R. Pollack and E. Welzl (1990): Weaving patterns of lines and line segments in space. Proc. SIGAL Conf. on Algorithms, Tokyo 1990. Lecture Notes in Comp. Sci. **450**, 439–446. Springer

Pach, J. and M. Sharir (1992): Repeated angles in the plane and related problems. J. Combinat. Theory ser. A. **59**, 12–22

Pach, J., W. Steiger and E. Szemerédi (1989): An upper bound on the number of planar k-sets. Proc. 30th Ann. IEEE Symp. Found. Comput. Sci.: 72–79. Also in: Discrete and Computat. Geometry **7** (1992) 109–123

Pach, J. and J. Törőcsik (1992): Some geometric applications of Dilworth's theorem. Discr. Computat. Geom., submitted

Paturi, R. and J. Simon (1984): Probalistic communication complexity. 25th Annual Symp. on Foundations of Comp. Sc.: 118–126

Perles, M. (1990): Personal communication

Reiterman, J., V. Rödl and E. Šiňajová (1989a): Embeddings of graphs in Euclidean spaces. Discr. Computat. Geom. **4**, 349–364

Reiterman, J., V. Rödl and E. Šiňajová (1989b): Geometrical embeddings of graphs. Discr. Math. **74**, 291–319

Roberts, F. S. (1969a): Indifference graphs. Proof Techniques in Graph Theory: 301–310 (F. Hararay, ed.). Academic Press, New York. [MR 40 #5488]

Roberts, F. S. (1969b): On the boxicity and cubicity of a graph. in Recent Progress in Combinatorics (W. T. Tutte, ed.): 301–310. Academic Press, New York. [MR 40 #5489]

Rosenfeld, M. (1991): Almost orthogonal lines in E^d. In: Applied Geometry and Discrete Mathematics, The Victor Klee Festschrift (P. Gritzmann and B. Sturmfels, eds.), DIMACS Series, Vol. 4, AMS-ACM, 489–492

Schneierman, E. R. and D. B. West (1983): The interval number of a planar graph: three intervals suffice. J. Combinat. Theory Ser. B **35**, 224–239. [MR 85e: 05101]

Schulte, E. (1987): Analogues of Steinitz's theorem about non-inscribable polytopes. In: Intuitive Geometry (K. Bőrőczky, G. Fejes Tóth, eds.) Colloq. Math. Soc. J. Bolyai **48**, 503–516

Sharir, M. (1991): On joins of lines in 3–space. Submitted to Combinatorica

Spencer, J., E. Szemerédi and W. Trotter Jr. (1984): Unit distances in the Euclidean plane. Graph Theory and Combinatorics, Proc. Conf. hon. P. Erdős , Cambridge 1983: 293–303. Academic Press, London. [Zbl 561.52008. MR 86m: 52015]

Steinitz, E. (1928): Über isoperimetrische Probleme bei konvexen Polyedern. J. Reine Angew. Math. **159**, 133–143

Sylvester, J. J. 1893. Mathematical Question 11851. Educational Times **46**, 156

Thurston, W. (1985): The Geometry and Topology of Three-Manifolds. Princeton Notes. Chapter 13

Trotter, W. T. (1979): A characterization of Roberts' inequality for boxicity, Discr. Math. **28**, 303–313. [MR 81a: 05118]

Trotter, W. T. and F. Harary (1979): On double and multiple interval graphs, J. Graph. Theory **3**, 205–211. [MR 81c: 05055]

Turán, P. (1954): On the theory of graphs. Colloq. Math. **3**, 19–30. [MR 15 – 976]

Vesztergombi, K. (1985): On the distribution of distances in finite sets in the plane. Discr. Math. **57**, 129–145. [Zbl 568.52014]

Vesztergombi, K. (1987): On large distances in planar sets. Discr. Math. **67**, 191–198. [Zbl 627.52009]

Woodall, D. R. (1973): Distances realized by sets covering the plane. J. Combinat. Theory Ser. A **14**, 187–200. [MR 46 #9868]

Wormald, N. (1979): A 2–chromatic graph with a special plane plane drawing. J. Austral. Math. Soc. A **28**, 1–8. [MR 80k: 05060]

Živaljevič, R. T. and S. T. Vrečica (1991): The colored Tverberg problem and complexes of injective functions. manuscript

Chapter XII. Set Theoretic Constructions in Euclidean Spaces

Péter Komjáth

Abstract. The most important methods of constructing, via the axiom of choice, various sets and decompositions in Euclidean spaces, omitting some geometrical configurations are surveyed.

0. Introduction

Since Mazurkiewicz' construction of a set on the plane hitting every line in exactly two points there have been several constructions producing paradoxical sets and decompositions in Euclidean spaces using set theoretical tools. As these theorems have been proved by remarkably similar methods and many of them remained unpublished for several years we decided to collect some typical arguments and results of the field.

We are mostly interested in those arguments which use more set theory, e.g. transfinite recursion or Skolem-type arguments. Though they are outside the scope of this survey, let me mention three beautiful and deep results which use only a little bit of set theory, like the axiom of choice or König's lemma on infinity.

Theorem 0.1 (The Banach-Tarski Paradox [3]). *In R^3, any two balls (of positive radius) are finitely equidecomposable.*

Theorem 0.2 (J. Mycielski [35]). *There is a set $E \subseteq R^3$ such that if D_0, D_1 are countable sets in R^3, then $(E - D_0) \cup D_1$ is congruent to E.*

Theorem 0.3 (M. Laczkovich [31]). *In R^2, a square and a disk of the same area are finitely equidecomposable.*

For an excellent discussion of these problems see [45].

The set theory to be used can be found in [24] or in the introductory chapters of [25] or [29]. We use the well-ordering theorem throughout as well as the identification of a cardinal with the least ordinal of that cardinality.

In Section 6 some real analysis is used, in Sections 5 and 6 we mention a few independence results.

Acknowledgements. This research has been partially supported by the Hungarian National Foundation for Scientific Research, Grants #1805 and 1908. I am grateful to János Pach, Fred Galvin, Jan Mycielski, and the referee for several remarks and suggestions.

1. Simple Transfinite Constructions

In this chapter we present some results which can be proved by simple transfinite recursion arguments. We start with a well-ordering of length continuum of the set of points, or lines of R^2, or R^3, etc., with all proper initial segments of this transfinite enumeration of size less than continuum. The theorem on transfinite recursion gives that it suffices to specify what to do at the αth step if we know the results of all previous steps. If we create only countably many objects at every steps, only less than continuum many objects have been selected in earlier steps, so, in some cases, we can proceed.

As usual in set theory, let 2^ω denote the cardinal continuum, as well as the least ordinal of size continuum.

Theorem 1.1 (folklore ?) R^3 *is the union of disjoint circles.*

Proof. Enumerate the points of R^3 as $\{p_\alpha : \alpha < 2^\omega\}$. By transfinite recursion on α we are going to find disjoint circles with the effect that eventually all points will be convered. Assume that we have reached the αth step, with all the circles, selected in previous steps, specified. If p_α is covered by them, we do nothing. If p_α is uncovered, we must find a circle C through p_α disjoint from the less than 2^ω many already selected circles. First, we select a plane P through p_α which is different from the planes of the circles already found. This is possible as 2^ω many planes go through p_α and only $< 2^\omega$ of them contain those circles. These circles intersect P in at most two points each, so once we decide to find the new circle in P, our task reduces to finding a circle in a plane passing through a certain point and missing $< 2^\omega$ other points. This is possible as there are 2^ω circles containing p_α but being otherwise mutually disjoint. Namely, those having the same line as the tangent line at p_α.   □

With similar constructions it is possible to decompose R^3 into the union of disjoint, unparallel lines, of R^3 with one (or even less than continuum many) points omitted into the union of disjoint lines, etc. In some cases, however, it *is* possible to give direct, choiceless constructions for these decompositions. For example, A. Szulkin [42] gave a simple, constructive proof for Theorem 1. 1. No such proof is known if we require the circles to be unit circles, a case easily handled by the above method. Jan Mycielski pointed out a simple decomposition of R^3 into unparallel lines. Decompose R^3 minus the z axis into suitable chosen hyperboloids, each formed by rotating a line around the z axis, with these lines having different angles with the xy plane. He points out that no such proof is known if we want the lines to intersect the unit ball.

Theorem 1.2 (Mazurkiewicz [33]). *There exists a set $A \subseteq R^2$ intersecting every line in exactly 2 points.*

Proof. Enumerate the lines of R^2 as $\{L_\alpha : \alpha < 2^\omega\}$. We are going to select 2 points on every line, by transfinite recursion. While picking those points we must make sure that

(a) in the steps *preceding* the αth we won't drop 3 or more points on L_α;
(b) in the steps *following* the αth we won't select more points on L_α (of which 2 points are already selected at the αth step).

To secure (a) we add the requirement that no 3 points of A be collinear. Assume that we have reached the αth step. Among the points selected in earlier steps, 2, 1, or 0 can be on L_α. In the first case we have nothing to do, we can pass to the next step. In the other case we have to select 1 or 2 points of L_α. Less than 2^ω points have already been selected. Condition (a) excludes at most one point of L_α for every pair of previously selected points. Likewise, (b) excludes $< 2^\omega$ points of L_α. It is possible to choose the needed 1 or 2 points, which are not ruled out, and so we can treat L_α.   □

We mention that there exist, by similar constructions, sets hitting every line (circle, ...) in 3 (4, 5, ...). An old unsolved problem is if a set with the above property exists which is Borel. Some other generalizations can be found in [34].

The following result was conjectured by H. Steinhaus.

Theorem 1.3 (Sierpiński [41]). *There exist sets A, $B \not\subseteq R^2$ such that $|\varphi(A) \cap B| = 1$ holds for every rigid motion φ of the plane.*

Proof. Let $\{\varphi_\alpha : \alpha < 2^\omega\}$ be an enumeration of the rigid motions. We are going to construct $A = \{a_\alpha : \alpha < 2^\omega\}$ and $B = \{b_\alpha : \alpha < 2^\omega\}$ with the intention that $b_\alpha = \varphi_\alpha(a_\alpha)$ is the only element of $\varphi_\alpha(A) \cap B$. As in the previous proof, we try to make sure that

(a) the points selected *before* the αth step may not produce 2 elements in $\varphi_\alpha(A) \cap B$;
(b) points chosen *after* the αth step may not add elements to $\varphi_\alpha(A) \cap B$.

If we require that
(a') no nonzero distance occurs both in A and B;
then (a) will be secured.

Assume that we have reached the αth step and $A_\alpha = \{a_\beta : \beta < \alpha\}$ and $B_\alpha = \{b_\beta : \beta < \alpha\}$ are given. If $\varphi_\alpha(A_\alpha) \cap B_\alpha$ is non-empty, it has just one element by (a), this element will be $b_\alpha = \varphi_\alpha(a_\alpha)$ and we can proceed. If $\varphi_\alpha(A_\alpha) \cap B_\alpha = \emptyset$, we have to find a $b = b_\alpha$ such that

(1.1) (for a') $d(b, x)$ $(x \in B_\alpha)$ is not a distance in A_α ;

(1.2) (for a') $d(\varphi_\alpha^{-1}(b), y) = d(b, \varphi_\alpha(y))$ $(y \in A_\alpha)$ is not a distance in B_α;

(1.3) (for a') $d(b, x) \neq d(\varphi_\alpha^{-1}(b), y) = d(b, \varphi_\alpha(y))$ for $x \in B_\alpha, y \in A_\alpha$;

(1.4) (for b) $\varphi_\beta(A_{\alpha+1}) \cap B_{\alpha+1} = \{b_\beta\}$ for $\beta < \alpha$.

Conditions (1.1), (1.2) exclude the points in the union of $< 2^\omega$ circles. As $\varphi_a(A_\alpha) \cap B_\alpha = \emptyset$, condition (1.3) excludes the union of $< 2^\omega$ many lines. The statement that a_α, b_α do not add elements to $\varphi_\beta(A) \cap B_\beta$ can be formulated as

(1.5) $b \notin \varphi_\beta(A_\alpha)$, $\varphi_\beta(a_\alpha) = \varphi_\beta \varphi_\alpha^{-1}(b) \notin B_\alpha$,

and

(1.6) $\varphi_\beta(a_\alpha) = \varphi_\beta(\varphi_\alpha^{-1}(b)) \neq b$,

i.e. b is not a fixed point of $\varphi_\beta \varphi_\alpha^{-1}$, this excludes a point or a line, and (1.5) excludes a set of size $\leq |\alpha|$. Together for all $\beta < \alpha$, we exclude the union of $< 2^\omega$ circles, lines, and at most $|\alpha|^2 < 2^\omega$ points. To conclude we need to show that $< 2^\omega$ many lines and circles cannot cover the plane. For this, take a line, not occurring among those lines. The allegedly covering lines/circles each meet this line in at most 2 points, so they cannot cover the line, let alone the plane. We can, therefore, select $b = b_\alpha$ and proceed with the construction.   □

St. Ulam asked if there exists a set $A \subseteq R^2$ intersecting each of its (proper) translated copies in a singleton. The following construction gives such a set.

Theorem 1.4 *If V is a vector space over Q, then there exists a set $A \subseteq V$ such that $|(A + a) \cap A| = 1$ for every $a \in V$, $a \neq 0$.*

Proof. Enumerate $V - \{0\}$ as $\{a_\alpha : \alpha < \kappa\}$. We are going to find b_α, $c_\alpha \in V$ such that $c_\alpha - b_\alpha = a_\alpha$ and no $\neq 0$ difference occurs twice. These conditions will ensure that if $A = \{b_\alpha, c_\alpha : \alpha < \kappa\}$ then A is as required. Assume that we have reached the αth step and $A_\alpha = \{b_\beta, c_\beta : \beta < \alpha\}$ has already been built up. If $A_\alpha \cap (A_\alpha + a_\alpha)$ is non-empty, it can only contain one element, and we have a unique choice for b_α, c_α. These elements are already in A_α, so we let $A_{\alpha+1} = A_\alpha$.

Assume that $A_\alpha \cap (A_\alpha + a_\alpha) = \emptyset$. We need to select a $b = b_\alpha$ such that

(1.7) $b - x, \; b + a_\alpha - x \quad (x \in A_\alpha)$ is not the difference between two elements of A_α ;

(1.8) $b - x \neq \pm a_\alpha, \; b + a_\alpha - x \neq \pm a_\alpha \quad (x \in A_\alpha)$,

and then we can let $b_\alpha = b$, $c_\alpha = b + a_\alpha$. The other possibilities, namely, that $a_\alpha = x - y$, or $b - x = b + a_\alpha - y$ for some $x, y \in A_\alpha$, are ruled out as $A_\alpha \cap (A_\alpha + a_\alpha) = \emptyset$, so (1.7) and (1.8) guarantee that no difference occurs twice. Conditions (1.7), (1.8) disqualify less than κ elements as b, so $b = b_\alpha$ and $c_\alpha = b_\alpha + a_\alpha$ can be selected. $\square$

We notice that it is impossible to extend the above result to the intersection of three copies.

Theorem 1.5 *If V is a vector space over Q, $A \subseteq V$, then there exist b, c V, different from zero and each other, such that $A \cap (A + b) \cap (A + c)$ is not a singleton.*

Proof. Assume that the statement is false. If there is a difference occurring three times, i.e. $x = b_1 - a_1 = b_2 - a_2 = b_3 - a_3$ for $a_1, a_2, a_3, b_1, b_2, b_3 \in A$, then $A \cap (A + (a_1 - a_2)) \cap (A + (a_1 - a_3))$ contains a_1 and $a_1 + x$, a contradiction.

So assume, that difference $c \neq 0$ occurs at most twice, $c = b_1 - a_1 = b_2 - a_2$ where b_2, a_2 may not exist. If d is an element of V, different from 0 and c, then for some $x \in V$ $\{x, x + c, x + d\} \subseteq A$, so x is a_1 or a_2. This gives that for every $d \in V - \{0, c\}$, $a_1 + d$ or $a_2 + d$ is in A, i.e. $(A + (-a_1)) \cup (A + (-a_2))$ covers $V - \{c\}$. If now a is any element different form $-a_1$, $-a_2$, then $A + a = ((A + a) \cap (A + (-a_1))) \cup ((A + a) \cap (A + (-a_2)))$ plus possibly one element. This gives that A has at most 5 elements, but then it certainly has two disjoint copies. $\square$

I have been unable to find these results in the literature, though I don't think I proved them first.

There is a dual question when we want that A and the translations of $V - A$ intersect in a small set.

Theorem 1.6 (St. Banach [2]). *If V is a vector space over Q of cardinality $\kappa > \omega$, then there exists a disjoint decomposition $V = A \cup B$, $|A| = |B| = \kappa$ such that $|(A + a) \cap B| < \kappa \; (a \in V)$. For $\kappa = \omega$ this is not true.*

Proof. Assume first that $\kappa > \omega$. Let $\{b_\alpha : \alpha < \kappa\}$ be a basis, for $v = r_1 b_{\alpha_1} + \ldots + r_n b_{\alpha_n}$ with $\alpha_1 < \ldots < \alpha_n$ put v into A iff α_n is an even ordinal. Given $a \in V$, for all but $< \kappa$ vectors x, the last co-ordinate of x and $a + x$ are the same, so $x \in A$ iff $a + x \in A$.

Assume that $\kappa = \omega$, and $V = A \cup B$ is a decomposition as in the statement. Fix a non-zero vector, a, from V. The different sets of the form $b + \mathbb{Z}a = \{\ldots, b - 2a, b - a, b, b + a, b + 2a, \ldots\}$ disjointly cover V. If a $b + \mathbb{Z}a$ meets both A and B, then it adds at least one element to $(A + a) \cap B$ or $(A + (-a)) \cap B$. There is, therefore, a $b + \mathbb{Z}a$, entirely in, say, A. Fix it, as $b' + \mathbb{Z}a$. For $b \in V$, as $((b - b') + A) \cap B$ is finite, $b + \mathbb{Z}a$ contains only finitely many elements from B. If $B \cap (b + \mathbb{Z}a)$ is non-empty, $b + \mathbb{Z}a$ contains an element from $((b - b') + A) \cap B$, so only for finitely many b does $b + \mathbb{Z}a \neq \emptyset$ hold, i.e., B is finite.   □

Theorem 1.7 (L. Trzeciakiewicz [43]). *If V is a vector space over Q, $V = A \cup B$, $A \cap B = \emptyset$, $|V| = \kappa > \lambda \geq \omega$, and $|A|, |B| \geq \lambda$, then there is an $a \in V$ such that $|(A + a) \cap B| \geq \lambda$.*

Proof. Select $A' \subseteq A$, $B' \subseteq B$ with $|A'| = |B'| = \lambda$. Choose λ^+ different vectors, $\{x_\alpha : \alpha < \lambda^+\} \subseteq V$. We are done, if either $A' + x_\alpha \subseteq B$ or $B' + x_\alpha \subseteq A$ for some $\alpha < \lambda^+$. In the other case, for every α there exist $a_\alpha \in A'$, $b_\alpha \in B'$ such that $a_\alpha + x_\alpha \in A$, $b_\alpha + x_\alpha \in B$. There is a set $X \subseteq \lambda^+$ of size λ^+ such that $a_\alpha = a$, $b_\alpha = b$ for $\alpha \in X$. Then $b + x_\alpha \in (A + (b - a)) \cap B$ so this latter set has at least λ^+ elements.   □

We mention that the last two theorems were later independently discovered by P. Lax and P. Erdős. Some extensions to groups are given in [36].

2. Closed sets or Better Well-Orderings

In the examples of Section 1 we used some conditions of finite character like non-collinearity, etc., which made sure that the construction can get stuck only in successor steps. In some other cases this does not work. Assume, for example, that one wants to color R^2 with countably many colors, with no two monocolored points spanning some rational, nonzero distance. By determining the colors along a transfininte recursion it may happen that a certain point cannot get any color, because for every color there is a point with that color, and of rational distance from the point. The rescue is that we are going to find a better transfinite enumeration of R^2 such that in that order no point is of rational distance of infinitely many earlier points. The technique for establishing the transfinite order is even more interesting and important.

The forthcoming arguments are based on the following statement. Assume that X is an infinite set, $f_0, f_1, \ldots$ are finitary functions, i.e. each with a finite arity, from X to X. A subset $Y \subseteq X$ is *closed*, if it is closed under the f_i's in the algebraic sense, that is, when f_i is n-ary, and $x_1, \ldots, x_n \in Y$ then $f_i(x_1, \ldots, x_n) \in Y$.

Lemma 2.1 (a) *If $Y \subseteq X$ is infinite, then there is a closed Z, $Y \subseteq Z \subseteq X$ with $|Z| = |Y|$.*

(b) If $\kappa = |X| > \omega$, then there exists a continuous, increasing decomposition of X into closed X_α's, i.e. $X = \bigcup\{X_\alpha : \alpha < \kappa\}$ such that $|X_\alpha| < \kappa$ for $\alpha < \kappa$.

Proof. (a) Define $Z = \mathrm{cl}(Y) = Y_0 \cup Y_1, \ldots$ where $Y_0 = Y$, and $Y_{t+1} = \{f_i(x_1, \ldots, x_n) : x_1, \ldots x_n \in Y_0 \cup \ldots \cup Y_t, f_i \ n-ary\}$. As the f_i's are finitary, Z is closed under them, and it is also easy to see that it is the smallest closed set containing Y. By induction on $t < \omega$, one can see that $|Y_t| = |Y|$, and so $|Z| = |Y|$.

(b) Enumerate X as $\{x_\alpha : \alpha < \kappa\}$ and take $X_\alpha = \mathrm{cl}(\{x_\beta : \beta < \alpha\})$.   □

By counting the f_i's by repetition, if needed, we can extend Lemma 2.1 to the case when f_i assigns a *countable* set to *every* finite set.

Theorem 2.2 (P. Erdős – A. Hajnal [18]). *There is a decomposition $R^2 = \bigcup\{S_i : i < \omega\}$ such that no S_i spans a nonzero, rational distance.*

Proof. We are going to show *by transfinite induction* on $\kappa = |X|$ that every $X \subseteq R^2$ can be so decomposed. The case $\kappa \leq \omega$ is obvious, we can put the points into different S_i's.

Assume that $\kappa > \omega$ and we know the above statement for every cardinal smaller than κ. We notice that if $x_0 \neq x_1$ are in R^2 then the set $\{y \in R^2 : d(x_0, y), d(x_1, y) \in Q\}$ is countable. This follows from the countability of Q and the fact that if $q_0, q_1 \in Q$ then there are at most 2 y's with $d(x_0, y) = q_0$ and $d(x_1, y) = q_1$. There is, therefore a function f_0 assigning these y's to x_0, x_1 and we can use the machinery just set up.

Let $X = \bigcup\{X_\alpha : \alpha < \kappa\}$ be a decomposition as in Lemma 2.1(b) with the X_α's closed under f_0. We are going to 'color' X_α, i.e. determine $X_\alpha \cap S_i(i = 0, 1, \ldots)$ by transfinite recursion on α. Put $Y_\alpha = X_{\alpha+1} - X_\alpha$ then $X = \bigcup\{Y_\alpha : \alpha < \kappa\}$ is a *disjoint* decomposition of X. As X_α is closed, no $y \in Y_\alpha$ is in rational distance from two or more points in X_α. Also, as $|Y_\alpha| \leq |X_{\alpha+1}| < \kappa$ we can, by the transfinite inductive hypothesis, color Y_α by countably many colors.

This coloration, however, can confront with the coloration of X_α, as for $y \in Y_\alpha \cap S_i$ there may be an $x \in X_\alpha$, with $d(x, y) \in Q$ which was also put into S_i. But, given $y \in Y_\alpha$, there can only be just *one* bad $x \in X_\alpha$, and so there can only be just one bad $i < \omega$. So, if $y \in Y_\alpha$ is colored by color n by the 'naive' coloration, assign *two* colors, $2n$ and $2n + 1$ to y. As we started with a coloration of Y_α, for every $y \in Y_\alpha$ we can use either of the colors assigned to it. Now select the one not confronting the color of an $x \in X_\alpha$.   □

It is possible to give a more transparent extension of Theorem 2.2.

Theorem 2.3 (Erdős – Hajnal). *There is a well-ordering, $<$, of R^2, such that for every $x \in R^2$, the set $W(x) = \{y < x : d(x, y) \in Q\}$ is finite.*

Before proving the result, we notice that it trivially implies Theorem 2.2. One can color R^2 as described in Theorem 2.2. by a transfinite recursion along $<$.

Proof. It suffices to show by induction on $|X|$ that every $X \subseteq R^2$ possesses such a well-ordering. For $|X| \leq \omega$, any ordering into type $\leq \omega$ will do. For $|X| > \omega$ take a decomposition as in the proof of Theorem 2.2, let $<_\alpha$ be a well-ordering of Y_α satisfying our inductive assumption and let $x < y$ if $x, y \in Y_\alpha$ for some α and $x <_\alpha y$ or if $x \in Y_\alpha$ and $y \in Y_\beta$ for some $\alpha < \beta$. Observe that if $x \in Y_\alpha$, $W(x)$ can be only one larger than the corresponding $W_\alpha(x)$ inside Y_α. □

Our following example is an interesting result of G. Gruenhage.
Let $\mathcal{L}$ denote the set of the straight lines of R^2.

Theorem 2.4 (Gruenhage [23]). *There exist functions F, G such that for every $x \in R^2$, $F(x)$ is a set of finitely many lines through x, for $L \in \mathcal{L}$, $G(L)$ is a finite subset of L, and if $x \in L$, then either $x \in G(L)$ of $L \in F(x)$.*

Proof. We are going to work with the two-sorted algebraic structure $(R^2, \mathcal{L})$. Let f_0, f_1 be the following functions. If $x_0 \neq x_1$ are in R^2, $f_0(x_0, x_1)$ is the line passing through them. If $L_0, L_1 \in \mathcal{L}$, $L_0 \neq L_1$, then $f_1(L_0, L_1)$ is the point in their intersection, if that intersection is nonempty. For the other values, f_0, f_1 can be arbitrary.

Again, by induction on the cardinality, we show for every $(X, \mathcal{F})$ $(X \subseteq R^2$, $\mathcal{F} \subseteq \mathcal{L})$ the existence of F, G as wanted. If $|X| + |\mathcal{F}| \leq \omega$, enumerate them as $X = \{x_0, x_1, \ldots\}$ and $\mathcal{L} = \{L_0, L_1, \ldots\}$ and put

$$L_i \in F(x_j), \text{ if } i \leq j \text{ and } x_j \in L_i \,;$$
$$x_i \in G(L_j), \text{ if } i < j \text{ and } x_i \in L_j \,.$$

It clearly works.

If $\kappa = |X| + |\mathcal{F}| > \omega$, let X_α, $\mathcal{F}_\alpha$ be the decomposition given by Lemma 2.1(b). Put $Y_\alpha = X_{\alpha+1} - X_\alpha$, $\mathcal{G}_\alpha = \mathcal{F}_{\alpha+1} - \mathcal{F}_\alpha$. As $(Y_\alpha, \mathcal{G}_\alpha)$ is of size $< \kappa$, by our inductive hypothesis there are appropriate F_α, G_α on $Y_\alpha, \mathcal{G}_\alpha$. If $x \in Y_\alpha$, as $\mathcal{F}_\alpha$ is closed under f_1, there can be at most one $L \in \mathcal{F}_\alpha$ such that $x \in L$. Take $F(x) = F_\alpha(x) \cup \{L\}$. If $L \in \mathcal{G}_\alpha$, as X_α is closed under f_0, there can be at most one $x \in X_\alpha$ with $x \in L$. Expand $G_\alpha(L)$ by this x to get $G(L)$. It is easy to check that F, G are as wanted. □

If 2^ω is small, the result can be sharpened.

Theorem 2.5 (F. Galvin, G. Gruenhage [22], see also [6]). *If $2^\omega = \omega_n$ $(n = 1, 2, \ldots)$, then there exist F, G as in Theorem 2.4 with $|F(x)| \leq n + 1$ for $x \in R^2$.*

Proof. It suffices to show that if $(X, \mathcal{F})$ is as in the proof of Theorem 2.4, and $|X| + |\mathcal{F}| \leq \omega_n$, then F, G exist with $|F(x)| \leq n + 1$ for $x \in X$. As in the induction from $n - 1$ to n we add only one element to F, it suffices to check the case $n = 0$. Modify the construction to

$$F(x_i) = \{L_j\} \text{ if } L_j \text{ is the first line containing } x_i,$$
$$G(L_j) = (L_0 \cap L_j) \cup \ldots \cup (L_{j-1} \cap L_j) \quad (\text{of size } \leq j).$$

If now $x_i \in L_j$ then either x_i is not in $L_0 \cup \ldots \cup L_{j-1}$ and so $F(x_i) = \{L_j\}$, or, in the other case $x_i \in G(L_j)$. $\qquad\square$

W. Sierpiński [37, 40] proved that CH (the continuum hypothesis) is equivalent to the existence of a decomposition $R^2 = A \cup B$ such that the intersection of A with every 'horizontal' line, as well as the intersection of B with every 'vertical' line is countable. R.O. Davies gave several variants and extensions of this result. We present some of them. As follows, let $\theta_0, \theta_1, \ldots$ be a finite or countable sequence of distinct planar directions, and let $\mathcal{L}_i$ be the set of those lines with direction θ_i.

Theorem 2.6 (R.O. Davies [13]) *If $\theta_0, \ldots, \theta_{n+1}$ are directions, $2^\omega \leq \omega_n$ then there is a decomposition $R^2 = S_0 \cup \ldots \cup S_{n+1}$ such that if $L \in \mathcal{L}_i$ then $|L \cap S_i| < \omega$.*

Proof. Let F, G be the functions given by Theorem 2.5. For $x \in R^2$, as $|F(x)| \leq n + 1$, there is an i such that $F(x) \cap \mathcal{L}_i = \emptyset$. Put then x into S_i. Obviously, if $L \in \mathcal{L}_i$, $x \in L - G(L)$, then $x \notin S_i$. $\qquad\square$

Theorem 2.7 (R.O. Davies [13], see also [6]). *If $2^\omega > \omega_n$, no decomposition as in Theorem 2.6 exists.*

Proof. Let e_i be the unit vector of direction θ_i. Assume that $R^2 = S_0 \cup \ldots \cup S_{n+1}$ is a decomposition as in Theorem 2.6. If $0 \leq i \leq n + 1$, and $x_j \in R$ $(0 \leq j \leq n + 1, j \neq i)$, then the line

$$\{x_0 e_0 + \ldots + x_{i-1} e_{i-1} + y e_i + \ldots + x_{n+1} e_{n+1} : y \in R\}$$

is in $\mathcal{L}_i$, so for only finitely many y's, these elements are in S_i. Let $F_i(x_0, \ldots, x_{i-1}, \ldots, x_{n+1})$ be the set of those y's. As $|R| > \omega_n$, we can choose $Z_i \subseteq R$ with $|Z_i| = \omega_i$ $(0 \leq i \leq n)$. Select $a_{n+1} \in R - \{F_{n+1}(x_0, \ldots, x_n) : x_i \in Z_i\}$, possible, as $|R| > |Z_0| \cdot \ldots \cdot |Z_n|$. Then, by reverse induction, choose

$$a_i \in Z_i - \{F_i(x_0, \ldots, x_{i-1}, a_{i+1}, \ldots, a_{n+1}) : x_0 \in Z_0, \ldots, x_{i-1} \in Z_{i-1}\}.$$

Finally, put $b = a_0 e_0 + \ldots + a_{n+1} e_{n+1}$. If $b \in S_i$, then $a_i \in F_i(a_0, \ldots, a_{i-1}, a_{i+1}, \ldots, a_{n+1})$, a contradiction. $\qquad\square$

Theorem 2.8 *If* $2^\omega \leq \omega_n$, $\theta_0, \ldots \theta_n$ *are directions, there exists a decomposition* $R^2 = S_0 \cup \ldots \cup S_n$ *such that* $|L \cap S_i| \leq \omega$ *for* $L \in \mathcal{L}_i$. *For* $2^\omega > \omega_n$ *this is false.*

Proof. Modify the proof of Theorem 2.5. to give $|F(x)| \leq n$ and $|G(L)| \leq \omega$. The first case is when $|X| + |\mathcal{F}| \leq \omega_1$, the appropriately modified proof works.

For the reverse direction, modify the proof of Theorem 2.7.   □

For infinitely may directions the following is known.

Theorem 2.9 (R.O. Davies [14]). *If* $\theta_0, \theta_1, \ldots$ *are distinct planar directions, then there is a decomposition* $R^2 = S_0 \cup S_1 \ldots$ *such that if* L *is a line of direction* θ_i *then* $|L \cap S_i| \leq 1$.

Proof. It suffices to confirm the existence of a well-ordering $<$ of R^2 such that if $x \in R^2$, then there are only finitely many lines through x of some directions θ_i which contain some earlier (by $<$) points. Then, put x into S_i for an i such that no L through x of direction θ_i has earlier points. This obviously works, we even didn't need that $<$ is *well*-ordering.

We show the existence of $<$ for every $X \subseteq R^2$, by induction on $|X|$. The case $|X| \leq \omega$ is easy, any well-ordering into type $\leq \omega$ will work. For $|X| = \kappa > \omega$ we can decompose X as $\cup\{X_\alpha : \alpha < \kappa\}$ such that $x \in X_\alpha$ provided that there exist $y, y' \in X_\alpha$, and L, L' of some directions θ_i, θ_j, with $y, x \in L$, $y', x \in L'$. This can be done, using Lemma 2.1 (b), by appropriate functions f_i. If $x \in Y_\alpha = Y_{\alpha+1} - X_\alpha$, x is contained in at most one line of direction θ_i (some i), having points in X_α. As $|Y_\alpha| < \kappa$, we have a well-ordering for Y_α, we can fuse them as in Theorem 2.3.   □

These results can be generalized to R^n. We mention the generalization of the last theorem. We notice that perhaps the best description of the method presented here is given in the paper containing the following result.

Theorem 2.10 (R.O. Davies [16]). *Let* $P_0, P_1, \ldots$ *be proper hyperplanes in* R^n *such that*

(*) $\quad P_t \cap P_{t+1} \cap \ldots \; = \{0\}$ *for all* $t < \omega$.

Then there is a decomposition $R^n = S_0 \cup S_1 \cup \ldots$ *such that* $|S_i \cap P| \leq 1$ *whenever* P *is parallel to* P_i. *Condition* (*) *is also necessary for such a decomposition.*

Proof. We can thin out $P_0, P_1, \ldots$ to a subsequence in which the intersection of any n members is $\{0\}$. We now proceed as in Theorem 2.9.

For the other direction assume that $P_{p-1} \cap P_p \cap \ldots \neq \emptyset$ for some $p < \omega$, and the statement still holds. Selecting lines from $P_0, P_1, \ldots P_{p-2}$ and $P_{p-1} \cap P_p \cap \ldots$ we get directions $\theta_0, \ldots, \theta_{p-1}$ and $R^n = S_0^* \cup \ldots S_{p-2}^* \cup S_{p-1}^*$

where $S_0^* = S_0, \ldots, S_{p-2}^* = S_{p-2}, S_{p-1}^* = S_{p-1} \cup S_p \cup \ldots$ and $|S_i^* \cap L| \leq 1$ if L is a line of direction θ_i. We can select vectors e_i of direction θ_i such that all the vectors in

$$X = \left\{ \sum_{i=0}^{p-1} k_i e_i : \quad 0 \leq k_i \leq p, \quad k_i \text{ integer} \right\}$$

are different. For example, if $|e_i| > p(|e_0| + \ldots + |e_{i-1}|)$. Then, for every $i < p$, S_i^* can cover at most $\frac{1}{p+1}|X|$ points of X, as no two can differ only at the i'th co-ordinate. But then $S_0^* \cup \ldots \cup S_{p-1}^*$ cannot cover X.   □

3. Extending the Coloring More Carefully

Having learnt the technique in Section 2 we seek for proofs requiring further tricks.

Theorem 3.1 (J. Ceder [8]). *There is a decomposition $R^2 = S_0 \cup S_1 \cup \ldots$ with no S_i spanning an equilateral triangle.*

Proof. Again, we are going to prove that every $X \subseteq R^2$ possesses such a decomposition. There is no problem if $|X| \leq \omega$.

For $|X| = \kappa > \omega$, let $X = \bigcup \{X_\alpha : \alpha < \kappa\}$ be a decomposition as in Lemma 2.1 by some functions $f_0, \ldots$ not yet fixed. It is reasonable to assume that $f_0(x, x\prime) = \{y, y'\}$ where y, y' are the two points forming equilateral triangles with x, x'. Let $S_0, S_1, \ldots$ be a decomposition which is good for every $Y_\alpha = X_{\alpha+1} - X_\alpha$ and assume that p, q, r from an equilateral triangle in S_i. Assume that α is the least ordinal that $p, q, r \in X_{\alpha+1}$, i.e. say $r \in X_{\alpha+1} - X_\alpha$, and $p, q, \in X_{\alpha+1}$. $p, q \in Y_\alpha$ is ruled out as p, q, r form a bad triangle. So assume that $p \in X_\alpha$. If $q \in X_\alpha$, then $r \in f_0(p,q) \subseteq X_\alpha$, a contradiction. So we are stuck with the situation that $p \in X_\alpha$, $q, r \in Y_\alpha$. If CH is assumed, $|Y_\alpha| \leq \omega$ so we can color Y_α with different colors, and so this case does not occur.

We call a pair $\{q, r\}$ *dangerous*, if there exists a $p \in X_\alpha$ as above, and observe that it suffices to show that the graph of dangerous edges on Y_α is $\leq \omega$-chromatic (see the graph theoretic notions in [7]). We show by adding more functions to the f_i's we can actually make it *two-chromatic*. For this, we must eliminate all odd circuits (see [7], p. 6), so assume that $\{q_0, q_1, \ldots, q_{2s}\}$ is one. By the definition of danger, there are $p_i \in X_\alpha$ ($0 \leq i \leq 2s$) such that q_i, q_{i+1}, p_i form an equilateral triangle. By another formulation there exist rotations φ_i by $\pm 60°$ around p_i such that q_0 is a fixed point of $\varphi_0 \varphi_1 \ldots \varphi_{2s}$. The composition of odd many rotations by $\pm 60°$ is a rotation again, so q_0 and therefore $q_1, \ldots, q_{2s}$ can be "computed" from $p_0, \ldots, p_{2s}$ and the signs of the rotations. If $f_{2s+1}(p_0, \ldots, p_{2s})$ contains them, then the dangerous graph will, in fact, be two-chromatic.   □

For completeness' sake we sketch Ceder's original proof. We start with another result.

Theorem 3.2 (R. Rado). *Every vector space V over Q is the union of countably many 3-element-arithmetic-progression free sets.*

Proof. Let $\{b_i : i \in I\}$ be a basis of V, with some ordered index-set I. (The axiom of choice ensures the existence of even a well-ordered basis.) Every $x \in V$, $x \neq 0$ can uniquely be written as

$$x = \sum_{t=1}^{n} r_t b_{i_t}$$

with $i_1 < \ldots < i_n$, $r_t \neq 0$, $r_t \in Q$. Put two x's into the same part if the two ordered strings $(r_1, \ldots, r_n)$ agree. Vector 0 forms a part itself.

Assume that $p + q = 2s$, and p, q, s are in the same part, the one with string $(r_1, \ldots, r_n)$. Let i be the least index which occurs in p, q, or s. Then the co-ordinates in this index are r_1 or 0. As $p + q = 2s$, this equality holds in this co-ordinate, so all are r_1. Continuing, we get that the next index is r_2 and so on. So $p = q = s$. □

A moment's reflection shows that this argument works for every vector space over a *countable* field f, if $\alpha p + (1 - \alpha)q = s$ is excluded, with some $\alpha \in f$, $\alpha \neq 0, 1$. But in R^2 p, q, s form an equilateral triangle if and only if the above equation holds with $f = Q(\sqrt{3}i)$ and $\alpha = \frac{1}{2} \pm \frac{\sqrt{3}}{2}i$.

This argument works if an $n + 2$-element configuration is excluded in R^n. P. Erdős conjectured that an equilateral triangle can be excluded in every R^n. Ceder's argument doesn't seem to work in higher dimensions. An argument similar to the one given in the proof of Theorem 3.1, but much more involved gives a decomposition of R^3 into countably many pieces none containing a regular tetrahedron. [26]

Added in Proof. This was recently extended to R^n by J. Schmerl.

P. Erdős further conjectured that even isosceles triangles can be omitted as in Theorem 3.1. We have only been able to show

Theorem 3.3 (P. Erdős – P. Komjáth [21]). *There is an $f : R^2 \to \omega$ such that if C is a circle with centre x, then $\{y \in C : f(y) = f(x)\}$ is finite.*

Proof. Similarly to Theorems 2.2 and 2.3 it suffices to show that there are a well-ordering, $<$, of R^2, and functions F, G such that for every circle C, $F(C)$ is a finite subset of C, for every $x \in R^2$, $G(x)$ is a finite subset of R^2, and if x is the centre of circle C, and $y \in C - F(C)$, then $x < y$ and $x \in G(y)$.

Once this is done, we can define $f(x)$ by a transfinite recursion along $<$, such that if $y < x$, $y \in G(x)$, then $f(x) \neq f(y)$. This suffices as if C is a circle with centre x and $y \in C - F(C)$ then $f(y) \neq f(x)$.

As in Theorem 2.4, we are working with two-sorted structures of type $(X, \mathcal{C})$ where $X \subseteq R^2$ and $\mathcal{C}$ is a set of circles. $(X, \mathcal{C})$ is *closed*, if

(3.1) if $C \in \mathcal{C}$ then the centre of C is in X;

(3.2) if $C_1 \neq C_2$ are in $\mathcal{C}$ then the points of $C_1 \cap C_2$ are in X;

(3.3) $C \in \mathcal{C}$ provided that three points of C are in X.

It is easy to find finitary functions $f_0, f_1, \ldots$ such that $(X, \mathcal{C})$ is closed in the above sense if and only if it is closed under those functions.

For $(X, \mathcal{C})$ countable, well-order X as $\{x_0, x_1, \ldots\}$ $\mathcal{C}$ as $\{C_0, C_1, \ldots\}$ and set

(3.4) $F(C_i) = \{x_j : j \leq t, \ x_j \in C_i, \ x_t \text{ is the centre of } C_i\}$;

(3.5) $G(x_i) = \{x_j : j < i\}$.

This clearly works.

If $|X| + |\mathcal{C}| = \kappa > \omega$, decompose $X = \bigcup\{X_\alpha : \alpha < \kappa\}$, $\mathcal{C} = \bigcup\{\mathcal{C}_\alpha : \alpha < \kappa\}$, take $<_\alpha$, F_α, G_α for $X_{\alpha+1} - X_\alpha$, $\mathcal{C}_{\alpha+1} - \mathcal{C}_\alpha$ and fuse the well-orders as in Theorem 2.3. If $x \in X_{\alpha+1} - X_\alpha$, by (3.2) there can be only one $C \in \mathcal{C}_\alpha$ with $x \in C$, and the centre of C, say y must be in X_α, we set $G(x) = G_\alpha(x) \cup \{y\}$. If $C \in \mathcal{C}_{\alpha+1} - \mathcal{C}_\alpha$, by (3.3) at most two points of C may be in X_α, say y_1 and y_2, and we can set $F(C) = F_\alpha(C) \cup \{y_1, y_2\}$. If now $C \in \mathcal{C}$ has centre x, and $y \in C - F(C)$, and $y \in X_{\alpha+1} - X_\alpha$, $C \in \mathcal{C}_{\alpha+1} - \mathcal{C}_\alpha$, then $x < y$ and $x \in G_\alpha(y) \subseteq G(y)$. If C occurs 'earlier' then $x \in G(y)$, by the definition. If C occurs 'later', then y is put into $F(C)$. $\square$

Another construction where a further trick is needed beyond the technique of Chap. 2 is the 3-dimensional analog of Theorem 2.2. One could guess that if x_0, x_1, $x_2 \in R^3$ are different points then the set $Z = \{y \in R^3 : d(x_i, y) \in Q, i < 3\}$ is countable, and the same proof goes through. This statement is, however, false. Namely, if x_0, x_1, x_2 are three appropriately selected points on a line, then Z contains a full circle. It is actually the disjoint union of countably many circles. As every circle is of size 2^ω we cannot add full circles to our closed X's. The idea is that we do not, we only reserve a rich set of colors for the points of the circle when it appears.

Theorem 3.4 (P. Komjáth [21]). *There is an $f : R^3 \to \omega$ such that $d(x, y) \notin Q$ when $f(x) = f(y)$ and $x \neq y$.*

Proof. A subset $E \subseteq R^3$ is called a *combinatorial line*, if there is a line L, and a point $p \notin L$ such that $E = \{x \in L : d(x, p) \in Q\}$. Clearly, $|E| = \omega$, and two distinct points are contained in only countably many combinatorial lines. This latter observation makes possible to find functions, such that Lemma 2.1 applies to get closed sets, where a set $X \subseteq R^3$ is *closed*, if

(3.6)   if E is a combinat. line, $|E \cap X| \geq 2$, then $E \subseteq X$;

(3.7)   if $x_0, x_1, x_2 \in X$ are not collinear, $d(y, x_i) \in Q$ $(i < 3)$, then $y \in X$.

We use the well-known notion of zero (one) density sets. A set $A \subseteq \omega$ is of density zero (one), if

(3.8) $$\lim_{n \to \infty} \frac{|A \cap \{0, 1, \ldots, n-1\}|}{n} = 0 \text{ (or 1)}.$$

We prove the following statement by transfinite induction on $|X|$. If X is closed, φ assigns a density one subset of ω to every point of X, then there is a mapping $f : X \to \omega$ such that $f(x) \in \varphi(x)$ $(x \in X)$, $d(x, y) \in Q$ implies $f(x) \neq f(y)$, and if $E \subseteq X$ is a combinatorial line, then $f(E)$ is of zero density. This clearly suffices, as the choice $X = R^3$, $\varphi(x) = \omega$ $(x \in X)$ gives the result.

If X is countable, a one-to-one function with $f(X)$ of zero density can easily be selected.

If $|X| = \kappa > \omega$, decompose $X = \bigcup \{X_\alpha : \alpha < \kappa\}$ into closed sets, and color $Y_\alpha = X_{\alpha+1} - X_\alpha$ by transfinite recursion on α. Assume that $f_\alpha = f|X_\alpha$ is already given. For $x \in Y_\alpha$, put

(3.9) $\varphi_\alpha(x) = \varphi(x) - \{f_\alpha(y) : y \in X_\alpha, \ d(x, y) \in Q\}$.

As X_α is closed, $\{y \in X_\alpha : d(x, y) \in Q\}$ is either of size ≤ 1 or is a combinatorial line. $\varphi_\alpha(x)$ is therefore, a set of density one. Applying the inductive hypothesis, we get $g_\alpha = f|Y_\alpha$ to Y_α, φ_α. It sufficies to show that if $E \subseteq X$ is a combinatorial line, then $f(E)$ is of zero density. There is an ordinal $\alpha < \kappa$, such that $|E \cap X_\alpha| \leq 1$ but $|E \cap X_{\alpha+1}| \geq 2$. Then, as $X_{\alpha+1}$ is closed, $E \subseteq X_{\alpha+1}$, and E is handled at the αth step.   □

In the above proof combinatorial lines and circles were regarded as small, resp. large sets in the sense that they had a zero resp. one density set of colors. In higher dimensions the same sets become large and small in the same time. Much more involved ideas are therefore used in the extension of this theorem to every R^n (see [27]).

4. The Use of the Continuum Hypothesis

The assumption of continuum hypothesis, i.e. that $|R| = \omega_1$ radically simplifies the situation. If, for example, one wants to color R^n with countably many colors omitting some monocolored configurations, we basicly have just one case 'of the induction'. Namely, when R^n is decomposed into countable X_α's. It is reasonable to require, that f, the promised coloring, be one-one on every Y_α. This already implies that every monochromatic configuration can have only one point per Y_α.

The following unpublished result of Kunen answered a question of P. Erdős.

Theorem 4.1 (K. Kunen). *If CH holds, then there is a coloring $f : R^2 \to \omega$ with no monochromatic triangle with rational, nonzero area.*

Proof. Let again $\mathcal{L}$ be the set of planar lines. We decompose, using Lemma 2.1(b), $(R^2, \mathcal{L})$ as $R^2 = \bigcup\{X_\alpha : \alpha < \omega_1\}, \mathcal{L} = \bigcup\{\mathcal{L}_\alpha : \alpha < \omega_1\}$, such that $|X_\alpha| + |\mathcal{L}_\alpha| = \omega$ and the structures $(X_\alpha, \mathcal{L}_\alpha)$ are closed in the following sense. Let $t(x, y, z)$ denote the area of the triangle with nodes x, y, z.

(4.1)   if $L_0, L_1 \in \mathcal{L}_\alpha$, $L_0 \neq L_1$, then $L_0 \cap L_1 \subseteq X_\alpha$;

(4.2)   if $x_0 \neq x_1$ are in X_α, then for every $r \in Q$, the line $\{y \in R^2 : t(x_0, x_1, y) = r\}$ is in $\mathcal{L}_\alpha$ (it is actually two lines if $r \neq 0$, but that doesn't matter);

(4.3)   if $L \in \mathcal{L}_\alpha$, then $|L \cap X_\alpha| = \omega$.

Certainly, we promise f to be one-one on each Y_α. We are going to construct $f|X_\alpha$ by transfinite recursion on α. To start, let $f|X_0$ be an arbitrary one-one function. If $f|X_\alpha$ is already defined, by (4.1), for every $y \in Y_\alpha$, there can be only one $L \in \mathcal{L}_\alpha$ passing through y. Then select $f(y)$ to be an element of $f(L \cap X_\alpha)$. This is possible by a diagonalization once we prove that $f(L \cap X_\alpha)$ is infinite. But the least α for which X_α contains 2 elements of L satisfies that $|L \cap X_\alpha| = \omega$, and on the infinitely many elements added by X_α, f is one-one.

Now assume that $t(x_0, x_1, x_2) = r \neq 0$, $r \in Q$, $f(x_0) = f(x_1) = f(x_2)$, $x_0, x_1, x_2 \in X_{\alpha+1}$, with α least. Let, for example $x_0, x_1 \in X_\alpha$, $x_2 \in Y_\alpha$. Then, by the choice of $f(x_2)$, there is a $y \in X_\alpha$, with $f(y) = f(x_2)$ and $t(x_0, x_1, y) = r$. Then, $\{x_0, x_1, y\}$ is monocolored, and as $r \neq 0$, it is a proper triangle, contradicting to the minimality of α. $\square$

It is probable that Theorem 4.1 holds without any assumption. This has not yet been proved. In the following proof we use a similar idea but with circles rather than lines.

Theorem 4.2 *If CH hold, then there is a coloring $f : R^2 \to \omega$ with no four distinct, monocolored points x, y, z, t such that $d(x, y) = d(z, t)$.*

Proof. Decompose $R^2 = \bigcup\{X_\alpha : \alpha < \omega_1\}$ such that

(4.4) $x \in X_\alpha$ provided that there are y, $y' \in X_\alpha$ such that $d(x, y)$, $d(x, y')$ occur as distances in X_α.

If we let $f_0(y, y', u_1, v_1, u_2, v_2)$ to be the set of those x's for which $d(x, y) = d(u_1, v_1)$ and $d(x, y') = d(u_2, v_2)$ then Lemma 2.1(b) will do the job needed.

To define the coloring on $X_{\alpha+1}$, for every $y \in Y_\alpha$ there can only be one $x \in X_\alpha$ such that with some $u, v \in X_\alpha$ u, v, x, y would form a forbidden quadruple, and this gives only one forbidden color for y. We can, therefore, easily extend f to Y in a one-one fashion.   □

Quite interestingly, the argument becomes considerably harder if we allow coincidences, i.e. if we exclude isosceles triangles. The reason is that under any closure construction it may happen that an element $y \in Y_\alpha$ may get no color, as there exist $x_i, x_i' \in X_\alpha$ such that $f(x_i) = f(x_i') = i$ and $d(x_i, y) = d(x_i', y)$. We must somehow exclude the possibility of this case.

Theorem 4.3 *If CH holds, there is a coloring $f : R^2 \to \omega$ with no monochromatic isosceles triangles.*

Proof. We are going to build the decomposition $(R^2, \mathcal{L}) = \bigcup \{(X_\alpha, \mathcal{L}_\alpha) : \alpha < \omega_1\}$ such that

(4.5)   if $L \neq L'$ are in $\mathcal{L}_\alpha$ then $L \cap L' \subseteq X_\alpha$;

(4.6)   if $x, x' \in X_\alpha$, $x \neq x'$, then their perpendicular bisector $H(x, x') = \{y : d(x, y) = d(x', y)\}$ is in $\mathcal{L}_\alpha$;

(4.7)   if $x \in X_\alpha$, $L \in \mathcal{L}_\alpha$, then x reflected through L, in short $R(x, L)$ is in X_α;

(4.8) $y \in X_\alpha$ provided there are $x, x' \in X_\alpha$, such that $d(x, y), d(x', y)$ occur in X_α.

We try to define $f : R^2 \to \omega$ in such a way that

(4.9) $f|Y_\alpha$ is one-one;

(4.10) if $L \in \mathcal{L}_{\alpha+1} - \mathcal{L}_\alpha$, then for all but finitely many $y \in Y_\alpha$, $f(y) \neq f(R(y, L))$;

and of course

(4.11) no monochromatic isosceles triangle is formed.

To get (4.9), (4.10), enumerate Y_α as $\{y_0, y_1, \ldots\}$, $\mathcal{L}_{\alpha+1} - \mathcal{L}_\alpha$ as $\{L_0, L_1, \ldots\}$ and selecting $f(y_i)$ let it differ from $f(R(y_i, L_0)), \ldots, f(R(y_i, L_i)), f(y_0), \ldots, f(y_{i-1})$. Can we also meet (4.11)? Assume that y_i forms an isosceles triangle with some $x, x' \in X_\alpha$. If $d(x, x') = d(x', y_i)$, then by (4.8) x' and so $f(x')$ is determined by y_i so we can easily exclude $f(x') = f(y_i)$. If $d(x, y_i) = d(x', y_i)$ then y_i is on $L = H(x, x')$. We are done unless there are infinitely many z_i, $z_i' \in X_\alpha$ such that $f(z_i) = f(z_i') = n_i$ and $H(z_i, z_i') = L$. As $f(z_i) = f(z_i')$, by (4.9), z_i, z_i' are not in the same Y_β. If $L \in \mathcal{L}_{\beta+1} - \mathcal{L}_\beta$, then $z_i \in X_\beta$, $z_i' \in Y_\beta$ or vice versa. But we have taken care of all but finitely many z_i, z_i' by (4.10).   □

The last two results give

Corollary 4.4 (R.O. Davies [15]). *If CH holds, then R^2 is the union of countably many pieces such that in no piece a distance occurs twice.*

Or, as P. Erdős usually expresses, any 4 points determine 6 distances. The corresponding result for R had been proved by Erdős and Kakutani [20]. The extension to every R^n, a long-standing conjecture of Erdős, was finally proved by K. Kunen [30].

It is worth noticing, that the negation of CH implies the negation of the statement, even for R. This was observed by P. Erdős, who deduced this from the following result of Erdős–Hajnal. (For the graph theory notions see (7).)

Lemma 4.5 (Erdős–Hajnal). *If the edges of the complete bipartite graph with bipartition classes of size ω_1, ω_2 are colored by ω colors, then there is a monochromatic 4-circuit.*

Proof. Assume that the statement does not hold. Let the bipartition classes be $A = \{a_\alpha : \alpha < \omega_1\}$ and $B = \{b_\beta : \beta < \omega_2\}$. For every pair $a_\alpha, a_{\alpha'}$ in A, and color $i < \omega$, there can be at most one b_β such that the color of $\{a_\alpha, b_\beta\}$, $\{a_{\alpha'}, b_\beta\}$ both are i as two b_β's would give a 4-circuit of color i.

For all i's and all pairs together there are at most ω_1 such β's, so we can select a β that is not assigned to any pair. But then the edges from a_α to b_β (all $\alpha < \omega_1$) must get different colors, a contradiction, since there are only countably many colors.   □

Theorem 4.6. *If $2^\omega > \omega_1$, then for every decomposition $R = S_0 \cup S_1 \cup \ldots$ there are x, y, z, t in some S_i with $x - y = z - t$.*

Proof. As $2^\omega \geq \omega_2 = \omega_2 + \omega_1$ there are linearly independent (over Q) reals $\{e_\alpha, f_\beta : \alpha < \omega_1, \beta < \omega_2\}$. Color the bipartite graph on $\{e_\alpha\}$, $\{f_\beta\}$ with the color of the edge between e_α and f_β being i if $e_\alpha + f_\beta \in S_i$. By Lemma 4.5, there are $\alpha_0 < \alpha_1$, $\beta_0 < \beta_1$ such that $x = e_{\alpha_0} + f_{\beta_0}$, $y = e_{\alpha_0} + f_{\beta_1}$, $z = e_{\alpha_1} + f_{\beta_0}$, $t = e_{\alpha_1} + f_{\beta_1}$ are in the same S_i. As $x - y = z - t = f_{\beta_0} - f_{\beta_1}$, we are done.   □

By the above proof, if CH fails we get a counter-example to Theorem 4.2 but not to Theorem 4.3. This suggests, as is conjectured by P. Erdős, that Theorem 4.3 is true without any additional hypothesis. It is true for R, as then we only need to exclude 3-member arithmetic progressions, and that can be done, see Theorem 3.2.

Another result where the assumption of CH is essential is the following.

Theorem 4.7 (P. Erdős–P. Komjáth [21]). *CH is equivalent to the existence of a coloring $f : R^2 \to \omega$ with no monochromatic right angled triangles.*

We omit the involved proof, only give the argument of the 'only if' part. Assume that $2^\omega \geq \omega_2$ and $R^2 = S_0 \cup S_1 \cup \ldots$. Let $\{E_\alpha : \alpha < \omega_1\}$ $\{F_\beta : \beta < \omega_2\}$ be lines parallel to the x, resp. y axis. Color a bipartite graph as in Lemma 4.5. as follows. If $E_\alpha \cap F_\beta$ is in S_i, color the edge between E_α, F_β by i. If now $E_{\alpha_0}, E_{\alpha_1}, F_{\beta_0}, F_{\beta_1}$ form an i-colored 4-circuit, then the points in $E_{\alpha_0} \cap F_{\beta_0}$ $E_{\alpha_0} \cap F_{\beta_1}$, $E_{\alpha_1} \cap F_{\beta_0}$ and $E_{\alpha_1} \cap F_{\beta_1}$ give a monochromatic *rectangle*.

5. The Infinite Dimensional Case

In this section we investigate the Hilbert space H, where the elements are infinite vectors of the type $x = (x_0, x_1, \ldots)$ with $x_i \in R$, $\sum x_i^2 < \infty$. The distance between $x = (x_0, x_i, \ldots)$ and $y = (y_0, y_1, \ldots)$ is

$$d(x, y) = \sqrt{\sum_{i=0}^{\infty} (x_i - y_i)^2} .$$

(It is usually denoted by ℓ_2.) On the problems about H research was done by P. Erdős, K. Kunen, L. Pósa, and the author. [17]

Most of the results proved in earlier sections fail for the case of H. For the non-extension of Theorem 2.2 we have the following statement

Theorem 5.1 (folklore). *In H, there exist 2^ω points such that the distance between any two is rational.*

Proof. Take a countable set of orthonormal vectors. Rather than indexing them by 0, $1, \ldots$, we index them by the finite 0–1 sequences, $a_\emptyset$, a_0, a_{00}, a_1, a_{01}, a_{10}, a_{11}, $\ldots$. If s is a finite 0–1 sequence, let $|s|$ denote its length. For every infinite 0–1 sequence f, let

$$v(f) = \sum_{i=0}^{\infty} \alpha_n a_{f|n}$$

for some $\alpha_n \in R$ to be determined later, where $f|n$ denotes the string of the first n elements of f. If f, g are 0–1 sequences with first difference at the n'th place, then, as the a's are orthogonal, unit vectors, $d^2(f, g) = 2 \sum_{i > n} \alpha_i^2$. Then, $d(f, g) = \frac{1}{2^n}$, if

$$\alpha_n^2 = \frac{1}{2 \cdot 4^{n-1}} - \frac{1}{2 \cdot 4^n} . \qquad \square$$

Notice that the construction gives 2^ω points determining only isosceles triangles, and that provides the strong failure of Theorems 3.2 and 4.3. For equilateral triangles we have the following.

Theorem 5.2 (a) (K. Kunen) *If $H = S_0 \cup S_1 \cup \ldots$ then some S_i contains infinitely many points which are equidistant.*

(b) *It is consistent that 2^ω is arbitrary large and H is the union of ω_1 sets each omitting equilateral triangles.*

(c) *If MA_κ holds, then every $X \subseteq H$ of size κ is the union of ω sets each omitting equilateral triangles.*

Proof. (a) Modify the construction of Theorem 5.1 to strings of natural numbers rather than 0–1 strings. A decomposition of H transfers to a decomposition of all $\omega \rightarrow \omega$ functions. By the Baire category theorem, there are infinitely many functions in the same S_i which agree up to a certain natural number and get different values there. If MA_κ holds, the same argument works with κ many pieces.

(b), (c) The partial ordering which adds an equilateral-triangle-free ω-decomposition by finite parts is *ccc* (see (29)), as H is separable.   □

Theorem 5.3 *It is consistent that 2^ω is arbitrarily large and there exists 2^ω points that among any ω_1 of them an equilateral triangle can be found. The same holds if CH is assumed.*

Proof. We have to modify the argument in Theorem 5.2 (a). This is done by the usual Luzin set constructions. Even an infinite dimensional simplex can be found.

6. Large Paradoxical Sets in Another Sense

The results in the previous sections on decompositions of Euclidean spaces have the corollary that some part in the decomposition must be large in some sense. In real analysis, the most important notions of smallness are measure zero and first category. We remind the reader that a set in R^n is of *first category*, if it is the union of countably many nowhere dense sets. If it is not of first category, it is called of *second category*.

It follows, therefore, from Theorem 2.2 that there exist second category or positive outer measure sets in R^2 omitting rational distances. It is, actually, quite easy to show that for any predetermined configuration there exist very large sets in R^n omitting it. A set $X \subseteq R^n$ is called of *full outer measure*, if $\lambda(G \cap X) = \lambda(G)$ for every ball G in R^n (λ denotes the Lebesgue outer measure). It is *categorically dense*, if $G \cap X$ is of second category for every nonempty open set G.

Theorem 6.1 *In R^n, there exist full measure/categorically dense sets omitting rational distances.*

Proof. Enumerate the positive measure/second category Borel sets as $\{B_\alpha : \alpha < 2^\omega\}$. It suffices to find an $a_\alpha \in B_\alpha$ such that $A = \{a_\alpha : \alpha < 2^\omega\}$ omits rational distances. If $\{a_\beta : \beta < \alpha\}$ are already selected, the points that may not be chosen as a_α are in the union of $< 2^\omega$ spheres. We must, therefore, show that B_α cannot be covered by $< 2^\omega$ many spheres. This is done by an argument similar to the one given in Theorem 1.3. If B_α is of positive measure, there is a straight line, L, such that $L \cap B_\alpha$ is of positive (linear) measure. (By the Fubini theorem.) Each sphere meets L in at most two points, so they cannot cover the positive measure set (which is of size 2^ω). The dual argument works for the category case.   □

The above theorem can be proved for the exclusion of any configuration, as collinear points (this was done by Sierpiński [38]), isosceles or right angled triangles, polyhedra of rational volumes, etc., etc.

One may notice that our results actually give more. Namely, e.g. if $A \subseteq R^2$ is a set of positive outer measure, then there exists, by Theorem 2.2, a positive outer measure subset $B \subseteq A$, omitting rational distances. Can we sharpen this to $\lambda(B) = \lambda(A)$? The above proof works if A is measurable, but in general it is hopeless, as the size of A may be smaller than continuum, so we cannot select elements in a process of length 2^ω. Of course, if CH or at least MA is assumed, the proof works nicely. But without extra hypotheses, we have not been able to prove the statement even for R. The following much weaker statement was observed by the author.

Theorem 6.2 *If $A \subseteq R$, then there is a $B \subseteq A$, $\lambda(B) = \lambda(A)$ with no points of difference 1.*

Proof. We use the theorem of Luzin that every set $A \subseteq R$ can be split into two disjoint subsets with outer measure $\lambda(A)$ ([32]). This implies, that if $A_1, \dots, A_n$ are given, there exist *disjoint* subsets $B_i \subseteq A_i$ with $\lambda(B_i) = \lambda(A_i)$. Given $A \subseteq R$, we let $A_n \subseteq [0,1)$ be such that $A_n + n = A \cap [n, n+1)$ for $n \in Z$. In ω many steps we can select $B_n \subseteq A_n$ such that $\lambda(B_n) = \lambda(A_n)$, and $B_n \cap B_{n+1} = \emptyset$, and then $B = \bigcup\{B_n : n \in Z\}$ works.   □

With a weak form of CH we do have the result for rational distances. Let (*) be the assumption that no cardinal $\leq 2^\omega$ is weakly inaccessible, i.e. a regular, limit cardinal.

Theorem 6.3 *If (*) is assumed, then every $A \subseteq R$ contains a $B \subseteq A$, with $\lambda(B) = \lambda(A)$ and omitting rational distances.*

Proof. We may assume that $\lambda(A) > 0$, as otherwise the empty set works.

Ulam and Sierpiński deduced from (*) that every set $A \subseteq R$ is the disjoint union of ω_1 many sets, each with outer measure $\lambda(A)$ [44, 39]. By the one-dimensional (actually, trivial) case of Theorem 2.2, there is a set $A^0 \subseteq A$ with $\lambda(A^0) > 0$, omitting rational distances.

Claim. There are an $A' \subseteq A$, and a Borel set $P \supseteq A'$ such that A' omits rational distances, $\lambda(P) = \lambda(A') > 0$, and

(6.1) $\lambda((A - P) \cap (A' + r)) = 0$ for $r \in Q$.

We first show that the Claim implies the Theorem. We are going to construct by transfinite induction A_α, P_α for $\alpha < \omega_1$. If A_β, P_β $(\beta < \alpha)$ are given, let

$$A^\alpha = A - \left[\bigcup_{\beta < \alpha} P_\beta \cup \bigcup_{\beta < \alpha} \bigcup_{r \in Q} (A_\alpha + r) \right].$$

If $\lambda(A^\alpha) = 0$, terminate the construction. If, however, $\lambda(A^\alpha) > 0$, apply the Claim, and get A_α, P_α as in the Claim, for A^α. Assume that the construction is terminated at $\gamma \leq \omega_1$. As the Borel sets $\{P_\alpha : \alpha < \gamma\}$ are disjoint and of positive measure, $\gamma < \omega_1$. Put $B = \bigcup\{A_\alpha : \alpha < \gamma\}$. As $\lambda(A^\gamma) = 0$, A is covered by $\bigcup\{P_\alpha : \alpha < \gamma\}$ and by a measure zero set. As the P_α's are disjoint, $\lambda(A) \leq \sum\{\lambda(P_\alpha) : \alpha < \gamma\}$, and also $\lambda(B) = \sum\{\lambda(A_\alpha) : \alpha < \gamma\} = \sum\{\lambda(P_\alpha) : \alpha < \gamma\}$, so $\lambda(B) = \lambda(A)$. Also, if $x \neq y$, $x, y \in B$, then either $x, y \in A_\alpha$ for some $\alpha < \gamma$, or $x \in A^\alpha$, $y \in A_\alpha$, or vice versa for some $\alpha < \gamma$, in either case, $d(x, y)$ is irrational.

Proof of Claim.

Let $A^0 \subseteq A$ be as above.

For $\alpha < \omega_1$ we are going to construct, by transfinite recursion A^α, P^α, Q^α, A_ξ^α $(\alpha < \xi < \omega_1)$ such that

(6.2) P^α is Borel, $\lambda(P^\alpha) > 0$, $Q^\alpha = \bigcup\{P^\beta : \beta < \alpha\}$, $P^\alpha \cap Q^\alpha = \emptyset$;

(6.3) $A^\alpha \subseteq P^\alpha$, $\lambda(A^\alpha) = \lambda(P^\alpha)$, $A^\alpha = \bigcup\{A_\xi^\alpha : \alpha < \xi\}$

(6.4) $\lambda(A_\xi^\alpha) = \lambda(A^\alpha)$ $(\alpha < \xi)$;

(6.5) $A^\alpha \cup A_\xi^\beta$ $(\beta < \alpha < \xi)$ has no rational distances.

Of course, already (6.2) carries the desired contradiction; there are no ω_1 disjoint sets of positive measure.

If these objects are given up to the αth step, then $A' = \bigcup\{A_\alpha^\beta : \beta < \alpha\}$ is a set with no rational distances, and is of full outer measure in Q^α. By our indirect assumption, there is an $r(\alpha) \in Q$, such that $\lambda((A' + r(\alpha)) \cap (A - Q^\alpha)) > 0$. Select $\beta(\alpha) < \alpha$ such that $\lambda((A_\alpha^{\beta(\alpha)} + r(\alpha)) \cap (A - Q^\alpha)) > 0$ and then put $A^\alpha = A_\alpha^{\beta(\alpha)} + r(\alpha)$, $P^\alpha \supseteq A^\alpha$ Borel, $P^\alpha \cap Q^\alpha = \emptyset$, $\lambda(P^\alpha) = \lambda(A^\alpha)$, then decompose A^α as needed in (6.3). To show (6.5), A^α has no rational distances, as it is a translated copy of $A_\alpha^{\beta(\alpha)}$. If $x \in A^\alpha$, $y \in A_\xi^\beta$, $x - y \in Q$, then there is a rational distance in $A_\alpha^{\beta(\alpha)} \cup A_\xi^\beta$, where $\xi > \alpha$ which contradicts (6.5).   □

Using Theorem 4.7 it is easy to show that, assuming CH, every second category planar set contains a second category subset omitting right angled triangles. It is hard not to conjecture at least that this statement holds without any additional hypothesis. And it doesn't.

Theorem 6.4 *If the existence of a measurable cardinal is consistent then it is consistent that there exists a second category set in R^2 not containing a second category subset which omits right angled triangles.*   □

The proof can be found in [28]. We only mention that a second category set $A \subseteq R$ is forced such that the graph of any $A \to A$ function is a first category subset of R^2.

References

[1]   F. Bagemihl, P. Erdős: Intersections of prescribed power, type, or measure. Fund. Math. **41** (1954) 57–67

[2]   St. Banach: Sur les transformations biunivoques. Fund. Math. **19** (1932) 10–16

[3]   St. Banach, A. Tarski: Sur la décomposition des ensembles de points en parties respectivement congruentes. Fund. Math. **6** (1924) 244–277

[4]   P. Bankston, R. Fox: Topological partitions of Euclidean space by spheres. Amer. Math. Monthly **92** (1985) 423–424

[5]   P. Bankston, R. McGovern: Topological partitions. General Topology and Appl. **10** (1979) 215–229

[6]   G.M. Bergman, E. Hrushovski: Identities of Cofinal Sublattices. Order **2** (1985) 173–191

[7]   B. Bollobás: Graph theory, An introductory course. Springer, 1979

[8]   J. Ceder: Finite subsets and countable decompositions of Euclidean spaces. Rev. Roum. Math. Pures et Appl. **14** (1969) 1247–1251

[9]   J.H. Conway, H.T. Croft: Covering a sphere with congruent great-circle arcs. Proc. Cambridge Philos. Soc. **60** (1964) 787–800

[10] R.O. Davies: Equivalence to the CH of a certain proposition of elementary plane geometry. Zeitschr. für Math. Logic und Grundlagen Math. **8** (1962) 109–111

[11] R.O. Davies: On a problem of Erdős concerning decomposition of the plane. Proc. Camb. Phil. Soc. **59** (1963) 33–36

[12] R.O. Davies: On a denumerable partition problem of Erdős. Proc. Camb. Phil. Soc. **59** (1963) 501–502

[13] R.O. Davies: The power of continuum and some propositions of plane geometry. Fund. Math. **52** (1963) 277–281

[14] R.O. Davies: Covering the plane with denumerably many curves. J. London Math. Soc. **38** (1963) 343–348

[15] R.O. Davies: Partitioning the plane into denumerably many sets without repeated distances. Math. Proc. Camb. Phil. Soc. **72** (1972) 179–183

[16] R.O. Davies: Covering space with denumerably many curves. Bull. London Math. Soc. **6** (1974) 189–190

[17] P. Erdős: Geometrical and set-theoretical properties of subsets of the Hilbert space. Math. Lapok **19** (1968) 255–258 (in Hungarian)

[18] P. Erdős: Problems and results in chromatic graph theory. In: Proof techniques in graph theory (ed. F. Harary). Academic Press, New York London 1969, pp. 27–35

[19] P. Erdős: Problems and results in combinatorial geometry, Disc. Geometry and Convexity. New York Acad. Sci. **440** (1985) 1–11

[20] P. Erdős, S. Kakutani: On non-denumerable graphs. Bull. Amer. Math. Soc. **49** (1943) 457–461

[21] P. Erdős, P. Komjáth: Countable decompositions of $\mathbb{R}^2$ and $\mathbb{R}^3$. Discrete and Comp. Geometry, **5** (1990) 325–331

[22] F. Galvin, G. Gruenhage: Plane geometry and the continuum hypothesis. Preprint

[23] G. Gruenhage: Covering properties on $X^2 - \Delta$, W-sets, and compact subsets of $\sum$-products. Topology and its Applications **17** (1984) 287–304

[24] K. Hrbacek, T. Jech: Introduction to set theory. Marcel Dekker, 1984

[25] T. Jech: Set Theory. Academic Press, 1978

[26] P. Komjáth: Tetrahedron free decomposition of R^3. Bull. London Math. Soc. **23** (1991) 116–120

[27] P. Komjáth: A decomposition theorem for R^n. To appear

[28] P. Komjáth: A second category set with only first category functions. Proc. Amer. Math. Soc. **112** (1991) 1129–1136

[29] K. Kunen: Set Theory. North-Holland, 1980

[30] K. Kunen: Partitioning Euclidean space. Math. Proc. Camb. Philos. Soc. **102** (1987) 379–383

[31] M. Laczkovich: Equidecomposability and discrepancy; a solution of Tarski's circle-squaring problem. Crelle Journal **404** (1990) 77–117

[32] N. Lusin: Sur la décomposition des ensembles. C. R. Acad. Sci. Paris (1934) 1674

[33] S. Mazurkiewicz: Sur un ensemble plan. Comptes Rendus Sci. et Lettres de Varsovie **7** (1914) 322–383

[34] A.W. Miller: Infinite combinatorics and definability. Annals of Pure and Applied Logic **41** (1989) 179–203

[35] J. Mycielski: About sets invariant with respect to denumerable changes. Fund. Math. **45** (1958) 296–305

[36] W.R. Scott, L.M. Sonneborn: Translations of infinite subsets of a group. Coll. Math. **10** (1963) 217–220

[37] W. Sierpiński: Sur un théorème équivalent à l'hypothèse du continu. Bull. Int. Acad. Sci. Cracovie **A** (1919) 1–3

[38] W. Sierpiński: Sur un problème concernant les ensembles mesurables superficiellement. Fund. Math. **1** (1920) 112–115

[39] W. Sierpiński: Sur une proprieté des ensembles linéaires quelconques. Fund. Math. **23** (1934) 125–134

[40] W. Sierpiński: Hypothèse du continu. Warszawa, 1934

[41] W. Sierpiński: Sur un problème de M. Steinhaus concernant les ensembles de points sur le plan. Fund. Math. **46** (1958) 191–194

[42] A. Szulkin: R^3 is the union of disjoint circles. Amer. Math. Monthly **90** (1983) 640–641

[43] L. Trzeciakiewicz: Remarque sur les translations des ensembles linéaires. Comptes Rendus de la Société des Sciences et des Lettres de Varsovie C1. III. **25** (1932) 63–65

[44] St. Ulam: Über gewisse Zerlegungen von Mengen. Fund. Math. **20** (1933) 221–223

[45] S. Wagon: The Banach-Tarski Paradox. Cambridge University Press, 1985

Author Index

Subject Index

Algorithms and Combinatorics

Editors: R. L. Graham, B. Korte, L. Lovász

Springer-Verlag
Berlin
Heidelberg
New York
London
Paris
Tokyo
Hong Kong
Barcelona
Budapest